STRATEGIC ORGANIZATIONAL DIAGNOSIS AND DESIGN

The Dynamics of Fit

Third Edition

STRATEGIC ORGANIZATIONAL DIAGNOSIS AND DESIGN

The Dynamics of Fit

Third Edition

Richard M. Burton
The Fuqua School of Business
Duke University
Durham, North Carolina

Børge Obel
Department of Management
Odense University
Odense, Denmark

Kluwer Academic Publishers
Boston/Dordrecht/London

Distributors for North, Central and South America:
Kluwer Academic Publishers
101 Philip Drive
Assinippi Park
Norwell, Massachusetts 02061 USA
Telephone (781) 871-6600
Fax (781) 871-9045
E-Mail: kluwer@wkap.com

Distributors for all other countries:
Kluwer Academic Publishers Group
Post Office Box 322
3300 AH Dordrecht, THE NETHERLANDS
Telephone 31 786 576 000
Fax 31 786 576 254
E-mail: services@wkap.nl

 Electronic Services <http://www.wkap.nl>

Library of Congress Cataloging-in-Publication Data

A C.I.P. Catalogue record for this book is available from the Library of Congress.

STRATEGIC ORGANIZATIONAL DIAGNOSIS AND DESIGN: The Dynamics of Fit, Third Edition, by Richard M. Burton and Børge Obel.

ISBN 1-4020-7684-3 (hardbound)
ISBN 1-4020-7685-1 (paperback)

Printed on acid-free paper.

Printed in the United States of America.

To Helmy H. BALIGH
Teacher and Friend

R.M.B.
B.O.

TABLE OF CONTENTS

CHAPTER 1 Diagnosis and Design 1
Introduction 1
What is an Organization?...2
An Information-Processing View of Organizations 5
    Neo-Information Processing as a Basis for Organizational
    Diagnosis and Design...9
Strategic Organizational Design 13
The Multi-contingency Diagnosis and Design Model.......................16
    Fit Criteria for Designing the Knowledge Base.........................18
    Contingency Fit...20
    Strategic Fit...21
    Design Fit..23
    Total Fit..24
Synthesis of Knowledge Bases...25
    Composing the Knowledge Base 30
The OrgCon 38
Validating the OrgCon...40
Summary..41

CHAPTER 2 What is an Organizational Design?.....................43
Introduction 43
Organizational Configurations..46
    Simple Configuration 47
    Functional Configuration...48
        A Functional Organization 50
    Divisional Configuration..54
        A Divisional Organization 56
    Matrix Configuration 60
        A Matrix Organization...62
    Ad Hoc Configuration 66
    Bureaucracies...67
    Virtual Network 69
    International Configurations...70
Organizational Complexity 73
    Horizontal Differentiation 74
    Vertical Differentiation 75
    Spatial Differentiation 77
Formalization..78

Centralization 80
Incentives 81
Coordination and Control 83
Summary 85

CHAPTER 3 Leadership and Management Style 87

Introduction 87
Leadership and Organization...91
Literature Review on Leadership 93
Dimensions and Typologies..99
Measuring Leadership Style 103
Leadership Style as a Contingency 107
    Describing a Leader 109
    A Leader's Effect on Structure...110
    Describing a Producer 113
    A Producer's Effect on Structure 114
    Describing an Entrepreneur...116
    An Entrepreneur's Effect on Structure...............................117
    Describing a Manager 119
    A Manager's Effect on Structure..121
    Interaction effects of Size and Leadership Style 124
Managing the Leader 125
Summary 126

CHAPTER 4 Organizational Climate 127

Introduction 127
Climate and Culture 131
Literature Review on Climate..137
Measuring and Categorizing Climate..141
Climate as a Contingency 145
    Describing a Group Climate 146
    The Group Climate Effects on Structure 148
    Describing the Developmental Climate 150
    Developmental Climate Effect on Structure 151
    Describing the Internal Process Climate 153
    Internal Process Climate Effects on Structure....................155
    Describing the Rational Goal Climate 157
    Rational Goal Climate Effects on Structure........................158
Managing the Climate 160
Summary 160

CHAPTER 5 Size and Skill Capabilities 163
Introduction 163
Literature Review on Size 165
 The Measurement of Size 166
 Size as Imperative 168
An Information-Processing Perspective on Size 168
Measuring Size and Skill Capability for Design Purposes 171
Size as a Contingency 174
 Size Effects on Complexity 175
 Size Effects on Centralization 176
 Size Effects on Formalization 178
 Size Effects on Configuration 179
 Size Effects on Incentives and Coordination and Control 180
Managing Size and Skill 181
Summary 183

CHAPTER 6 The Environment 185
Introduction 185
Literature Review on Environment 189
 Measures of the Environment 189
 The Environment-Structure Relationship 195
Describing the Environment 198
Environment as a Contingency 205
 Environmental Effects on Formalization, Centralization
 and Complexity 206
 Environmental Effects on Configuration, Coordination,
 Media Richness and Incentives 215
Operationalizations of the Environmental Measures 224
National culture 227
Managing the Environment 231
Summary 233

Chapter 7 Technology 237
Introduction 237
Literature Review on Technology 240
Measuring Technology 249
Technology as a Contingency 254
 Technology Effects on Formalization 254
 Technology Effects on Centralization 256
 Technology Effects on Organizational Complexity 257

Technology Effects on Configuration 259
Technology Effects on Coordination and Control...............................260
Information Technology 262
Managing Technology...267
Summary...270

CHAPTER 8 Strategy...271
Introduction 271
Literature Review on Strategy...276
Structure Follows Strategy 276
The Counter Proposition: Strategy Follows Structure 278
Strategy and Structure; Fit...279
Measuring and Categorizing Strategy...284
Strategy as a Contingency 291
Describing a Prospector 294
Prospector Effects on Structure...296
Describing a Defender 297
Defender Effects on Structure 299
Describing an Analyzer without Innovation 300
Analyzer without Innovation Effects on Structure...........................302
Describing an Analyzer with Innovation 304
Analyzer with Innovation Effects on Structure.................................305
Reactor Strategy...307
International Dimensions...308
Choosing the Right Strategy 309
Summary...310

Chapter 9 Diagnosis and Misfits...311
Introduction 311
Literature Review on Misfits..312
Strategic Misfits..314
Environment-Strategy...315
Technology-Environment...317
Climate-Environment 317
Climate-Strategy..319
Technology-Strategy 320
Climate -Technology...320
Technology-Leadership Style 321
Climate - Leadership Style 322
Leadership Style - Environment 323
Leadership Style – Size Skill 323

Contingency Misfits 324
   Environment Misfits 324
   Strategy Misfits 326
   Technology Misfits 328
   Size Skill Misfits 329
   Climate Misfits 330
   Leadership Style Misfits..331
More Specific Misfits 333
   Environment - Strategy Misfits..333
   Technology - Environment Misfits 335
   Climate - Environment Misfits...336
   Climate - Strategy Misfits...337
   Strategy - Technology Misfits 337
   Climate - Technology Misfits 338
   Technology - Size Skill Misfits 339
   Climate - Leadership Style Misfits 340
   Leadership Style - Environment - Size Misfits........................340
   Leadership Style - Strategy Misfits 341
   Environment Misfits 342
   Strategy Misfits 343
   Climate Misfits 344
   Leadership Style Misfits..344
Summary..345

CHAPTER 10 Organizational Design Fit................................347
Introduction 347
Design Parameter Fit 354
Design Synthesis 358
   The Simple Configuration 359
   The Functional Configuration 360
   Divisional Configuration..362
   Matrix Configuration 363
   Ad Hoc Configuration 365
   Virtual Network 366
   The Professional Bureaucracy 367
   The Machine Bureaucracy...368
   Organizational Complexity...369
   Formalization 373
   Centralization..376
   Coordination and Control...380
   Media Richness and Incentives 381
Summary..382

CHAPTER 11 The Dynamics of the Change Process 385
Introduction 385
Diagnosis and Design 386
Lifecycle Management: Diagnosis and Design 388
The Dynamics of Diagnosis and Design 393
 Misfits: Where Do They Come From? 394
 External Changes: Evolutionary and Revolutionary 396
 Internal Changes: Intended Misfit Creation 397
The Mechanisms of Misfits..399
How to Fix Misfits 403
How Fast is Fast Enough 407
Design is Exploration and Exploitation....................................411
Design is Learning 415
Summary..418

References..421
Subject Index 435

CD: Teaching Material and OrgCon
Cases:
 Oticon (Dorthe Døjbak and Mikael Søndergaard)
  The Top Management Group that got Fired ... by Itself
  The Spaghetti that became Revolutionary
  What Happens if the Spaghetti gets Cold?
  Scouts are More Attractive
 ABB Electromechanical Meters (David R. Thomason)
 Duke University Press (Kathy Brodeur)
 GTE (Louis Grilli)
 Bluestone Group. Inc. (Hyun-Chan Cho, Hajime Kitamur and Ting-Hsuan Wu)
 Joy Builders (Dorthe Døjbak)
 Meki (Dorthe Døjbak)
 Dandairy Ltd. (Jens Krag)
 Seeds Control, Ltd. (Jens Kragh)
OrgCon and Training Course Installation Files

Preface to the Third Edition

In this third edition, we have made a number of updates and changes - both in the text and in the OrgCon software.

First to the reader, we repeat a note from the preface in the second edition as it is even more relevant here:

There are two very contrasting approaches to reading this book and learning about organizational design. The more traditional approach is to read the book, and then use the OrgCon on cases and applications. The second approach is to begin with the OrgCon software and only examine the book as you find it helpful. Which approach is better? It is your choice, not ours. In our experience, students in organizational design prefer to start with the OrgCon and a case, rather than with the book itself. Readers who have more background in organization theory and design usually examine the book first. We have tried to write the book so that it can serve both as a reference and an integrated presentation.

There are numerous changes in the third edition. The literature review in each chapter has been updated. The information processing approach is strengthened and applied more comprehensively as the theoretical underpinnings. Throughout we have rewritten the text beyond normal editing in an attempt to make the presentation clearer and easier to read. In addition, we have made the following substantive changes:

- Introduced virtual network configurations in Chapter 2 and then used it throughout.
- Enhanced the discussion of incentives throughout.
- Leadership style in Chapter 3 is now a two dimensional construct: delegation and uncertainty avoidance.
- National culture is included as part of the organization's environment in Chapter 6.
- Diagnosis and misfit management in a new Chapter 9 are developed in greater depth and based more firmly upon empirical studies and the literature.
- Design and contingency fit are revised discussions in Chapter 10.
- Change management and the dynamics of design is a new Chapter 11.

- OrgCon has been revised to include all of these changes and to include new insights from the literature and our ongoing validation.
- OrgCon now includes a "Trainer" for the first time user.
- OrgCon has a comprehensive HELP which gives definitions and illustrations for every question and concept.
- And, many new cases have been added and included on the CD along with the OrgCon.

We made these changes for both types of readers mentioned above.

As with previous editions, we are very fortunate to have the support of many friends and colleagues who have contributed in innumerable ways to make this revision possible. Pernille Dissing Sørensen developed the "Trainer" module in the OrgCon. Aparna Venkatraman read, ciritized, suggested changes on numerous drafts. Peter Obel is an expert programmer and makes the OrgCon technically sophtisticated and user friendly. Nadine Burton further developed the HELP version for the OrgCon. Mike Roach has been our technical consultant and in-class assistant. Mona Andersen is our technical editor for the text; she has been tireless in her insistence on getting it right. For many years, Thorkild Jørgensen of Aarhus University has offered a critique and many suggestions from his use of the text and OrgCon with his students. Ray Levitt of Stanford University and his students have offered very helpful suggestions. Perhaps, the most fundamental critique is that of Carl R. Jones of the Naval Postgraduate School; he always asks tough questions.

We would like to thank Dorthe Døjbak Håkonsson and Jens Krag for their permission to include the new cases on the CD. These cases have been translated from Danish to English by Axel Moos.

We also want to thank our colleagues at Duke University and University of Southern Denmark for discussions and comments. All have made this revision better and we are very appreciative.

Our home institutions - The Fuqua School of Business, Duke University and the University of Southern Denmark - have been very supportive. At Kluwer, Gary Folven is a most supportive editor so we can concentrate on the book itself.

Again, Helmy Baligh reminds of what is important.

Rich Burton and Borge Obel

Preface to the Second Edition

We will not repeat our preface discussion from the first edition. Here we only add some new comments:

- a note to the reader and user,
- changes in the book and the Organizational Consultant (OrgCon),

and,

- our thanks to the many individuals who have contributed critically to this venture, read and reviewed the book, contributed chapters and cases, and similarly used and critiqued the OrgCon.

For the reader, there are two very contrasting approaches to reading this book and learning about organizational design. The more traditional approach is to read the book, and then use the OrgCon on cases and applications. The second approach is to begin with the OrgCon software and only examine the book as you find it helpful. Which approach is better? It is your choice, not ours. In our experience, students in organizational design prefer to start with the OrgCon and a case, rather than with the book itself. Readers who have more background in organization theory and design usually examine the book first. We have tried to write the book so that it can serve both as a reference and an integrated presentation.

We made a number of changes in this edition. We have revised the OrgCon and put it in a Windows frame. It is much friendlier and you can spend your time focusing on the cases and learning about organizational design. The OrgCon is contained on the accompanying CD-ROM - here you will also find a number of data input files for cases.

In the book itself, we have kept the main ideas and the basic approach. But, there are significant changes. We have added a new chapter on climate and incorporated climate into the multidimensional contingency theory. Climate helps determine what the organization should be, and how it should be designed. The addition of climate to the model makes it more nearly complete and adds considerably to our understanding of the organization and its design.

The synthesis Chapter 8 on diagnosis and design in the first edition has been totally rewritten as Chapter 9 here. We have eliminated a good deal of the earlier redundancy and spelled out more clearly what we mean by design. It is not an easy concept. Design is both a product and a process. Design is organizational lifecycle management and dynamic. Design can be viewed as March's concepts of exploration and exploitation. And finally, design is organizational learning: it is a never ending process of experience, error and refinement.

We have added capsules on new organizational forms and a number of other topics. They are snippets which are related and relevant, but may not fit the flow and the order of the book. Norman Davies, in his EUROPE: A HISTORY, 1996, called them "capsules." His capsules: cross the boundaries of the chapters, illustrate curiosities, whimsies, and inconsequential side-streams, glimpses of "new methods, new disciplines, and new fields" of recent research. Our capsules are similar in spirit: the research support is weak, most of support is conjectural or assertive; a popular new word as delayering or concept as hypercompetition; or small case stories. We hope the reader will find them interesting.

The Windows 95 version of OrgCon has been a multi-person, multi-location project. In Odense, Jan Mikkelsen has been our fundamental programmer. Roxanne Zolin, Stanford University, planned and managed the project as well as adding critical programming elements. In Omaha, Miriam Zolin developed the help-part. Peter Obel has been our expert for all the things that go beep in the night as well as fixing problems when others gave up. The computer team, Bjarne Nielsen and Jan Pedersen at Odense have as usual provided invaluable 24 hours a day support.

Our friends have made this venture fun; even when we could not laugh, they did for us. We have been very fortunate to have their support and help. Starling D. Hunter III, Mikael Søndergaard, and Dorthe Døjbak permitted us to include research chapters; we are very pleased. Starling in Chapter 11 demonstrates how to use publicly available information to create the input data for an OrgCon analysis, and then the implication of the results. Mikael and Dorthe in Chapter 12 developed a very interesting case: Oticon, the spaghetti organization. This book is a research book. It is not an easy read. Yet, a few colleagues have used it as a textbook and provided very valuable feedback. We would like to thank Ray Levitt, Thorkild

Jørgensen, Gerry DeSanctis, Bo Eriksen, Carl Jones, and Henrik Bendix. Several individuals have read and critiqued our effort: Pat Keith, Stan Rifkin, Ray Levitt, Jan Thomsen, Starling Hunter, Dorthe Døjbak, Torbjørn Knudsen, and Jeanette Larsen. Arie Lewin, Duke University, is our special critic; he has the ability to coach and make the end result better. We are very grateful.

The Fuqua School of Business, Duke University and The School of Business and Economics, Odense University, our home institutions, have provided much support. Part of the revision was written at Stanford University, where Obel was a Research Scholar at Cife and Scancor, Stanford University. Ray Levitt has been a most gracious host. We also like to thank Jim March and the many researchers at Scancor for the many question they asked, which forced us to rethink and sharpen our arguments. We express our thanks to Stanford University. Our work has been generously supported by The Danish Social Science Research Council.

At Odense University, Mona Andersen made disarray into wonderful text. She made the impossible sccm normal for this edition as she did for the first one.

At Kluwer, Zack Rolnick has made it all possible. The Kluwer team has responded at critical times to make it happen.
As ever, Helmy Baligh is our anchor who keeps us focused on what is important.

We would like to thank them all.

Rich Burton and Børge Obel

Preface to the First Edition

Organizational design is a normative science with the goal of prescribing how an organization should be structured in order to function effectively and efficiently. Organizational theory is a positive science that states our understanding about how the world operates and contrasts that understanding with a view of how the world could possibly operate. It provides the theoretical underpinnings for organizational design. In this book, we attempt to construct an approach for diagnosing and designing organizations built on a knowledge base of organizational theory.

Organizational design is a young field that incorporates many concepts and approaches. In organizational design literature to date, there seems to have been only two ways of doing things in this field - either to be so general and so simple that the various interpretations do not yield practical design implications, or to be so detailed and specific that generalization to other situations is almost impossible. We attempt here to strike a balance - and offer an approach that is applicable to a broad range of situations.

In our view, organizational theory exists as a large body of related languages, definitions, hypotheses, analyses, and conclusions. Our knowledge is vast, diverse, somewhat inconsistent, and generally unconnected. Yet there is an underlying core of knowledge that can be used for analytical purposes. Creating this balanced approach requires that the knowledge be distilled and augmented to produce a set of clear and consistent design rules that can be used to recommend what the organization's design should be.

In organizational theory we have attempted to find simplicity. The search for a dominant contingency or imperative has led to a parade of paradigms which examine elements of technology, size, environmental, and strategy paradigms. Moreover, little effort has been made to remove the simplicity of these paradigms and put the various pieces together in a reasonable fashion. Schoonhoven (1981) lists a number of problems with the contingency view of organizational design, including lack of clarity, lack of understanding about interactions, and a lack of acknowledgment of the functional forms of the interactions. Others (Donaldson, 1982, 1987; Schreyogg, 1980; Pennings, 1987) have debated whether a contingency view was appropriate or not. The difficulties come from many sources. In many studies the distinction between description and prescription is not at all clear. In many cases a silent shift is made from the description to the prescription. Indeed, we must incorporate knowledge that comes from descriptive studies. But we want to prescribe. We want to recommend what the organizational design should be.

Designing an organization is no simple task. From Mintzberg's five classes of structure, Mintzberg (1979), one can generate over a million design alternatives. How does one search such a design space? To illustrate, let us assume that an organizational structure can be defined as functional, divisional, or matrix and as centralized or not and formalized or not. There are then 3×2×2 = 12 possible designs from which to choose. The number of choices grows nonlinearly as the number of organizational dimensions approaches useful realistic proportions. Let us add the choice that the organization could have high specialization or not. The number of choices is now 2×12 = 24. In a contingency theory of organization, these design alternatives would need to be evaluated for all possible conditions. In the simple case of ten contingency dimensions and only two values allowed for each, we would have to evaluate twenty four classes of structures under 1,024 different conditions. When the dimensions and number of variables approaches a useful size for analytical purposes, the choice and evaluation of organizational design alternatives becomes an enormous problem. There is no way students, teachers, consultants, and managers can handle such a complex choice situation without some order and systematic approach.

This book is accompanied by Organizational Consultant–a computer based expert system that operationalizes the theoretical propositions developed in this book. Organizational Consultant is a program that applies the underlying theory to deal with millions of design possibilities in a systematic and comprehensive manner to diagnose an organization and offer recommendations for its design, configuration, complexity, formalization, centralization, media richness, committees, meetings, and liaison relations.

The book is divided into two parts. Part 1 consisting of the first 8 chapters is the theory part, where the multiple contingency model is developed. One special feature of Part 1 is its detailed analysis of Scandinavian Airlines Systems (SAS) from 1950 to today. SAS has undergone a number of changes in its situation and design during this period, and we analyze those changes to provide insight into organizational design. Other business cases are included to illustrate various concepts and to relate the theory to practice.

Part 2 is a manual/case part consisting of Chapters 9-13. Part 2 describes how the Organizational Consultant operates and how it can be used to analyze cases.

This part also contains seven cases specifically developed to be used with Organizational Consultant. For Part 2, Nancy Keeshan is a coauthor.

Although Part 1, Part 2, and Organizational Consultant can be considered as a unit, each is self-contained. For instance, Organizational Consultant can also be used with other textbooks in organizational theory such as Robbins (1990), Daft, (1992), and Mintzberg, (1984).

Organizational Consultant is a teaching tool in itself. Explanation and help functions contain many parts of the theory. Extensive use of Organizational Consultant can teach the student many issues of the theory.

The book and software have been developed over some years. We have used both the book and software in many courses both at The Fuqua School of Business and Odense University. We would like to thank our students for valuable comments and critique. Many students developed cases from their own organization as part of the courses and we were allowed to include some of these cases in Part 2 of this book. We would like to thank Steve Hulme, Louis Grilli, David Thomason, Kathy Brodeur, Barbara Johnson, and Kelly Leovic. Ray Levitt at Stanford also used both the book and Organizational Consultant in his course. Ray's many comments helped improve the final version. Additionally, he convinced his students Hyun-Chan Cho, Hajime Kitarnura, and Tin-Hsuan Wu to prepare a case for Part 2. We are very grateful for his effort. Many individuals have worked with us as teaching assistants, programmers, and research assistants. We would like to thank Berit Jensen, Claus Bo Jørgensen, Mogens Kjær Harregaard, Kim Madsen, Mads Haugård, Bo Hegedal and Søren Hjortkjær. Our colleagues Caroll Stephens, Arie Lewin, Tom Naylor, Mikael Søndergaard, Bjarne Nielsen, and Henrik Bendix offered comments, critique, and help in various ways. A number of executives allowed us to test Organizational Consultant in their organization. We are grateful to Don Namm, Glenn Weingarth, Don Melick, Benny Mortensen, Hans Jørgen Hansen, Jørn Henrik Petersen, Sven-Erik Petersen, and Arne Fredens. We would also like to thank the reviewers and staff from Kluwer. Their comments and help is highly appreciated. Mona Andersen has worked with us in all these years. She has typed and organized several versions as well as prepared this final one. It has been a pleasure to work with Mona.

We would to thank the Danish Social Science Research Council, Duke University, and Odense University for the support they have provided to make this research possible.

And finally, and most importantly, Helmy H. Baligh has been our fellow traveler on many ventures. He is coauthor on a number of base papers for this book and has been our most supportive and most critical colleague– thus the dedication.

Richard Burton and Børge Obel

CHAPTER 1

Diagnosis and Design

INTRODUCTION

On Saturday, January 30, 1993, the Herald Sun, the local newspaper in Durham, North Carolina, reported that IBM was facing problems. IBM had in the previous week announced a cut in dividends and planned to replace its chairman, John Akers. IBM's profit went from a $6 billion profit in 1990 to $2.8 billion and $4.9 billion losses in 1991 and 1992 respectively. Peter Lieu, a computer analyst at Furman Selz, was quoted as saying that "John Akers inherited a mess and the mess is a highly centralized organization with virtually no delegation of responsibility." The problems facing IBM were compared to the situation at AT&T, which after losing $1.23 billion in 1988, turned profitable in 1990 with a reported 1992 fourth-quarter profit of $1 billion. The success of AT&T was attributed to a decentralized management style introduced by the late James Olson and continued under the new leadership by Robert Allen. Allen cut staff by 5 percent and wrote off $6.7 billion in old analog technology. He diversified and brought in a new management. Meanwhile, IBM was struggling with its old management style and was accused of "failing to 'obsolete' its own products quickly enough." IBM's problems had arisen because of a new competitive situation in all of its markets-a change away from the use of mainframe computers and toward workstations and networks. Additionally, competition in the personal computer market had changed with declining growth and confusion about which operating system would take the lead. Further, IBM also "had some bad luck in the form of the global recession, which cut into international profits." This story points to a key issue important for business success. Situations changed in the past and they will change in the future. Competition, technology, and economic conditions change over time, and firms have to adjust to these changes. Adjustments depend on the management and its style, which is the basis for selecting a strategy that will lead the firm to success in its new conditions. Finally, the firm needs an organization and an organizational structure that will enable it to carry out its selected strategy. The proper fit among the firm's strategic factors: leadership style, climate, size, environment, technology, strategy, and the firm's structure is a necessary condition for a business success. However, it is not sufficient nor a guarantee for success.

IBM and AT&T are two very large corporations, but the same arguments can be made for small and medium-size firms as well. AT&T and IBM struggled to find a way to fight back. It has not been easy for them to find the appropriate response, even with large staffs of experts and numerous consultants. Small and medium-size firms may have equal difficulties finding the proper fit. The story also shows that change is not a new phenomenon, and it is as important today with as it was in 1993. Today we may have even more and more rapid changes.

This book summarizes our knowledge on how to diagnose a firm's situation and then design an efficient and effective organization to meet changing conditions. Strategic organizational diagnosis and design includes evaluating the strategic position as well as the composition of organizational units, reporting relationships among units, and other structural relationships. It also includes the design of organization-wide methods, procedures, and work technologies (Nadler and Tushman, 1988) as well as the organization's incentive system.

WHAT IS AN ORGANIZATION?

Organizations are everywhere and central to our lives. We are part of many organizations-where we live, where we work, where we shop, and where we do many other activities. IBM, AT&T, Scandinavian Airlines System (SAS), General Motors, University of Southern Denmark, The Fuqua School of Business, and a two-person pizza shop - all are organizations. Organizations are such a part of our lives that we seldom reflect on exactly what we mean when we speak of an organization.

So what is an organization? Some of the definitions found in the literature are presented below:

- "Organizations are social units (or human groupings) deliberately constructed and reconstructed to seek specific goals" (Etzioni, 1964).
- "The purpose of organizations is to exploit the fact that many (virtually all) decisions require the participation of many individuals for their effectiveness ... Organizations are a means of achieving the benefit of collective action in the situations in which the price system fails" (Arrow, 1974).
- "Organizations are social entities that are goal-directed, deliberately structured activity systems with an identifiable boundary" (Daft, 1992).
- "An organization is a consciously coordinated social entity, with a relatively identifiable boundary, which functions on a relatively continuously basis to achieve a common goal or a set of goals" (Robbins, 1990).

The definitions have much in common and explain the concept of what an organization is.

An organization is a social entity. It exists for and is made up of individuals (Arrow, 1974, p. 33) Corporations, small businesses, political parties, and religious institutions are examples. An organization is more than a simple collection of people, such as shoppers at a supermarket who are there for their own individual purposes. An organization has goals and it exists for a purpose. Its members share this purpose for the most part. Arrow (1974, p. 33) states, "The purpose of organization is to exploit the fact that many (virtually all) decisions require the participation of many individuals for their effectiveness." Individuals fulfill their own goals by participating in the organization. Organizations have activities; they sell groceries, make automobiles, sell computers, elect councils, and help the poor. Organizations have boundaries; some individuals or activities are inside, and others are outside. The outside is called the environment, which is an important consideration for the design of the inside. Today with the use of networks, virtual teams, and strategic alliances, it may be difficult to tell when something is inside or outside the organization. Finally, an organization is deliberately constructed; it is an artificial entity, it is designed.

Organizations are designed to do something-undertake activities to accomplish goals. Additionally, members of an organization benefit by being members. They gain from cooperation because individuals have different talents and individuals' efficiency usually improves with specialization (Arrow, 1974). Organizations emerge when cooperation and specialization cannot be obtained through a market and price system but activities have to occur within an organization. A fundamental issue in designing an organization is to group small activities together so that goals are realized or conversely, to take a large task and break it into smaller tasks. In manufacturing an automobile, a large task, thousands of smaller tasks must be grouped to accomplish the larger task.

The organizational design problem is both how to put the smaller tasks together and how to take a large task and break it down into appropriate smaller tasks. Whichever view is adopted, the smaller activities must be coordinated to accomplish the larger task and to realize the organization's goal. Coordination is putting the pieces together. Without coordination, we do not have an organization but only a collection of separate activities. Coordination, then, introduces the need for information and information exchange in order to reduce the uncertainty associated with the tasks; otherwise the required coordination cannot be obtained. Generally, we use information processing as an integrating concept in the design of organizations (Nadler et al., 1988).

As discussed in the next section, various criteria may be applied to help us select among possible designs. Three criteria are paramount in organizational design-effectiveness, efficiency, and viability-an organization should be designed to meet these criteria:

- Effectiveness: An organization is effective if it realizes its purpose and accomplishes its goals.
- Efficiency: An organization is efficient if it utilizes the least amount of resources necessary to obtain its products or services.
- Viability: An organization is viable if it exists over a long period of time.

Effectiveness is contrasted with efficiency. Effectiveness is doing the right thing; efficiency is doing it right. Usually, effectiveness does not incorporate efficiency: that is, an organization can accomplish its goals but be quite inefficient in its use of resources. An efficient organization uses its resources well, but may not accomplish its goal well. Efficiency has its focus on the internal working of the organization while effectiveness addresses the organization's positioning vis a vis the environment. We want to design organizations that are both effective and efficient, as both are likely to be important for viability or long-term survivability. However, an organization may survive for many years and not be particularly effective or efficient. The U.S. government is relatively long-lived; effective in many of its activities, but not known for its efficiency. These criteria represent different concepts that are desirable and provide general guidance in selecting appropriate organizational configurations and organizational properties.

AN INFORMATION-PROCESSING VIEW OF ORGANIZATIONS

We live in a world where information, information exchange, and information processing are of paramount importance. Companies are linked to their customers and suppliers by numerous information networks. Telephone, fax, electronic mail, and the internet are modern ways of communicating. The information revolution (Business Week, June 13, 1994) is changing the world and how it works. The development of cellular phones, multimedia systems, and interactive television systems affects the way people and organizations operate and work (OR/MS Today, June 1994, pp. 20-27).

These new media for information exchange may be viewed as a new phase of automation similar in importance as the industrial revolution in the beginning of the 20th century. Consider two examples: first, information and travel, and then information and small firm management. Preparing for a trip may involve buying an airplane ticket, reserving a hotel room, renting a car, and communicating with the friends that you will visiting. Finding the way from the airport to your friends requires a map, and it would also be nice to know the weather conditions so you can bring the appropriate clothes. This trip preparation is information intensive and requires extensive information processing by numerous individuals and organizations. Precise planning and coordination are required to assure you have the services at the moment you want them. It is interesting that your communication with the organizations to plan your trip can be done without talking to anyone. It has been automated using e-mails and web-based systems. The automation has moved from automation inside the organization to include customers and suppliers. Once you are on the trip you meet and interact with many individuals: airport security agents, flight attendants, and your friends. If you plan another trip, the various organizations may remember who you are and your preferences. The next communication will be even more efficient. Thus, the organization remembers to be more efficient and effective.

Consider the case of a small lumber firm where the CEO coordinated activities by reading the mail in the morning when orders arrived. He processed the orders and issued directions for the day. He was the central figure coordinating activities in an environment in which orders and specifications had to be in writing. He balanced the risk and centralized the information and decision-making in order for his firm to be effective and efficient. The introduction of the fax machine changed how this firm operated. With the fax, orders now came at any given time a day. The demand for quick response grew because customers wanted offers immediately by fax. Because modern fax systems allow customers to send requests to more than one firm, competition was enhanced and became fierce. The lumber firm's old centralized decision procedure no longer worked; decisions had to be delegated, and rules for handling orders and balancing the risk had to be developed. New decision-making procedures had to be learned. Changing the information flow created a need for learning new decision-making procedures and changed the design of the organization.

An organization also processes information in order to coordinate and control its activities in the face of uncertainty where uncertainty is an incomplete description of the world (Arrow, 1974, p. 34). By processing information, it observes what is happening, analyzes and makes choices about what to do, and communicates the above to its members. Information processing is a way to view organizations and their designs. Information "channels can be created or abandoned. Their capacities and the types of signals to be transmitted over them are subject to choice, a choice based on a comparison of benefits and costs" (Arrow, 1974). Both information systems and individuals possess a capacity to process information, but "this capacity is not, however, unlimited and the scarcity of information-handling ability is an essential feature for the understanding of both individual and organizational behavior" (Arrow, 1974). Work involves information processing; individuals are information and knowledge based workers. They talk, read, write, calculate and analyze. Various media are available - from pens and face to face conversation to word processors and video meetings. Innovations in information technology change both the organization's demand for information processing and its capacity for processing information but not the basic nature of work.

The basic design problem is to create an organizational design that matches the demand for information processing with the information processing capacity. Galbraith (1973; 1974), in a seminal work, presented the organizational design problem as an information-processing problem: "the greater the uncertainty of the task, the greater the amount of information that has to be processed between decision makers" (Galbraith, 1974, p. 10). The task uncertainty can arise from the technology and environment (Thompson, 1967) as well as other sources. If the information processing demand comes from many routine and predictable tasks, then formalization in form of rules and programs can increase the number of tasks that can be handled. When there are uncertainties associated with the tasks, then information processing is referred up the hierarchy to a level where an overall perspective exists. This is the traditional exception-based hierarchical decision-making. Such hierarchical decision-making can handle only a limited amount of uncertainty. If the uncertainty exceeds the capacity of the hierarchy, then targets or goals have to be set for the various tasks, making them somewhat independent. The coordination has moved from a procedure orientation to a results orientation. Following March and Simon (1958), the mechanistic model increases information processing capability to obtain integration of interdependent activities by coordination using rules or programs, by hierarchical information processing, and by coordination by targets or goals. These alternative information processing activities are incomplete and costly, but can add value in effectiveness and efficiency of the organization (Arrow, 1974).

Organizations can either reduce their need for information processing or increase their capacity to process information (Galbraith, 1974). The need for information processing can be reduced by increasing slack resources: just-in-time (JIT) inventory requires precise coordination; buffer inventory is an alternative. Buffer inventory replaces the need to process the information required for JIT. Information processing needs can be reduced by creating self-contained tasks; for example, a two product firm can create two self-contained single-product divisions that need not communicate. Of course, this strategy of reducing the need for information processing may incur high opportunity costs from loss of coordination. Buffer inventory is expensive, and JIT eliminates that cost but may increase other costs. JIT increases the likelihood of not having raw materials available when they are needed, which incurs an opportunity loss. Single-product divisions may ignore interdependencies in production or marketing, which may be costly in lost opportunities. Thus, reducing information needs must be balanced with the returns to coordinating the activities.

An alternative approach for the organization is to increase its capacity to process information. In a hierarchical organization, the hierarchical processing of information can be increased by investment in a vertical information system. The demand for information processing capacity that arises from uncertainty frees the organization to be able to react to unforeseen events. In a highly specialized organization, it may be difficult to have all specialized groups react in a coordinated manner. An information system may increase the speed and amount of information that can be exchanged. The introduction of satellites, information computer networks, the Internet and integrated CAD-CAM systems can increase the information processing capacity of the organization. The type of feedback and amount of information may require face-to-face communications. Information processing capacity also can be increased by creating lateral relations. Direct contact, liaison roles, task forces and permanent teams are examples of strategies that will increase information processing capacity.

The development of new information technology, theories of organizational learning and knowledge management require a revisit of Galbraith's strategies. Interactive information networks, multimedia systems, and generally the speed and amount of information that can be processed should be taken into account. When Galbraith developed his strategies, a liaison role was required; now a multimedia interactive information system may be the answer.

Following Arrow (1974), we argue that the definition of information is qualitative. This means that information also is related to the value of the information, issues of ethics, etc. The cost of information and information channels is an important consideration in designing effective and efficient organizations. Organizations are increasingly information processing entities; both the information processing demands and the capacity have increased recently as the cost of information processing has decreased, i.e., as information processing becomes less expensive, more information is used and alternative mechanisms are reduced. There has been a reduction in slack resources, perhaps a slight increase in self-contained units, a large investment in information systems, and a large increase in lateral relations, which has led to "leaner and meaner" organizations: less inventory, less equipment, and fewer employees, particularly middle managers. Those who remain use information much more quickly and efficiently. This introduces the issue of information of the human resources in the organization. Many organizations have invested in the technical side of knowledge management systems without getting the benefits often because the human side was neglected. Here we introduce a model that specifically incorporates both the management and the employees in the information processing model.

The information processing view of organization thus underlies many of the arguments in this book. It gives managerial substance and relevance to the contingency theory propositions in strategic organizational design.

Neo-Information Processing as a Basis for Organizational Diagnosis and Design

The information processing view of organization is well established, and has been the basis for further development in the generation of organizational models (Carley, 1995). Burton and Obel (1984) simulated the decentralized firm as an information processing model to investigate the effects on alternative information and decision-making structures. The Virtual Design Team (VDT) (Jin and Levitt, 1996) is an information processing model of a project organization where individual actors analyze, communicate and make decisions. The actors have different information processing skills and information channels as well as different communication media. Organizational behavior can then be simulated to see the effect of deployment of organizational resources, team building, and organizational structure. The original VDT approach modeled large but relatively routine projects, and has recently been extended to model semi-routine fast-pace-project organizations (Thomsen, 1998). In these models information processing is described over time based on the individual's information processing capacity. Additionally, the fit between the information processing capacity and information processing demand over time is measured. This is the micro view of the information processing approach to organizational design for a specific set of organizations.

Knowledge Management and Organizational Learning

In a recent issue of *Organization Science* (2002) devoted to "Knowledge, Knowing and Organizations," Birkinshaw, Novel and Ridderstråle (2002, p. 274) show strong association between dimensions of knowledge and organization structure - partial support for the "fit" hypothesis in contingency theory.

Knowledge management is an organizational design issue to improve the effectiveness and efficiency of an organization and its people by sharing knowledge and information. Argote and Ingram (2000) argue that knowledge transfer can be the basis for competitive advantage. In order to share the information, it is codified, stored and communicated, interpreted and applied. This is the continuing creation of a knowledge base and its application by the organization. It is an information processing task.

Knowledge management involves a number of difficult tasks. The codification of knowledge is a significant issue. Tacit and implicit knowledge by definition can not be codified; but we spent a good deal of effort to transform the valuable tacit knowledge into explicit knowledge; this requires a common language among the users (Arrow, 1974). With a common language and understanding we share the knowledge. The storage and communications require a classification scheme for retrieval so that individuals who do not know, but want to know can find the relevant knowledge. Having the knowledge in a coded form is not sufficient; applying it to a real situation also requires tacit knowledge and interpretation. Knowledge management includes the knowledge base, but involves a deeper understanding including the value of the knowledge in the particular organizational context.

The design of a knowledge management system is an organizational design problem where the focus is directly on information and how information will be processed and utilized by the organizational members. Our model suggests that the development and design of a knowledge management system should not only include the information issues of coding, storing and communication, retrieval and interpretation, but also the goals and strategic issues of management style, climate, size, environment, strategy and technology and their relation to the organizational form and properties. Knowledge management is the task of the organization, and not a technical add on. The CKO (chief knowledge officer) focuses on the design of the organization and the application of knowledge.

Organizational learning is the improvement in the knowledge base of the organization - the continuing update of what we know and how to apply it in the organization. Huber (1990) uses knowledge and information interchangeably; it is to know how to do something that broadens the possible ranges of behavior. Organizational learning involves four constructs: knowledge acquisition, information distribution, information interpretation and organizational memory. We can then distinguish between knowledge and learning. Knowledge is what we know; learning is adding to and changing what we know. Learning is the dynamic or change process for knowledge. Argote (1999) measures the organizational learning using classical learning curves to measure increases in efficiency. Organizational learning is then realized when the organization has new knowledge about how to do something and implements that process. Knowledge management is developing the knowledge base of the organization; organizational learning is improving that knowledge base for increased value to the organization. Knowledge management is the continual improvement of the information processing capacity and capability of the organization to realize its goals for being effective and efficient. The design of a knowledge management system is the design of an information processing organization, which includes the total design of the organization.

In this book we take a macro view of the information processing approach to organizational design - a neo-information processing view - yet build on the ideas information processing of the individual actors, or agents in the organization. Here the information processing demand on the organization is compared over time to the information processing capacity of the organization. The unit of analysis is the organizational unit. Additionally, we posit that differences between the information processing demand and capacity may arise due to more general phenomena than uncertainty as presented by Galbraith (1973; 1974). Differences between the information processing demand and capacity may arise due to uncertainty, but they may also arise due to strategic choices. If an organization decides to change its overall policy for meeting customer orders from a one week guaranteed delivery to delivery within 48 hours, the information processing demand is changed dramatically. To meet such a demand may require changes in the information structure and organizational structure, and different human capacities may be needed. An advanced information processing system may be necessary, formalization of procedures may be installed, limitations in the type of orders accepted may be decided, and more delegation to better trained individuals may be needed.

The neo-information processing view is a frame to interpret research findings whether they were generated from this basis or not. Thus, we shall consistently interpret the positive science research literature on the organization through information processing lenses. We then utilize the research literature to support and enhance the information-processing model of organization for design purposes. Our goal is not to simply restate and review the literature as it exists; our goal is to interpret the vast knowledge base of strategic organization theory for our own purposes: organizational diagnosis and design utilizing an information processing view of organization. This is the neo-information processing model of organizational design.

To develop a neo-information processing model of organization, we posit that the basic work of an organization can be seen as information processing: observing, transmitting, analyzing, understanding, deciding, storing and taking action for implementation. In the literature these issues may be labeled with other words like learning, tacit vs. explicit knowledge, knowledge management, and data mining.

Information is the nerve system of organization and that is our focus. We assume:

- Individuals are rationally bounded (March and Simon, 1958).
- Information is costly to gather, transmit, store and analyze (Arrow, 1974).
- Information activities are not performed with perfect reliability; information is then scarce, imperfect, and associated with uncertainty; and thus, coordination is problematic (Galbraith, 1974).
- Individuals have self-interest as well as organizational concerns. (Cyert and March, 1963).

These assumptions are the basis for an information processing view of organization and are incorporated in the March and Simon's (1958) classic work on organization, Galbraith's (1974), organizational design framework, Arrow's 1974 limits on organization and finding the balance of costs and value of information, Williamson's

(1975) transactions economics cost paradigm, and Nadler and Tushman's (1988) information processing model of organization, to name a few. Information processing provides an interpretative framework for describing the world around us, the essence of positive science. Further, the mechanisms of organization for design purposes are information processing choices.

The information processing view is the basis of our thinking about organization. For example, the observation that a large organization tends to be decentralized can logically be supported from the information processing view. From an information processing view, we argue that large organizations need to decentralize in order to deal with the information processing demands and mitigate for possible top management information overload. Decentralization may also be decided from motivational reasons. Better-motivated employees have higher information processing capacity. We integrate such views in our model by introducing organizational climate dimensions as well as incentives.

STRATEGIC ORGANIZATIONAL DESIGN

The goals and mission of the organization are the basis for the specification of what the organization should do. If the goal is to run a profitable gourmet restaurant, then its overall style and the type of food and wine it serves are what the organization does. The specific tasks by which the food is prepared and served depend on the structure. From the goals, missions and the particular environment a strategy has to be found. The strategy could be to have a wide variety of different choices on the menu. But variety also could be obtained by having only a narrow choice each night and changing the choice every night or every week. The strategy affects the way the restaurant is run, which affects structure, decision-making, standardization, and so on. Some specificity about what the organization does is the basis for the design, but the level of specificity depends on the particular organization. The next step in the design of an organization is to specify the values for factors that affect the choice of the right organizational structure and then have decision rules that can be applied.

Generally, the point of departure for the strategic organizational design is the goals and mission of the organizational unit. The organizational unit may be an independent owner-run organization, a corporation, or a subsidiary or department. In each case, the design process begins by asking what the goals of the organization are. From there on, the contextual situation can be specified. The level of detail is the choice of the designer. It requires a specification of the unit of analysis, the boundary of the organization as the interface between what is inside the organization and what is outside, i.e., the environment.

The unit of analysis: inside, boundary and outside of the organization

For a strategic organizational diagnosis and design it is very important to be clear about the choice of the unit of analysis - i.e., the organization, the boundary and the environment. In a large corporation, the unit of analysis could be the whole corporation or a department within the corporation. The corporation and its department are different units of analysis, different boundaries and different environments.

Traditionally, employees are inside and individuals who are not employees are outside. Later in Chapter 5 on size, we define the size of the organization as the number of employees, which also sets the boundary of the organization as employees as part of the organization and everyone else as outside the boundary. The boundary also sets "what" is inside and outside the organization, e.g., owned assets are inside the boundary of the organization. This is the accounting convention as well as the legal notion. March and Simon (1958, p. 89-90) note the legal boundary as fundamental, but also acknowledge its limitation for managerial purposes. The point is clear: the boundary of the organization is not given for all time and purpose, but can be set to meet a particular need or purpose. Traditionally, the managerial prerogatives of authority and responsibility followed the property rights for the organization. The ownership property "line" marked what was outside and what was inside and therefore what could be managed. The Porter (1980) model of the firm within an industry is an example where the boundary between the firm and the market is well defined.

Baligh and Burton (1982) argue that the managerial boundary of the organization and the legal boundary are normally not the same. The managerial boundary is a choice for management and should be thought of in information processing terms. Advertising is a direct intervention by management beyond the legal boundaries of the firm. (Automobile companies gather information and take action for automobile dealers, whether they own the dealership or not.) The managerial information boundary for the modern organization goes "outside" in many ways: gather environmental information, advertising, influence buyers, influence government and policy, monitor new product and process technologies, monitor and direct supplier.

New forms of organization, such as joint ventures, strategic alliances, virtual organizations and outsourcing, all share one common aspect; they break down the traditional boundaries and the simple notions of property rights and the associated management prerogatives and responsibilities. Later in Chapters 2 and 8, new forms will be discussed in more detail. In short, the new forms have complex boundaries of ownership, but more importantly, of managerial prerogative and complexity. Older notions of authority, responsibility, command and control break down and call for new attitudes and concepts. However, it is important that the unit of analysis be specified for the strategic organizational diagnosis and design. The unit of analysis should be determined from a managerial point of view, which may or may not coincide with the legal boundaries.

The strategic organizational design specifies the organizational groupings-the units in the organization and the relationships among the units. The organization wide information, measurement, control, and incentive systems are part of the strategic organizational design.

Our main goal is to design organizations that are effective and/or efficient for good performance. Fit is a matching process where the organization can be matched with the environment or vice versa. A good fit means better performance (Donaldson, 1987). Burton, Lauridsen and Obel (2002) demonstrated that those organizations that fit perform better. For example, as we discuss in more detail in Chapter 6, an organization may choose a flexible structure that fits a turbulent environment, or it may manipulate its environment to reduce the turbulence. We also have seen a number of smaller companies that deliberately decide not to grow because growth would require a more elaborate decentralized structure. In those cases, the CEO did not want to give up the position of influence and therefore controlled the situation so that growth would not occur.

This book develops a comprehensive theory of strategic organizational diagnosis and design. Our basic view is that an organization is an information-processing entity; our model is a synthesis of many partial theories of organizational design. Organizational theory is a positive science that focuses on understanding organizations. It is a multi-disciplinary science, in which separate disciplines have created their own distinct questions, hypotheses, methodologies, and conclusions. These various views do not necessarily fit together into a comprehensive and consistent view, nor do they necessarily offer recommendations on how organizations should be designed. It is not our purpose in this book to review the organization theory; many excellent books do that job (Hall, 1991; Scott, 1998). Our purpose is to develop an information processing approach to organizational design, which incorporates the research and experiential knowledge from organization theory and design. We have argued in various ways and sought strong support for our statements. We scrutinize the organizational theory literature and analyze organizations to develop a knowledge base to help students learn about the fundamentals of organizational design and help practitioners design effective and efficient organizations.

THE MULTI-CONTINGENCY DIAGNOSIS AND DESIGN MODEL

Organizational diagnosis and design is a normative science that focuses on creating an organization to obtain given goals. Design "is concerned with how things ought to be, with devising structures to attain goals" (Simon, 1981). Organizational design depends on organizational theory; its prescriptive purpose complements the descriptive function of the positive theory. The model developed in this book takes its point of departure in an information-processing organizational design framework. This framework is used to synthesize what we know from organizational theory. It tries to evaluate the normative implications of descriptive analyses from organization theory.

Figure 1.1. Organizational Context

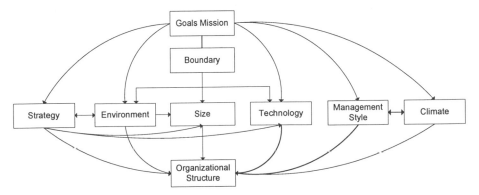

Organizational mission and goals form the basis for organizational design. The goals and mission determine the boundary of the organization relative to its environment. Goals and mission influence the choice of technology as well as size. These choices depend on the strategy. The strategy usually has the closest link to the goals and mission. These relationships are summarized in Figure 1.1.

The first step in the organizational design process is to establish the contextual basis from the organization's goals and mission. From this basis the multiple contingency relationships have to be established. The multiple contingency model states that the organizational structure depends on multiple dimensions in the contextual situation. It was obvious from the story about IBM and AT&T that environment, management style, technology, size, and strategy play an important role for the choice of a new organizational structure.

Figure 1.2. The Multi-Contingency Diagnosis and Design Model

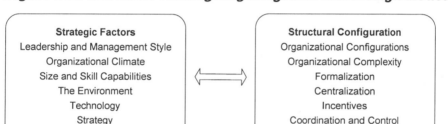

The multi-contingency concepts and relations are summarized in Figure 1.2. The diagnostic factors are on the left, and the design recommendation possibilities are on the right. An effective and efficient organizational design provides a good fit between the diagnostic factors and design properties and the structure of the organization. Additionally, we look at the multi-dimensional contingency models in a dynamic perspective, where the contextual factors and the organizational structure have to be aligned as changes in one or the other happen over time.

The contingency theory of organization is the basis for the knowledge base in our model. That knowledge base is gleaned from many literature sources. The knowledge from the literature has been translated into meaningful "if-then" rules and composed into a set of consistent rules. For example the rule that states, "If the organization is large, then decentralization should be high" is a consistent rule. Overall, there are more than 500 rules in the knowledge base as implemented in OrgCon, gleaned from a broad spectrum of the contingency theory literature. As depicted in Figure 1.2, these if-then rules relate the contingency factors of size, technology, strategy, environment, climate, ownership, and management preferences to the organizational design recommendations.

The if-then rules in the knowledge base should be a clear translation of our knowledge using consistent definitions of terms. The rules must also provide an integration of the many partial theories. We have chosen to use the structural properties shown in Figure 1.2 and then translate the recommendations given in the literature into these concepts using the neo-information processing perspective as a means in the translation. This provides a basis for the development of a multiple-contingency model. Ketchen et all (1997) found in a meta analysis of 40 contingency studies that the studies with the broadest set of variables had the highest predictive power. It is thus important to develop a comprehensive and relevant set of variables that should constitute a multiple-contingency model. We have followed similar attempts to combine knowledge from the literature, such as Daft (1992) and Robbins (1990). In section 1.6 this process is discussed in more detail.

Fit Criteria for Designing the Knowledge Base

Good fit leads to high performance organizations. Fit is an organizing concept for the creation and development of the knowledge base, i.e., the diagnosis and design rules. The development of the multicontingency model for organizational design rests upon the proposition that a fit among the "patterns of relevant contextual, structural, and strategic factors" will yield better performance than misfit will (Doty et al., 1993). Our model builds on Miles and Snow strategic fit model (Miles and Snow, 1978), Mintzberg's structuring model (Mintzberg, 1979), Galbraith's star model (Galbraith, 1973; Galbraith, 1995), Nadler and Tushman's congruence model (Nadler and Tushman, 1984), Miller's environment, strategy and structure model (Miller, 1987), and Meyer, et al's configurational model, among others (Meyer et al., 1993). There is a large literature that tests alignment and fit relations for a few variables, whereas our model is a comprehensive model involving many variables.

The development of a multi-dimensional model form requires a synthesis and integration of concepts and ideas to create definitions. The challenge is to create a knowledge base system that utilizes known theory for a given situation to suggest appropriate organizational design recommendations. To meet all of these goals, the knowledge base must fit together across a number of dimensions (Burton et al., 2002).

Here, we present four fit criteria that are necessary in order to obtain a useable system (Drazin and Van de Ven, 1985; Khandwalla, 1973; Miller, 1992). In Figure 1.3, the multi-contingency organizational diagnosis and design model is presented pictorially. This picture is a convenient means to illustrate the four fit criteria needed for a useable organizational design consultant. Those criteria are:

- Contingency fit.
- Strategic fit.
- Design fit.
- Total design fit.

In Chapter 9 we present a number of misfit statements. Misfit and fit are complementary concepts; each implicitly implies the other. Donaldson (2001) indicates that a "misfit produces a negative effect on organizational performance, structural change to a fitting structure is not usually immediate" (p.15). He discusses misfit as a condition that calls for the organization to move back to fit (p. 165). Fit recommendations are triggered by misfit observations. For example, a fit recommendation that the organization should be decentralized is operative only if it is not currently decentralized. Fit recommendations are "should be" statements; misfit statements are "should not be." Other terms have been used for misfit: deviation, misalignment, incongruence, out of kilter, incompatibility, gap and so on.

The fit - misfit concepts are fundamental in design. Alexander (1964), an architect, focuses on the design process, "the process of achieving good fit between two entities as a negative process of neutralizing the incongruities, or irritants, or forces, which cause misfit...though equivalent from a logical point of view, from a phenomenological and practical point of view, they (fit and misfit) are different" (p. 24). He states the design problem, "the form is the solution to the problem: the context defines the problem" (p. 15) and further, "The form is a part of the world over which we have control, and which we decide to shape while leaving the rest of the world as it is...Fitness is a relation of mutual acceptability between these two" (p.19).

Contingency Fit

Each if-then contingency proposition must be consistent with the neo-information processing framework and also be consistent with the contingency theory and represent our knowledge well.

Figure 1.3. Strategic Organizational Diagnosis and Design Fit

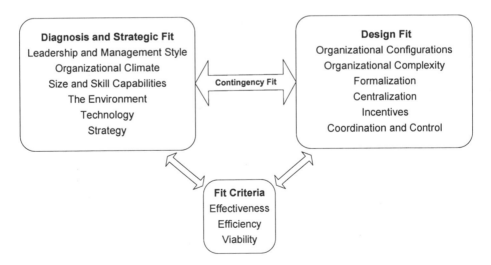

A rule that states, "If the organization is large, then the structure should be centralized" is not consistent with what we know about effective and efficient operations and should not be a part of the knowledge base. Further, these design propositions must be consistent for each contingency across all design parameters. This is labeled contingency fit in Figure 1.3. The contingency fit criteria can largely be achieved through careful attention to the contingency theory literature and translation of that knowledge into appropriate if-then statements. Most empirical research has emphasized how to obtain contingency fit for one or more contingency factors. The studies reviewed in Chapters 3 through 8 show that contingency fit is important for the performance of the firm.

Strategic Fit

The strategic fit requires that the organization's situation is internally consistent. That is, the strategic facts, which give answers to the "if" questions, must make sense together. For example, a dynamic and uncertain environment and a routine technology do not fit. There is no recommended design for this situation, and it is identified as a strategic misfit. Something must be changed. Here, the routine technology may be changed to a less routine technology. This misfit says that we cannot recommend an organizational design that can keep a routine technology viable in the face of a dynamic environment. The literature on strategy is abundant with research on this issue (Quinn et al., 1988).

The Scandinavian Airline System (SAS)

The Scandinavian Airline System (SAS) is used as an illustration in each chapter in this book beginning with an introduction to the company and its history. SAS is the result of a joint Scandinavian cooperation on air traffic. It was created in 1951 as strategic alliance between three national airlines. With the SAS agreement, the three Scandinavian airlines gained a stronger negotiating position in bilateral agreements, which was necessary due to the 1944 Chicago agreement. Copenhagen's position gave SAS access to the American market and later to the Japanese market (Buraas, 1972).

The International Air Transport Association (IATA), which was the air companies' own organization, was established in 1946. Its main task was to coordinate ticket prices on a global scale. Through membership in IATA, SAS could consider the prices fixed during the period between two IATA meetings (Buraas, 1972). Those cooperating in IATA focused on keeping prices high. SAS thrived on technical competence; pilots, technicians, and system managers were its heroes (Hofstede, 1997). Consequently, during the 1950s and 1960s, SAS offered its passengers modern equipment, which was frequently replaced.

This was SAS from its establishment in February 1951 until the mid 1970s. Then, in the mid-70s, the oil crisis struck a blow at the entire airline business. Known oil reserves were expected to last only until the end of the century. Oil prices that had followed normal price increases now rose explosively. In the short run, the world economy was so severely affected that the market for air traffic changed fundamentally. The turbulence continued with deregulation, which introduced competition among airlines, first in the United States in 1978 and then later in Europe.

SAS unexpectedly found itself in a completely new competitive situation for the first time since 1951 (Obel, 1986). Operating costs increased sharply, but revenues did not. With a radically changed environment, a substantially worse economy, and a stagnant market, SAS had to start thinking along new lines. However, the large, bureaucratic organization structure could not easily be turned around. Cost consciousness was not enough, and the board of directors recognized the need for a results oriented company, and a new CEO was hired.

When Jan Carlzon left the Swedish Linjeflyg and became president of the airline company SAS in 1980, a new era began. Carlzon changed SAS from a product oriented to a market oriented organization. Carlzon's goal was for customers to choose SAS because SAS was the best alternative. This worked well for some years until new problems emerged due to a failure in a diversification strategy.

Jan Stenberg became the president in 1994 and was faced with new financial problems. He developed a new strategy with a defined set of core activities and, at the same time, SAS was allowed to become broader as the company became a member of the STAR alliance that includes airlines such as Lufthansa and United.

In 2000 Jørgen Lindegaard replaced Stenberg. In Lindegaard's first year in office, he faced several crises including an EU cartel case, the U.S terrorist attacks on September 11, 2001, and a crash in Milan, where all the passengers were killed. The bilateral 1944 Chicago agreement - one of the reasons for creating SAS - was expected to be changed within the EU in the autumn of 2002. Consequently, another phase of deregulation was facing SAS. SAS has thus gone through a number of strategic and organizational changes. Throughout the book, we will return to discuss SAS in more detail and relate is development to the multi-contingency model. You may also get much more information about SAS vision, strategy, organization, and financial results as well as historical milestones from the SAS homepage: www.scandinavian.net.

   1.   Create four scenarios in OrgCon (Carlzon-a, Carlzon-b, Stenberg,

and Lindegaard)
2. Fill as much information as you can. You will get more information in subsequent chapters.

Strategic misfits have to be dealt with. Fit needs to be established. Strategic misfits may appear due to changes in the environment and thus be exogenous to the organization. Misfits may also appear because of management decisions. A manager may change the size and create misfits, which then have to be adjusted to.

The control of strategic fits and misfits is a key to organizational success. In the organizational life cycle model where the organization evolves from the small start-up organization to the multi-market-multi-product organization, the organization has to pass through many faces of misfits. Only those organizations that manage to create the proper misfits and then resolve these will be successful. In Chapters 9 and 11 we return to the discussion on how to manage misfits over time in a world where change is the norm.

Design Fit

For each design property and form, the recommendation set of if-then propositions must fit and be in balance. For example, a design recommendation that the organization should be decentralized can be driven by a number of contingencies. Leadership style, climate, size, environment, technology, and strategy all may suggest decentralization; the recommendation is strong. However, it is more likely that there are design propositions that suggest decentralization and others that suggest centralization. Here, the design propositions must be in proper relative balance to obtain an appropriate recommendation. For example, the leadership style may lead to centralization and the organizational size to decentralization. The issue is to weigh the two contrasting recommendations. The certainty factor (cf) approach[1] helps access the design fit. If the recommended design recommendation has a high certainty factor, it is likely that there is a design fit. Also, the cf's on all possible recommendations can be inspected. If there are many different recommendations all with high

---

[1] The certainty factor principle will be discussed in a later section.

cfs, then the possibility for a design misfit exists. The organizational theory literature itself includes hypotheses on such possible misfits. For example, if formalization is low, then incentives should not be based on process and application of standards, and procedures. Khandwalla (1973) showed that design would lead to higher performance.

The concept of design fit is associated with the notion of equifinality (Doty et al., 1993). The strategic contingency factors may recommend multiple acceptable properties of the organizational structure but not all combinations are acceptable. For example, it may be recommended that centralization should be high or low and that formalization also should be high or low. However, only two of the four possible combinations may be acceptable. Most likely only an organizational structure with a high centralization and a low formalization or an organizational structure with a low centralization and a high formalization will be acceptable; but high centralization and high formalization may work.

Total Fit

Total fit is the most demanding of all. It requires that the contingency fit, the strategic fit, and the design fit criteria have all been met. And now, the total fit criterion requires that the design recommendations fit together internally and, more importantly, fit the actual strategic situation. Miller (1992) discusses the problems of simultaneously obtaining contingency fit and design fit. The problems he encounters are partially due to different definitions of concepts in the various contingency paradigms. A problem that is solved in the model developed in this book. However, even with a strategic fit and a contingency fit, design fit may not be present. Each contingency relationship may lead to more than one design recommendation. A balanced combination then has to be chosen to obtain total design fit. Note also that total design fit is unobtainable if a serious strategic misfit exists. If the organization has put itself in a bad position, then no appropriate organizational design exists. This suggests a sequence in the way misfits are treated, as we will discuss in Chapters 9 and 11.

SYNTHESIS OF KNOWLEDGE BASES

The organizational design model follows the "if-then" format of production rules in expert systems. Since most knowledge in organizational theory and design can be stated as simple if-then propositions, it seemed natural to adopt such a representation. This book is an expert system in itself; the rules developed are put in a form that makes it relatively easy to analyze and diagnose organizational design problems.

To illustrate the procedure and model the next section takes the well-known models of Duncan (1979) and Perrow (1967) and restates them as rule-based expert systems. The issues of translation and composition are discussed, and a contingency theory rule-based knowledge base is briefly presented. Several partial theories of the contingency imperatives of size, strategy, technology, and so on are given. A comprehensive treatment of the various paradigms and their composition is given in the next chapters. The knotty problem of composing the several partial theories into a consistent and comprehensive knowledge base within an information-processing framework is analyzed. Unfortunately, the literature is rather silent on how to put the pieces together. Finally, we discuss the implementation of our rules in the OrgCon expert system.

Duncan (1979) and Perrow (1967)use different organizational theory models that can be translated into if-then rules. First, we demonstrate the translation of each into a knowledge base. Second, we consider how to put the separate knowledge bases together into a combined knowledge base. This illustrates the complexity of composing a knowledge base from a large, diverse, and sometimes inconsistent organizational theory literature. We begin by stating Duncan's (1979) model of environmental contingencies as a rule-based knowledge base.

Duncan's knowledge base can be stated as the following six if-then rules:

- If environmental complexity is simple
  and environmental change is static,
  then organizational structure is functional.

- If environmental complexity is simple
  and environmental change is dynamic,
  then organizational structure is mixed functional.
- If environmental complexity is complex
  and environmental segmentation is present
  and environmental change is static,
  then organizational structure is decentralized.
- If environmental complexity is complex
  and environmental segmentation is present
  and environmental change is dynamic,
  then organizational structure is mixed decentralized.
- If environment complexity is complex
  and environmental segmentation is absent
  and environmental change is static,
  then organizational structure is functional.
- If environmental complexity is complex
  and environmental segmentation is absent
  and environmental change is dynamic,
  then organizational structure is mixed functional.

Duncan himself presents the equivalent of the above rule-based knowledge base in the form of a decision tree rather than if-then statements. This knowledge base is in fact a small expert system. Consider the goal of this expert system; it is to recommend an organizational design that is stated as either functional, mixed functional, decentralized, or mixed decentralized. These are the possible answers to complete the statement "then the organizational structure is" as given in the six rules above. The recommended structure is thus dependent or contingent on the environment, as stated in the "if" part of each rule. Duncan describes the environment along three dimensions: complexity, change, and segmentation. Environmental complexity is either simple or complex, environmental change is either static or dynamic, and environmental segmentation is either segmentable or not. An organization can find itself in 2×2×2 = 8 possible environments. The knowledge base is the set of six if-then rules that links the environmental condition with the recommended organizational design, which is contingent on the organization's environment.

To illustrate Duncan's expert system, consider an airline company in the 1960s with a static and simple environment. Using the if-then rules in the knowledge base, we find that the recommended structure is functional. Today the environment is complex, probably not segmentable, and certainly dynamic. The recommended structure is now a mixed functional structure-functional with lateral activities. But if the environment were segmentable, then the structure should be mixed decentralized - decentralized with lateral activities. The recommended structure is dependent on the environment and is relatively sensitive to shifts in the environment; as the environment changes, the recommended structure changes as well.

We now turn to a Perrow's (1967) model of technology contingencies. Like Duncan's model, it can also be translated into a knowledge base of if-then rules. The goal is to recommend an organizational design from the choices: routine, engineering, craft, or nonroutine. The organization's technology has two dimensions: task variability, which can be routine or high, and problem analyzability, which can be ill defined or analyzable. There are then 2×2 = 4 possible technologies, which are the facts for the expert system.

Perrow's knowledge base can be stated as the following four statements:

- If task variability is routine and
  problem analyzability is ill defined,
  then organizational structure is craft.
- If task variability is high and
  problem analyzability is ill-defined,
  then organizational structure is nonroutine.
- If task variability is routine and
  problem analyzability is analyzable,
  then organizational structure is routine.
- If task variability is high and
  problem analyzability is analyzable,
  then organizational structure is engineering.

Perrow developed this knowledge base by describing and analyzing the real world. Here, these propositions serve as normative statements as to how organizations ought to be structured in order to become efficient. This approach takes the knowledge of positive science and uses it in a normative system that recommends how organizations should be structured. To take the step from a descriptive form to a normative form presumes a causal relation and further it is a desired relation that will yield a better-performing organization. To justify such steps, we have tried to make sure that the relation is found generally in the literature and not restricted to single study and that an argument for the causal relationship could be found using the information processing view of organization.

Above, Duncan's (1979) and Perrow's (1967) contingency models were translated into separate knowledge bases. In Duncan's system, the organizational design is dependent on the organization's environment; in Perrow's system, the organizational design is dependent on the technology. Each system is self-contained and independent. Yet the reader must somehow feel that each one is incomplete and limited. Can the two systems be put together so that the composed system is more complete and more practical?

Both the environment and the technology are important contingencies in design (Daft, 1992; Robbins, 1990). But the process of putting the two systems together is not at all obvious. Each approach has its own vocabulary and definitions, and the relations among the definitions are not clear. How do Duncan's functional and decentralized structures combine with Perrow's routine, engineering, craft, and nonroutine structures? Are Duncan's four organizational designs simply different names for Perrow's four designs? For example, is a functional organization the same as a routine organization, or are the two sets of recommendations independent?

Creating a comprehensive knowledge base is a design problem itself; it is a synthesis of the knowledge from many different sources. How do we put the pieces together so that the resulting design is a mechanism that meets its purpose?

Taken separately, Duncan and Perrow's systems individually are logically complete. Consider the nature of the Duncan system. There are six if-then statements. Each is stated without equivocation. For example, consider this:

- If environmental complexity is simple
  and environmental change is static,
  then organizational structure is functional.

This statement depends on two antecedents: environmental complexity and environmental change. Either the environmental complexity is simple, or it is not. And either the environmental change is static, or it is not. If both antecedents are factually met, then the organizational structure is functional or, interpreted normatively, should be functional. There are no maybes, probabilities, or equivocalities of any kind. If the two conditions are met, then the result is certain. If either condition is not met, then the statement does not apply; it is not "fired."

Duncan's six statements are identifiable; each applies or it does not, and it applies with certainty. Further, the six antecedent statements are mutually exclusive, so no two statements can be fired in a given situation. There cannot be a conflicting design recommendation. And further, assuming any environment can be categorized into one of the six probable environments, the Duncan system is comprehensive.

In brief, Duncan assigns all possible environments to six mutually exclusive categories, each of which has a certain design recommendation. Duncan says nothing about technology, size, managerial style or strategy. It is implicit that these contingent factors can take on any value. That is, Duncan's design recommendations do not depend on whether the size is small or large, the technology is divisible or not, the managerial style is hands-on or not, or the strategy is prospector or not. These contingencies are not relevant. Despite the reasonableness of Duncan's statement, there is much evidence that his system is normatively incomplete.

Similarly for Perrow, technology is categorized into one of our four mutually exclusive categories, each of which has a certain design recommendation. There is no consideration of size, strategy, managerial preference, or Duncan's consideration of environment. Perrow's approach, too, must be incomplete. Both Duncan and Perrow might argue that design was not their purpose. Each was describing the world: testing hypotheses in the best of positive science tradition. But our goal is to go beyond positive science and to use its knowledge to create a practical aid for designing organizations.

The synthesis of Duncan's and Perrow's knowledge bases can be simple or enormously complex. First, the simple approach is to consider the two systems as independent. That is, for any organization, the environment can be categorized according to Duncan and the technology according to Perrow. Then, a design recommendation could be a simple combination such as Duncan's recommendation of a functional organization plus Perrow's routine organization. But do these fit together? What does it mean? We also could have a functional organization and a craft organization. There are 6×4 = 24 possible design outcomes-each of which must be realizable. As for each system separately, these recommended designs are stated with complete certainty.

The second approach involves dependency and integration. One kind of dependency could be definitional. Are Duncan's organizational designs simply different words for Perrow's concepts, or are they different aspects of organization, as assumed above? Let us assume they are different aspects of organization but are not independent. Consider again Duncan's if-then proposition:

- If environmental complexity is simple
  and environmental change is static,
  then organizational structure is functional.

Let us assume that this statement is compatible with Perrow's routine organization but not compatible with a nonroutine organization. One approach is to simply add a third antecedent to Duncan's statement; this approach becomes unwieldy very quickly. A more fruitful approach is to recognize that Duncan's statement is not true universally for other conditions that may exist. That is, we hold the statement with a certainty less than total. But, by how much?

Unfortunately, the literature does not give excellent guidance, and there is considerable incompleteness and ambiguity about the strength of these dependencies. Our knowledge base relies heavily on the positive science literature and empirical studies that usually involve hypothesis testing: analysis of variance models, multiregression models, and so on. The explained variance is usually not high, and this is a major clue that the models explain only a small amount of the phenomena under study. Composing a consistent and comprehensive knowledge base from the empirical literature is difficult and requires additional validation.

Composing the Knowledge Base

The construction of a knowledge base for an expert system is a statement of our knowledge about organizational theory, diagnosis and design. The rules are chosen and put together to enhance the effectiveness and/or the efficiency of the organization. The Duncan and Perrow rule statements are clear and can be applied one by one, but they have yet to be composed into a consistent and coherent set knowledge basc. Each is individually clear, but can we put them together into a consistent and coherent system to obtain total fit? It is not obvious how to compose the two independent systems into an integrated expert system, even if we have resolved the problem of concept definitions. The composition of the rules is essential to the development of the knowledge base. There are a number of ways to put the pieces together. In the area of expert systems, fuzzy logic, Bayesian theory, and certainty factors (cfs) have been used to combine facts from a number of production rules (Harmon and King, 1985; O'Leary, 1996).

In our development we have used the certainty factor principle, which worked well in our OrgCon expert system. This is not to say that other systems would not function well. We have presented only the production rules in pure form in Chapters 3 through 8, that is, without associating certainty factors to the rules. In Chapter 10 we discuss how to combine the rules both formally and informally.

We now illustrate the use of the certainty factor principle to compose the knowledge base. We capture its essence with a few examples. The rules are meant to illustrate how we translated the knowledge from the literature into if-then rules for the knowledge base. We begin with size as a contingency factor and then consider other contingencies. In Chapters 3 through 8 we state more generally all the propositions that have formed the basis for our rules.

It is generally accepted that the size of the organization affects its structure. The literature is replete with support (Blau and Schoenherr, 1971; Pugh et al., 1969) and counter argument (Aldrich, 1972). We have taken this idea and translated it into rules that state that the sizc of the organization should affect its structure. To illustrate, one rule in our knowledge base is:

- If organizational size is large,
  then formalization should be high (cf 20).

The size factor is then a fact. An organization is large, or it is not. If it is large, then the recommendation is that the formalization of the organization should be high with written and well-defined jobs, regulations, work standardization and relatively less freedom on the job for employees. The certainty factor of 20 is a qualifier in the statement (certainty factor measures the degree of belief one has in the statement and can range from -100 to 100). A cf 20 implies that the statement has a relatively weak effect but cannot be ignored. A stronger statement would increase the certainty factor, and cf 100 would be total certainty. Negative certainty factors reflect disbelief. Hall, Hass, and Johnson (1967) argue that size is important but it is not the only determinant of structure; hence, we choose a relatively weak effect. Size does not only affect formalization but also organizational complexity, centralization, configuration, and so on, which are incorporated in the knowledge base.

The technology is another determinant of the appropriate structure. As discussed earlier, Perrow (1967) investigated the relation between technology and structure. We include the effect of the technology on the structure. To illustrate, one rule is given:

- If technology is routine,
  then organizational formalization should be high (cf 20).

Formalization implies organizational rules and how they are followed in the organization. Robbins (1990) indicates that there is empirical evidence but it is qualified. Hence, the certainty factor is 20. The goal is to recommend the appropriate level of formalization for the organization.

The organization's strategy is another determinant of organizational design. In 1962, Chandler stated his now famous proposition that "structure follows strategy." In our knowledge base, we use Miles and Snow's (1978) hypotheses about strategy and structure. The organization's strategy can be categorized as: prospector, analyzer, defender, or reactor. A prospector's "domain is usually broad and in a continuous state of development." To illustrate, one rule from the knowledge base is:

- If the strategy is prospector,
  then centralization should be low (cf 20).

The rationale for the rule statement is that the prospector who has a large number of diverse activities requires a decentralized structure; if not, organizational activity tends to slow down because of decision bottlenecks. The statement is qualified since strategy is not the only determinant of the structure; hence, the certainty factor is relatively low, as in previous examples.

The environment is an additional determinant of organizational design. Duncan (1972) provided early empirical evidence. For our knowledge base, one rule is:

- If the environmental uncertainty is low,
  then the centralization should be high (cf 20).

Duncan's model is stated differently but provides support (Duncan, 1979). Robbins (1990)argues that centralization is possible when the environment is stable, since there is time to process requisite information; but again, it is a qualified statement.

Environmental hostility (Robbins, 1990) is also a determinant of the structure. Greater hostility requires greater centralization of the organization. To illustrate, one rule is:

- If the environmental hostility is extreme,
  then the centralization should be high (cf 40).

We are more certain that an extremely hostile environment calls for a unified effort; hence, the certainty factor is 40.

In the contingency framework, is management style important? Child (1972) argues that the environment and technology leave some discretion for managers; they have some choices concerning the organizational design.

We include rules that relate management's leadership style as a determinant of the structure:

- If the leadership style follows that of a manager,
  then centralization should be high (cf 40).

That is, this particular management style indicates greater centralization. A certainty factor of 40 indicates some qualification, but is a strong recommendation.

In the above rules, we have included the influence of size, technology, strategy, environment, and management style on the structure: formalization, centralization, and so on. Many of the other rule statements are similar in form.

Compound rules, which have two or more contingent conditions required, are also part of the knowledge base. This, in part, takes care of some interaction effects (Pennings, 1987; Schoonhoven,

1981). To illustrate, one rule requires both high organizational complexity and nonroutine technology:

- If the organizational complexity is high
  and the technology is not routine,
  then the horizontal differentiation should be high (cf 60).

If both conditions are met, then horizontal differentiation should be high with considerable confidence - a certainty factor of 60. (The Duncan and Perrow knowledge bases, as represented earlier, contain only compound rule statements.) The interaction effect may also be modeled by relationships between various structural dimensions.

In our knowledge base, we also include rules that caution the user. This is a strategic misfit. To illustrate, one rule is:

- If the strategy is prospector
  and the technology is routine,
  then there is a strategic misfit.

Intuitively, one can judge that a prospector strategy and a routine technology are not compatible. This is a diagnostic statement, but does not give an explicit design recommendation.

These rules are representative of the 450-plus rules in the knowledge base. The rules have been translated from the literature into a set of if-then statements. The if-then rules are statements that specify when strategic and contingency fit is obtained. It also contains rules on misfits which focus on diagnosis.

We now turn to the composition of these rules when the contingency factor influences have to be balanced and the conflict in influences resolved. The certainty factors are used to compare each statement and summarize our understanding and knowledge. The resulting recommendation states how design fit can be obtained and the strengths of the recommendation. A high certainty factor specifies that design fit has been obtained. A low certainty factor specifies that there may be design misfits. Consider two knowledge-base statements:

- If size is large,
  then decentralization is high (cf 30).

- If the strategy is prospector,
  then decentralization is high (cf 20).

Now let us assume that both antecedents are true - that is, the size is "large" and the strategy is "prospector," then the rule for combining the conclusion on the structure is given by:

- Decentralization should be high (cf 44),

where the calculation follows the rules from the MYCIN concept (M4, 1991). Here, we have $30 + (100 - 30) \times 20/100 = 30 + 14 = 44$[2].

The resulting conclusion on decentralization is stronger than either statement would be alone. These two rules concerning size and strategy have a positive relation with decentralization. Each contingency factor adds to the conclusion, but neither contingency factor is sufficient to make a certain conclusion or recommendation. Greater confirmatory information leads to stronger certainty of the recommendations. This is consistent with the concept that additional confirmatory information has positive marginal effect.

We also may have knowledge that tells when a relationship is not positive. For example, one rule might state, "If structural complexity is low, then decentralization should not be high." We can incorporate the negation by using a negative certainty factor:

- If structural complexity is low,
  then decentralization should be high (-cf 30).

By combining all three statements, we find that decentralization is high (cf 20). Thus, positive and negative effects can be appropriately included. The negative certainty factors are used to obtain design fit and thus sort multiple recommendations.

In this way, we are combining the various contingency theories into a unified and consistent statement of our knowledge. The literature is rather silent on the appropriate combinations (although there is considerable evidence for the separate contingency factors to determine the appropriate structure). Validation of the knowledge base requires a continuing revision through application in real situations and cases, and we discuss this issue later in this chapter.

---

[2] Note $CF = CF1 + (100 - CF1) \times CF2/100$

It is important here to note that the sequence in which the information is collected and applied does not change the results. If we first know that decentralization should be high (cf 20) and then later learn from another rule that decentralization should be high (cf 30), then the calculation is as follows:

$$20 + (100 - 20) \times 30/100 = 20 + 24 = 44.$$

The result is the same, independent of the order. It is, however, interesting to note that the marginal contribution to results does depend on the sequence. In the first situation the rule that stated that decentralization should be high (cf 30) contributed with cf 30. It dropped to 24 in the second case. This can be interpreted as a decreasing marginal utility of information. If you do not know anything about the organizational structure, then just getting some information is of high value. But if you already know from other contingencies that the centralization should be high with a certainty factor of 80, then the most likely recommendation will be high centralization no matter how much more you learn. Therefore, the importance of the extra confirmatory information has a lower value.

From an information processing point of view, each contingency factor demands a degree of information processing capacity. The environment may demand a high information processing capacity, while technology demands low information processing capacity. The result is not necessarily a medium demand for information processing. There is a requirement for high information processing capacity related to the environment and a low capacity related to technology. Medium information processing capacity in both areas may be both too little and too much. The certainty factor principle will find the dominant demand for information processing but also summarize demand in directions other than the dominant one. The expert system approach also lets you explain why a particular result was composed, giving the full background for evaluation. The certainty factor principle has proven to be an efficient tool for diagnosis and design. A trace of rules that are fired to provide a recommendation will show where the demand for information processing capacity originates. In fact the expert system can be designed without any use of certainty factors. The expert system can then find the rules that apply to the particular input and then sort these in groups that lead to similar conclusions. The final balancing can then be done by the individual based on his or her knowledge or experience or view of the particular situation. In our expert system, OrgCon, we have, however, used certainty factors, and that has proven useful and beneficial.

Our model can be used without complete information about the situation. But the more facts that are known, the more likely it is that the knowledge base will provide a recommendation that has a high certainty factor. A high certainty factor indicates that many contingencies pull in the same direction and that little conflict about the demand for information processing exists. A low certainty factor can have multiple causes. A low certainty factor may come from lack of facts or from facts that give conflicting recommendations. The lack of facts can either be contingencies where you have no facts or contingencies where you are unsure about the value. You may think that the environment is uncertain, but you may want to discount the effect of this information by assigning a low certainty factor to the premise in the rule. A certainty factor assigned to a fact in a rule is given in parenthesis as shown in the rule below:

- If the environmental uncertainty is low (cf 20),
  then centralization should be high (cf 20).

The above rule will result in a recommendation that centralization should be high (cf $(20 \times 20)/100 = 4$).

In most cases complete information is not available. As shown above, there is diminishing marginal value of additional information. In many cases it may be too costly to obtain complete information, even if it were possible.

In summarizing, contingency theory suggests that an appropriate organizational design is contingent or dependent on factors as size, strategy, technology, environment, and leadership (Pennings, 1987). There is a large literature (e.g., (Daft, 1992; Robbins, 1990)) that supports this view. Yet the composition and integration into a comprehensive contingency theory of organization is still ad hoc. In our model such a view can be incorporated by assigning a certainty factor to a fact. In the OrgCon system this is part of the input to the expert system. The certainty factor value associated with the fact then changes the strength of that contingency factor. If we assume that size is large with a certainty factor 50, the result from the above statement about the relationship between size and decentralization would change from a recommendation that decentralization should be high (cf 30) to a recommendation that decentralization should be high (cf 15 (30×50/100)). The certainty factor associated with a fact can therefore either represent a belief about the importance of the contingency factor, or it can represent that there are uncertainties about the true value of the contingency factor and therefore the strength is diminished.

We now turn to the strength of these design rules. A larger certainty factor indicates a stronger belief in the design recommendation; a smaller certainty factor indicates a weaker recommendation. A negative certainty factor represents disbelief. A certainty factor of 100 or -100 is a recommendation and zero is no recommendation. There are many factors that go into assigning a certainty factor. A certainty factor of 60 or greater is a very strong statement. Between 30 and 60, the recommendation is a strong recommendation but not mandatory and is not sufficient by itself to be a strong recommendation. Thus, other contingencies must further support to obtain a strong recommendation. Above 80, the recommendation is almost certain (Harmon et al., 1985).

The statement of the if-then rules and the composition of these is the specification of the way relationships are modeled and as shown in Figure 1.2. The certainty factor principle is used to specify the strength of the statement and the composition of the rules. A high certainty factor represents a fit - both a contingency fit and a design fit. The rules individually and together help us diagnose problems and then offer recommendations for design to increase the effectiveness and/or the efficiency of the organization.

THE OrgCon

So far, we have discussed only the creation of the rule based knowledge base that represents our knowledge of the theory of organizational diagnosis and design. We have implemented the above model in the OrgCon[3] expert system, which aids in the diagnosis and design of organizations.

OrgCon analyzes the current organizational structure using the many facts supplied by the user related to the functioning of the or-

---

[3] OrgCon is accompanying this book and can be fond on the enclosed CD. The OrgCon is also available from the website: www.ecomerc.com.

ganization. The strategic situation itself is analyzed and possible strategic misfits are given. The structure is then described in terms of the configuration and its properties. Based on the input, the system recommends the configuration and structural properties that give a good fit with the specified situation. Finally, the current and prescribed organizational structures are compared and possible recommendations given. The system allows the user to change input values and rerun the consultation, thereby providing a way to perform sensitivity analyses and comparisons of scenarios.

Analyzing organizations involves four steps:

- Specify the organization's situation in terms of the particular organization analyzed.
- Transform the specification under step 1 into general concepts usable for the design purpose.
- Provide general recommendations based on the specifications under step 2.
- Translate the general recommendations to specific recommendations for the particular organization.

This book is about the knowledge we have to move from step 2 to step 3. It shows how general knowledge from a number of sources can help guide the particular organization in its choice of a proper organizational design.

The expert system provides decision support in moving from step 1 to step 4. Decision support for the other steps can also be created, but are not discussed in this book. The expert system does not make any decisions. It provides a methodology and a structure in moving from step 1 to step 4 using the existing knowledge on organizational theory intelligently.

The OrgCon knowledge base system is in included with this book on the accompanying CD. The CD contains the OrgCon system, a training course, sample input files, case write-ups, and the Adobe© reader. The Adobe© reader is needed to read the case files. Run "setup" from the accompanying CD and the installation program will guide you through the installation. Activate OrgCon by clicking on OrgCon8. For further instructions run the training program. Read the Readme.txt file and the registration guide on the CD.

VALIDATING THE OrgCon[4]

The first working version of OrgCon was the result of literature models and new theorizing only. Once running, the system produced designs that were used to modify and improve it. Briefly stated, the basic knowledge base and its rules were developed by the authors independently, who then compared their rules. They used the organization theory literature extensively as well as their own theoretic work and models. The result was a set of diagnostic and prescriptive if-then rules mapping facts into diagnosis and design recommendations. These rules were reviewed for accuracy, consistency, and completeness by the authors. The creation of design rules is relatively straightforward. Crafting a consistent and coherent set of rules is an ongoing and difficult job, but analyzing the knowledge-base weights is an even more formidable task. There are two ways to consider the base weights, or certainty factors, attached to the if-then rules. These factors may refer to the relative strength of the various contingency statements made in the literature and the authors' models. Consider these two statements: "If size is large, then decentralization is high" and "If the strategy is analyzer, then decentralization is medium." The issue is to assign certainty factors to each of these rules to reflect their relative strengths or importance to achieving effectiveness and/or efficiency goals of the organization - both separately and collectively. The knowledge-base weight may refer to the strengths of the arguments that support the rule - that is, the quality and correctness of the rule. It is this latter component that is the object of an in situ test. Testing the inference engine was not a major issue. These issues were considered using cases and working with executives and students.

The validation process used in OrgCon relies on information obtained from cases, consultation with executives, discussion with experts, and executive M.B.A. student assignments. Over the years several thousand students have used the system. The validation process is shown in Figure 1.4.

---

[4] This section is based on Baligh, Burton, and Obel (1996)

Figure 1.4. Validation Process

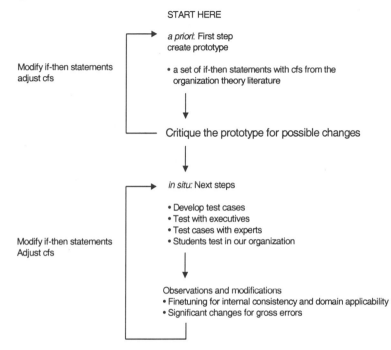

START HERE

Modify if-then statements
adjust cfs

a priori: First step
create prototype

• a set of if-then statements with cfs from the
  organization theory literature

Critique the prototype for possible changes

in situ: Next steps

• Develop test cases
• Test with executives
• Test cases with experts
• Students test in our organization

Modify if-then statements
Adjust cfs

Observations and modifications
• Finetuning for internal consistency and domain applicability
• Significant changes for gross errors

In the process of validation, we made small and large changes in the knowledge base rules, in the ease of use, and in the domain of applicability of OrgCon. Through application, issues are raised, but resolution depends on theory and its appropriate incorporation into OrgCon. It also forced us to reconsider the theory, thus necessitating further theory development. The validation process included three different types of tests: test cases, executives, and executive MBA students. More information about the actual validation can be found in Baligh, Burton, and Obel (1996).

SUMMARY

An information processing view of organization provides the basis for developing a normative approach for organizational diagnosis and design and the creation of a knowledge-base expert system to help us both learn about organizational design and make recommendations about what the organizational design should be - how to help make organizations effective and/or efficient. For the normative approach, we have utilized the positive science organization contingency theory as a source of knowledge within the design framework. The model is pictured in Figure 1.2 with the strategic factors - leadership and management style, organizational climate, size and skill capabilities, the environment, technology and strategy - on the left and the design elements - configuration, organizational complexity, formalization, centralization, incentives and coordination and control - on the right.

Four concepts of fit - strategic, contingency, design, and total - were introduced to construct the model and design organizations as shown in Figure 1.3. The model has been validated and implemented as the expert system OrgCon, which has proven to be a useful tool in teaching and consulting.

CHAPTER 2

What is an Organizational Design?

INTRODUCTION

When you examine the design of an organization for evaluating its efficiency and effectiveness, you can look at it in a number of ways. You may gather information about what the organization actually does - the goods and services it provides. You may be told who the boss is and who makes the decisions. You may be shown an organizational chart. Many companies also have explicitly stated objectives that drive their strategy. This information tells you about the design of the organization, the way the organization is put together, who does what, and who talks with whom.

The Organizational Design of SAS

The structure of Scandinavian Airline System (SAS) has evolved greatly due to market changes, new technologies, and new strategies. As discussed previously, 1975 was a significant turning point when the environment changed. Economic conditions for the airline industry since 1950 are summarized below.

1950-1975	1975-Present
Stable and rapidly growing market	Stagnant market
Slowly rising oil prices	Rapidly rising oil prices
Protectionist aviation policies	Liberalization of aviation
Cartel formation and market sharing	Competition, including price
Stable profitability	Mostly deficits

Despite the fact that general economic conditions changed around 1975, SAS did not change its strategy until 1981. In 1981, SAS found itself with declining productivity and overcapacity. Its image in terms of service and punctuality was declining. From 1975 onwards, the market became more competitive. The market was more segmented, and new entrants tried to exploit niches. There was an emphasis on both costs and services. Generally, the situation had shifted from an orientation to operations to an orientation to markets - a truly dramatic shift. Uncertainty in the environment was greatly increased, requiring greater information-processing capacity.

However, it was not until 1981 that SAS realized that its old strategies did not work in the new situation. SAS had operated from 1975 to 1981 with significant situational and contingency misfits, and contrary to previous periods of stable profit, losses were substantial.

In order to solve the crisis, the board brought in new management in 1981 and a new strategy was immediately developed for SAS. The main strategy was to make SAS "the businessman's airline", focusing on service and functionality.

Changes in the strategy were followed by a new organization. The old organization was a functional structure with a high degree of centralization and formalization. There was an emphasis on rules and programs for coordination with an authority and control hierarchy to ensure proper implementation and deal with exceptions, which were kept to a minimum.

From the mid eighties and onwards, deregulation increased the number of competitors, and price competition became more intense than ever. In 1980, SAS was badly suited to take on price competition because of its high fixed costs. As a means to differentiate its services SAS introduced the total traveling concept, offering customer's a total traveling package which, in addition to the plane ticket, included ground transportation, hotel bookings, etc. The total traveling concept necessitated wide spanning investments in many non-airline businesses. In 1986 SAS reorganized its businesses into five independent business units: the airline, SAS Service Partner, SAS International Hotels, SAS Leisure and SAS distribution. The rationale was that each of these businesses faced very different strategic demands and, therefore, was required to have its own management team to allow more aggressive business development. The new SAS organization was further decentralized and attempted to create relatively independent units that did not require a great deal of coordination. The SAS airline and SAS International hotels were treated as relatively independent businesses-that is, self-contained tasks. SAS had thus backed away from an earlier strategy of integrating the airline and hotel business into a single identity for the customer.

In April 1994, Jan Stenberg, former CEO of the Swedish Ericsson, succeeded Jan Carlzon. With the appointment of Stenberg, a new strategy was formulated. From the very beginning, Stenberg carried through a severe rationalization program, including the selling off of many of the company's side businesses, among others, Diners Club Nordic franchise, SAS Leisure, SAS Service Partner and several hotels. Contrary to Carlzon, Stenberg aimed at satisfying customers' demands for frequent departures by making alliances, rather than mergers with foreign airline companies. To carry through his rationalization program, Stenberg found it necessary to increase the level of centralization of the company.

The deregulations as well as the many strategic alliances increased SAS's information processing requirements. SAS was recentralized to initiate the changes needed to reduce costs, but at the same time, the company was dependent on the innovativeness and adaptability enabled through decentralization. Although SAS was made leaner and more centralized with the appointment of Stenberg, the structure and operations of SAS's three parent companies remained unchanged. SAS could still be described as a mixture of a divisional and functional structure. The lateral communication of the functional structure allows for the informal communication necessary for innovation and adaptation to changes in customers' wishes and demands. The divisional structure and its higher level of centralization allow the top management to control expenditures and assure the company's efficiency. As a further means to coordinate the actions of top management and front line personnel, the company is making increased use of professional managers. The role of professional managers is to have the necessary professional insight to understand top management's decisions, and effectively communicate the implications to front line personnel. At the same time, they communicate front line individuals' ideas as well as market trends to the top management.

When Jørgen Lindegaard became the president of SAS in 2001, he announced ideas on expansion. However, the September 11 event in 2001 and the SARS epidemic in 2003 reduced the demand for aviation services. Tougher competition and reductions were the result. Plans for reducing the number of employees by more than 5000 were made. At the same time, SAS bought a number of smaller European airlines as Braathens and Spanair. The SAS group was consolidated into a divisional configuration with 5 divisions: Scandinavian airlines, Subsidiary and Affiliated Airlines, Airline Support Business, Airline Related Business, and finally Hotels. The idea of the reorganization was also to put more decision responsibility on the divisions with results oriented incentives (SAS 2002 annual report).

1. In the different periods mentioned above, describe the organizational structure (configuration, centralization, complexity, formalization, coordination and control, and incentives).
2. What changes were made going from one period to the next?

In Chapter 1 we argued that for design purposes we want to understand how the organization uses information to make decisions, coordinate its activities, and implement what it wants to do. In this chapter, we will discuss a number of concepts that can be used to describe an organization for design purposes - the design side of Figure 1.2.

An organizational design is the specification of *configuration, complexity, formalization, centralization, incentives and coordination and control mechanisms*. We will review these concepts in this chapter. The specific measurement and operationalization can be seen in the way it implemented in the OrgCon system. Specifically, the help section in OrgCon provides such details.

ORGANIZATIONAL CONFIGURATIONS

In describing an organization, many people begin with the organizational configuration. The configuration is represented most frequently as an organizational chart. The configuration specifies the general principle for dividing work, breaking tasks into subtasks and coordinating activities. Thus, the configuration specifies the overall units that are the basis for making decisions and communicating with each other. The type of information that flows within an organization depends in part on the configuration. There are a large number of organizational configurations, and new ones emerge from time to time. The following basic configurations are the most common:

- *Simple*: A flat hierarchy and a singular head for coordination, communications, control and decision-making.
- *Functional*: Unit grouping by functional specialization (productions, marketing, finance, human resources, and so on).
- *Divisional*: Self-contained unit groupings into somewhat autonomous units coordinated by a headquarters unit (product, customer, or geographical grouping, including multi-national).
- *Matrix*: A structure that assigns specialists from functional departments to work on one or more interdisciplinary teams that are led by project leaders. Permanent product teams are also possible. A dual hierarchy (such as function and projects) manages the same activities and individuals at the same time. A three-dimensional matrix of product, function, and country is common in multi-national organizations.
- *Ad hoc*: High horizontal differentiation, low vertical differentiation, (high spatial differentiation), low formalization, decentralization, and great flexibility and responsiveness.
- *Bureaucracy:* Highly routine operating tasks, much formalized rules and regulations, tasks grouped into functional departments, centralized authority, decision-making follows the chain of command, and an elaborate administrative structure with sharp distinction between line and staff. The standard form is called a *machine bureaucracy*. If it has highly skilled professionals, high complexity, decentralization, and internal professional standards it is called a *professional bureaucracy*.

- *Virtual network*: An organization with a high spatial separation of a significant number of employees/units and an asynchronous mode of communication within the organization.
- *International configurations*: Special versions of the functional, divisional, and matrix configurations to consider international issues.

The specification of a configuration does not prescribe in detail how the organization's activities are divided. There are many ways to create a divisional or a functional structure. The number of functions or divisions and the basis for their differentiation determines the structure. Further, the first four categories can be combined with the last four categories. A functional configuration may be a machine bureaucracy or not; it may be virtual or not; and it may be international or not. These issues are discussed in later capsules and chapters. Next, we review each basic configuration more closely.

Simple Configuration

The simple configuration consists of a top manager and individuals. There may be little functional specialization and no well-defined departmental structure with departmental heads. Decision-making, coordination and control are usually performed by the top manager. Thus the center of the information flow is normally located within a single individual. In Figure 2.1, a simple configuration is pictured as an organization chart. Individual names are used to highlight the relations among these individuals, but the organization usually lacks task specificity, so functional activities are omitted. Normally, George, Jane, and Jens do what Peter tells them to do. In this configuration, the leader is dominant, and the number of vertical levels seldom exceeds two.

Small owner-run companies often choose this configuration in their early stages. This is the time when the founder wants to be involved in all the activities of the organization. The simple configuration has both, strengths and weaknesses. Decision-making and control rest with the top manager. Peter can personally coordinate all activities and assume that these activities meet his purpose for the organization. Peter assigns the activities or tasks directly to George,

Jane, and Jens and they have to report back to him. Each is expected to complete those activities that Peter assigns to each. Specific task assignments may evolve, but Peter can change them at will.

Figure 2.1. A Simple Configuration

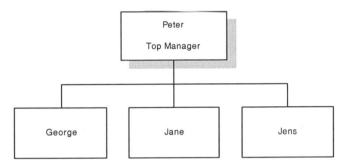

The weakness of the simple configuration is its heavy reliance on the top manager. He or she determines what to do and how to do it. If he or she decides well, then the firm succeeds; if not, it fails. There is little redundancy or fallback. This simple configuration is limited by the information processing capacity of the top manager. This limitation includes both the capacity to transmit and receive information as well as the capacity to make decisions based on the available information. As the firm grows in size or complexity of the tasks, the top manager can become overloaded and may simply be unable to cope with the information demands. This leads to a need for task specificity, i.e., who is doing what on a regular basis, so that the top manager needs to process less information. But that information must be specified in a form, that contains (or permits) the required coordination among the specialized tasks or activities.

Functional Configuration

The functional configuration has more vertical levels and more horizontal specialization than a simple configuration. There is a well-defined departmental structure. The departments are created based on the functional specialization in the organization. A distinction between line and staff departments is often made. For example, in a manufacturing firm, the line departments could be based on purchasing, production, and sales while the staff units could be financial, accounting, and human resources departments. Line departments are directly involved in the production of the firms' products and/or services while the staff departments are support functions. The functional organization is the most common configuration. Organizations that start out with a simple configuration may eventually change to a functional configuration. See Figure 2.2 for a prototypical functional organization.

Figure 2.2. A Prototypical Functional Configuration

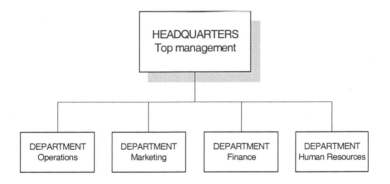

In a simple configuration, the top manager is involved in decision-making and therefore the coordination and control issues are simple; coordination and control are much more complicated in a functional configuration. Horizontal specialization creates distinct departments. Each manager of a particular department may see his department as the primary reason for the organization's competitive advantage, which can create conflict. However, horizontal specialization can be the basis for highly efficient use of resources, provided their use is coordinated. The dominant information flows tend to follow the hierarchy. The type of information that flows within the hierarchy is specified by the design of the functions. In the functional configuration shown in Figure 2.2, information related to coordinating production and marketing has to flow at the top level in the hierarchy. Delivery times, sales mix, productions schedules are examples of this type of information. Thus, the top management needs to be involved not only in strategic issues but, to a great extent, also in the tactical and operational issues.

A Functional Organization

Mack Trucks, Inc. has a classic functional configuration. Its top management and several functions are shown in Figure 2.3.

Figure 2.3. Mack Trucks, Inc.: The Functional Configuration

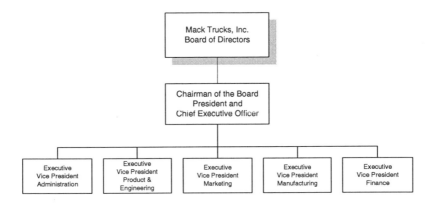

Source: *Conference Board Chart Collection* (1991).

Mack Trucks is an old company that builds top-of-the-line heavy-duty trucks in the United States. Mack has a small market share, which has not changed significantly for many years. It has made few technological innovations. Over time, truck design does change, but it evolves slowly. Mack trucks are more expensive than most competitive trucks, and their quality and reliability are high. Mack Trucks maintains a niche that it has developed over many years.

Functions: How many and what are they called?

The traditional functional organization has a number of departments, which have descriptive names: purchasing, manufacturing, sales, finance and personnel (see Figure 2.2). In terms of product flow, the raw materials are procured, and then manufactured into a product, which is sold to customers. The financial department handles the finances of the firm and the personnel department takes care of the employees of the firm. The organizing principle is that of function where efficiencies are realized through specialization of task. Each specialized function does its own part of the total and passes it on to the next function, where finance controls the resource issues and personnel deals with the employees.

In most functional organizations, there are approximately five functional departments. Mack Trucks, Inc. has five departments: administration, engineering, marketing, manufacturing and finance as shown in Figure 2.3. Here, administration includes personnel. The departmental names may vary from organization to organization, but some form of manufacturing or operations, marketing or sales and finance or accounting is typically present. These names are typical for product manufacturers. In Table 2.1, corporations in industries ranging from brewing to newspaper publishing have very similar functional department names. For engineering firms, typical functional departments are research and development, engineering and operations. Sales, personnel and finance functions are usually included at the firm headquarters. However, for service firms, the names and functions can vary widely. An advertising agency may have functions such as layout, creation, graphics, client relations, etc. The main principle is that tasks and individuals are grouped by specialized function where there are efficiencies to be realized, and further there is little substitution across functions, even though there is a workflow which requires coordination.

How many functions should there be? Five, plus or minus one or two is typical. There is no fixed number, although, most organizations do have a common set of functions, which must be performed as suggested above. Of course, accounting can either stand as a functional department or it can be included in finance or administration; accounting can also be outsourced. There is considerable variation. Similarly, most functions can be renamed and subsumed in others. The result is that there is no set number of functions.

The balance is that the efficiency of specialization should be realized, which suggests there should be at least a few functions. However, the more functions there are, the greater the difficulty of coordination among the functions. For most functional organizations, five, plus or minus one or two, functional departments, seems to work well.

The functional configuration remains the most prevalent in many industries. It is found in manufacturing, brewing, steel, pharmaceutical, telecommunications, transportation, public service, and hospitals, to name a few. Table 2.1 contains the name of several organizations, together with their functional departments. The functional department names vary among these organizations.

The coordination of these functions at the minimum requires that the products and their quantities should be the same for each of the functions. The firm should not sell fifty trucks, manufacture 100, and buy supplies for 150. Similarly, it should not make 100 if the market demands 150 at the going price. This coordination requirement is simply stated, but it is very complicated in reality. If there is one standard product, the above coordination description is sufficient. If there are hundreds or thousands of customized products, coordination requires enormous amounts of information. Those information flows usually are hierarchical, as shown in Figure 2.4, and the decision-making for coordination is done by the headquarters units or top management. Many functional based organizations use integrated IT-systems like SAP, Navision and Movex to increase the information processing capacity.

Table 2.1. Functional Configurations

Consolidated Edison of New York (2002)	Administration, Central Services Finance, Law, Public Affairs, Operation (Gas, Electric, Central)
Coors Brewing Company (1998)	Administration and Finance, Corporate Affairs, Engineering and Construction, Human Resources and Communication. Marketing, Plant Operations and Technology, Sales
Fiat Auto (1999)	Environmental and industrial policies, Logistics, Marketing and Sales, Manufacturing, Personnel and Organization, Purchasing, Quality, Research and Development
General Semiconductors, Inc (1999)	Corporate development, E-business, Finance, Information technology, General counsel, Operations
Mott's Group (1999)	Manufacturing, New business development, Manufacturing, Sales and marketing, Strategic planning, Supply chain

Source: *Conference Board Chart Collection.*

Figure 2.4. Product and Information Flow in the Functional Configuration

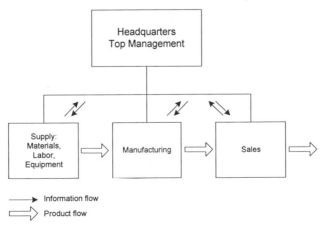

Galbraith's information processing strategies emphasizes standard operating procedures or coordination by rules or programs (Galbraith, 1973). Standardized products and standardized production processes reduce the need for information processing. As standardization decreases and customization increases, there is a need for greater information processing. The rationale for the functional configuration lies in the efficiencies of specialization; it is more efficient to have one group of individuals manufacture a product and another group to sell it than for everyone to do both functions. Different skills and interests are required, and not everyone can be excellent at both. This is the rationale for specialization-or the partitioning of the overall task into functional subtasks. The coordination issue becomes the essential short-time management problem-keeping the product flow in balance. In the longer run, management must match the functional capabilities with market demand for the product such that the firm is profitable and a return can be realized for the owners. This is a minimal requirement for a viable firm.

Coordination approaches vary across firms. In the above discussion, the headquarters, or top manager, balances the product, or work flows. This is central planning at the top. However, any one function could also serve in this role. For example, the sales function could forecast sales and then tell manufacturing how many of the product to make. Alternatively, manufacturing could set the production quantities and tell sales to sell them at the best price possible. These leader-follower coordination approaches are also used in functional configured firms.

Functional configurations have limitations. As the number of products and their customization grows, the coordination problem can become overwhelming as information processing needs escalate. Functional configurations also tend to lack flexibility and innovation, for several reasons. First, the organization may be information overloaded. This may be the case if changes in the environment make constant coordination across the functional specialization necessary. Second, no one may be directly responsible for innovation-particularly of new products. Both sales and manufacturing focus on today's products. Manufacturing may focus on cost reduction, but usually not on new products. Third, the functional incentives do not lead to innovation. Sales tries to maximize sales and revenue, and manufacturing tries to minimize costs. Both incentives are reasonable and easy to measure but may not lead to longer-run profitability and viability. The functional specialist's contribution to longer-term profitability is difficult to measure and hence, to reward. Additionally, product innovation itself requires coordination of the functional specialties. A new product has to please both manufacturing and sales.

In brief, the functionally configured firm does well in a stable environment that requires little change with a known and stable technology, that offers efficiencies in specialization, and that allows coordination to be obtained with reasonable information processing requirements.

Divisional Configuration

The divisional configuration is characterized by organizational subunits based on a grouping of products, markets, or customers. The units are relatively autonomous contrary to the units in the functional configuration. The aim in designing a divisional configuration is to minimize the interdependency of the subunits. Coordination of these subunits-called divisions-is very different from coordination in the functional configuration. In a pure divisional form the top management is not involved in operational and tactical issues, but concentrates on the overall strategic decisions (Chandler, 1962; Williamson, 1975). If top management engages in issues that are not of a strategic nature, the configuration is called a corrupt divisional configuration. The divisions themselves can have any configuration, but very commonly, they have a functional configuration as shown in Figure 2.5.

There are a number of different bases for the divisional configurations. Markets by product, geography, or customer can form the basis for a divisional grouping. The grouping also can be a mixture. However, most often the divisions are based on products or product groups. Usually a divisional configuration is best for a large organization where the relationship between the divisions is not very involved. Staff departments may exist both at the top and at the divisional level. This means that in many cases there is a mixture of divisionalization of the line activities and a functional basis of the support functions.

Figure 2.5. The Divisional Configuration

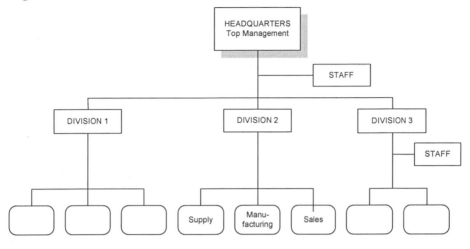

The divisions may be configured differently, as simple and functional configurations, for example. They also can be different businesses, such as automobiles and hamburgers. However, they all may also be alike. A retail chain with exactly the same type of store on each location is one such example. This particular version of the divisional configuration is called the carbon-copy divisional configuration (Mintzberg, 1979). The carbon-copy structure is often used in retail and service businesses when economies of scale are not important, either due to geographical dispersion or type of business. Retail chains and banks have identical shops and branches at different locations. In franchising setups, the carbon-copy divisional configuration is also often used.

Carbon copies may exist at the same location. The Danish advertising agency Bergsoe reorganized from a functional structure into a carbon-copy organization. As the number of projects increased, the horizontal coordination across departments in a regular functional organization became too difficult to handle. Bergsoe tried to solve the problem by installing a highly sophisticated computerized project scheduling system, which did not work. Projects did not meet deadlines, and people started to bypass the scheduling system. The situation grew worse. The solution was to break the company into small self-contained groups that had all the specialists needed to handle a project (recall one of Galbraith's approaches). Eight such groups were created, and all were located in the same office. To eliminate competition among the groups, a scheme for allocating projects and sharing profit was developed. Most of the eight groups were structured the same way. However, some were run as an adhocracy while others were simple or functional organizations. But they all did the same thing: no product specialization was allowed. For a given customer, each division dealt with all needs ranging from video production to advertising campaigns in newspapers and magazines. The reorganization changed both the information-processing demand and capacity.

A Divisional Organization

General Electric Company (GE) has a classic divisional configuration where each division is in a different business. The top management and the several divisions are shown in Figure 2.6. The several divisions report to the corporate executive office.

Figure 2.6. The Divisional Configuration of General Electric

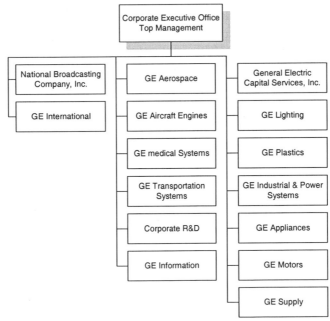

Source: *Conference Board Chart Collection* (1991)

The divisions are product or service based. GE is a very large corporation that develops, manufactures, and sells many different products and some services, such as broadcasting and finance. Over the last decade, GE has bought and sold many divisions. This configuration facilitates such actions.

The divisional configuration is the second most prevalent configuration. It has become more common as firms have become more product and customer focused.

The divisional configuration is also found across a broad range of industries: aircraft, heavy manufacturing, electronics, pharmaceutical and chemicals, construction, and consulting. Table 2.2 contains the name of several organizations, together with their divisions.

Table 2.2. Divisional Configurations

Bank of America (1998)	Global asset management, Retail banking, Wholesale banking
Bayer (2002)	Additives and rubber, Animal health, Biologicals, Coatings and colorings, Consumer care products, Diagnostics, Fibers, Industrial chemicals, Pharma, Plastics, Specialty metals,
Caterpillar Inc. (1998)	Building Construction Products, Caterpillar Overseas, Component Products, Construction and Mining Products, Engine, Financial Products, Human Services, Legal, Logistics and Products, North America, Asia-Pacific-Latin, America Commercial Divisions, Shin Caterpillar Mitsubishi Ltd., Solar Turbine, Technical Services
IT Group (2002)	Capital finance group, Commercial finance group, Strategic finance group
Hewlett Packard (2000)	Computing systems, Embedded and personal systems, Finance and administration, HP labs, Human resources, Imaging and printing systems, Strategy and corporate operations
Henkel KGaA (2001)	Adhesives, Cosmetics, Detergents/Household Cleaners, Finance, Hygiene Cleaners,
KPMG U.S (1997)	Assurance, Consulting, Financial services, Healthcare and life sciences, Manufacturing, Public services, Retail and distribution, Tax

Source: *Conference Board Chart Collection*

The divisional configuration is represented generally in Figure 2.7. The fundamental nature of the divisional configuration is the partitioning of the overall task into relatively autonomous units. Each divisional unit is self-contained in its operations. As discussed previously, the General Electric divisions have very few, if any, relations between the divisions, and each division is in a separate business. There are no flows of goods or information required among the divisions; each buys and sells into separate markets. However, all divisions are tied to the headquarters or top management through information, financial issues, and policy matters.

Figure 2.7. Product and Information Flow in the Divisional Configuration

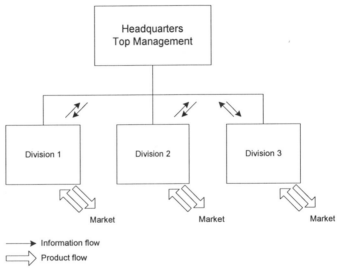

There are no coordination issues among the divisions. The aircraft engine division and the lighting division of GE do not need to coordinate production levels or sales plans. Consequently, as the number of decisions grows, the additional coordination requirements involve only relations between the division and top management.

Information processing requirements between the divisions and top management can be limited to financial matters and general policy issues, including strategic planning. The information processing demands of the divisional configuration are quite limited and additional information processing within a division due to increased products or customization does not increase the information processing demands at the divisional-corporate level. Consequently, the divisional configuration can grow by adding divisions or grow within any division without information processing difficulties.

Divisions: How many and what kind?

The variation in the number of divisions is large. There can be either very few or a large number. The basis for the divisions can be products, markets, customers or geographical location. The main principle is to have a close connection with the outside or environment of the organization. A customer division is organized around the buyer. Even the product organization is focused on the outside market and who will buy the product. Geographical or regional divisions are oriented to the local population, buyers and markets.

As shown in Figure 2.6, General Electric has some ten divisions. These divisions are quite varied in their foci from jet engines and financial services to television broadcasting. They are connected by a common ownership and financial outcome, but their processes, products or services and customers can be largely unconnected with each other. The coordination among the divisions is financial and strategic, but usually not operational. The GE headquarters can "manage" a large number of divisions as the information processing that is required at the headquarters is focused on finance and does not involve detail of operations or customers. The GE divisions are largely product based: lighting, plastics, engines, motors, etc. Some are service based as transportation, finance, and broadcasting.

Product or service is but one basis for a divisional configuration. Henkel KgaA is product based, but it is also customer based at the same time as shown in Table 2.2. Chemical products and cleaners are industrial products aimed at industrial users; cosmetics and household cleaners are household products aimed at the individual customers. Each requires a different focus, particularly for marketing and sales. Many service organizations will have a customer oriented divisional configuration. Banks and other financial institutions will have corporate divisions and personal divisions, sometimes these divisions will be called "client" divisions. The goal is to focus first on the customer and his/her needs and preferences, and secondly on the service and how it is provided. Bank of America as shown in Table 2.2 is an example of a service company, which matches its organization to the service and the type of customer.

Geography is a frequent division configuration rationale. Many international corporations, such as Philips, are organized by region or country. The country provides a local focus on the customer and the local customer needs as well as legal and other factors, which may be particular to the region or country. Recently, a number of country divisional European corporations have reorganized by product or customer within the EU region.

The divisional configuration gives a first order external focus to the organization. The focus can be the product and its market, the customer and his needs, or the region and its uniqueness.

The rationale of the divisional configuration depends on the efficiency of the coordination processes and rather minimal information processing demands.

Referring to Galbraith's strategic alternatives, the divisional configuration creates the divisions as relatively self-contained units. Each can operate independent of the other divisions without creating conflicts or loss of opportunity.

Additionally, each division can focus on its customers, markets, and products; specialization is by customer or product. It is an effective organization, as the division tends to do the right thing through its responsiveness to customers. However, the divisional configuration has limitations; most interdivisional efficiencies are overlooked and lost. Divisions can duplicate costly developments in research, underutilize or duplicate costly capital investments, and generally, lose opportunities of process specialization. Further, potential economics of scale can be lost. However, the financial performance of each division can be assessed using standard accounting measures of income and profits, and thus a performance-based incentive system can be put in place.

The divisional configuration can be quite flexible (Galunic and Eisenhardt, 1996). It depends on each division to pursue its market and technological opportunities. The division will normally focus on variations of its existing customers, markets, and products. New opportunities beyond the existing divisional activities are top management's responsibility.

The great advantage of the divisional configuration is the direct focus of each division on its customers, markets, products, and technology with measured performance goals. Each division is its own business.

Matrix Configuration

The three basic configurations discussed so far have been hierarchical. The matrix structure introduces a dual-hierarchy configuration; it incorporates the essential functional and divisional configuration on an organization simultaneously.

The matrix configuration can take many variations. One of the more usual ones is shown in Figure 2.8. Here the basis for the matrix configuration is a regular functional form with departments as described previously in this chapter. Together with the functional structure, the organization also has a project-based configuration. That is, a project manager who has responsibility for the project heads each project in the organization. Normally, he or she has to request the resources to carry out the project from the various departments. Each department head is responsible for ensuring that the resources are used as efficiently as possible.

Projects may be temporary, or they may be based on products, programs, or customers for a longer term. At the limit, a matrix configuration with only one department is called a *project configuration.*

Figure 2.8. A Matrix Configuration

	FUNCTION 1	FUNCTION 2	FUNCTION 3	FUNCTION 4
Project 1				
Project 2				
Product 1				
Product 2				
Customer 1				
Customer 2				

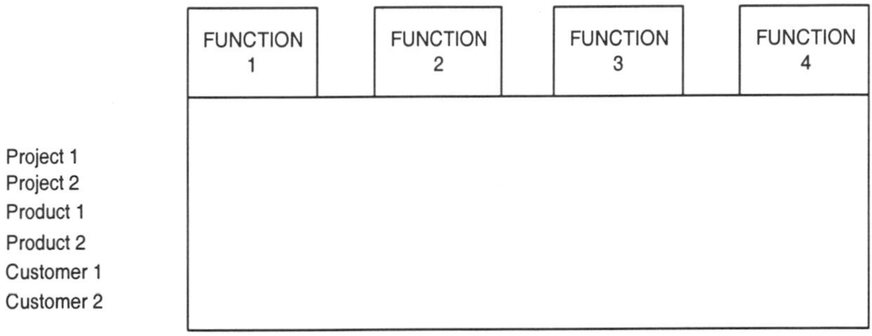

Matrix organizations are widely used in corporations and public organizations. Early use of the matrix form occurred in the production of airplanes and aerospace projects. The functions were research, development, engineering, assembly and testing. The projects were individual airplanes, rocket engines, guidance systems, and so on. Since then, the matrix organization (although the name matrix may not be used) has become quite common. The most common form is to have a functional organization with a project or product management. In multi-national corporations, the three-dimensional matrix of function, product, and region (country) is commonly used.

The matrix configuration addresses coordination issues in a functional organization when the coordination requirements are so high that the regular functional configuration is ineffective and the interdependencies between products are so many that a divisional configuration is not an efficient configuration. The goal is to obtain functional specialization and efficiency as well as project focus to realize an end objective effectively.

A Matrix Organization

Michael Allen Company is a service organization-a management consulting company. The functional dimension is the managing consultant specialization, and the divisional dimension is the practice client management. The service deliveries are various projects, and the matrix operations are obtained through project management-that is, a partner who is responsible for a given client provides services through project management. See Figure 2.9 (note that the functional and project dimensions are switched in Figures 2.8 and 2.9).

Figure 2.9. The Michael Allen Matrix Organization (1990)

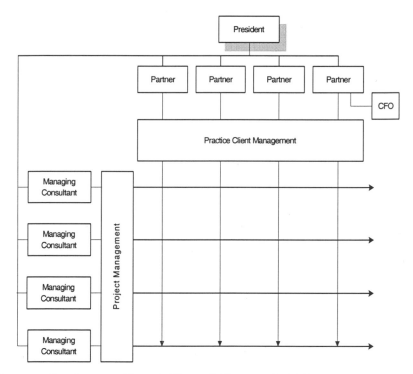

Source: *Conference Board Chart Collection* (1991).

Michael Allen is one of the few companies to state explicitly that its configuration is a matrix configuration. Many matrix configurations are not evident from the organization chart but can be inferred or can be observed. Nonetheless, matrix configurations are wide-spread and rather common, if not for the organization as a whole, for the various divisions or even departments.

Table 2.3. Matrix Organizations

Company	Function or Division	Supervision
American Bankers Association (1998)	Administration, Corporate planning, Finance, Information services, Investor relations, Legal	By project - Utilities, Auto Industry
A.T. Kearney (2001)	Global industry practices and Business Development, Global services and strategy	Various regions
Belgacom Communications (1999)	Carrier, Information and Network Operations, General	By customer type - Business, corporate, Multimedia and infohighways, Residential
Booz Allen Hamilton (2001)	Worldwide commercial business, Worldwide technical business	By market teams - aerospace, automobiles, defense, financial services, energy, health, information technology, organizations and management
Kirk Tyson International Ltd. (1998)	Client development, Consulting, European business development, Finance and administration, Key account management, Marketing, New ventures, Research	Project directors and managers
Nestle (2002)	Finance and control, Marketing, Production, Technical, R&D	By region - Americas, Asia, Europe, Oceania and Africa
Royal Dutch (2002)	Chemicals, Exploration and Production, Gas, Metals, Manufacturing, Marketing, Oil products, Transport and Trading.	Various regions

Source: *Conference Board Chart Collection*

For many organizations, matrix configuration and associated activities are suggested by the organization charts as both functions and divisions are equally highlighted. Table 2.3 shows some of those organizations, together with the two dimensions of the matrix. The dual hierarchy is the essential characteristic of the matrix configuration; there is a simultaneous explicit managerial focus on both the divisions (project, customer, program, and product) and the function (supply, manufacturing, sales). Each function will have a manager, and each division will have a manager.

The coordination in a matrix configuration is very complicated and requires large amounts of management time. To examine the coordination issues, consider the implications of changing one element in the matrix. Let us assume that the manufacturing phase of Product 1 is moved from June to July. What are other required changes? First, for Product 1 all subsequent activities must be moved back one month: the delivery time and sales efforts, for example, must be adjusted. But these adjustments are only the beginning; since Product 1 manufacturing time has changed, all other manufacturing efforts must be adjusted. Similarly, since Product 1 sales effort is adjusted, all other sales efforts must be adjusted. In a matrix configuration, the adjustment of one element may cause adjustments for all other elements. Sometimes, this is called the *jello effect*: you touch it, and it moves everywhere. There is a ripple effect throughout the matrix. It is straightforward to extrapolate that the coordination is difficult, requiring a great amount of information. The advantage is that these adjustments can be made and uncertainties can be managed.

The rationale for the matrix configuration is its focus on the customer, product, program, or markets, its ability to adjust to uncertainties in a timely fashion, and its ability to use source functional resources efficiently. The goal is to capture the effectiveness of the division as well as the efficiency of the functional configuration under uncertainty.

ABB: A Matrix Organization

The matrix configuration is often used for the small organization or project. ABB is a matrix organization, and it is very large. ABB has some 250,000 employees, revenues in excess of US$25B and operates in most industrial countries around the world. How does the ABB matrix work?

ABB's matrix has two dimensions: a local or country dimension and a business area dimension. ABB wants to focus on being a local company and at the same time global to realize the economies of scale and knowledge across borders. A local ABB company develops its own product strategy, deals with labor unions, markets its products, deals with the government, assures local capital and banking relations, and generally manages a total business. The business area managers are responsible and make decisions for product strategies throughout the whole of ABB. At the intersection, there are some 1100 local companies, whose presidents have two bosses. The matrix is to realize the shared knowledge from around the world and take advantage of global markets for given products along the one dimension. Along the other dimension, ABB is local in each country or region. ABB has purposeful internal contradictions, which must be managed: global and local, small and large, and decentralized and centralized. The two matrix dimensions focus on local organization and global product lines and are the organizational means to manage these counter forces. The individual local ABB companies can be small (there are 1200 of them) and yet it is a huge company.

Decision-making is also a contradiction, as decisions are made at the lowest level, but decisions cannot be made at variance with the global product strategy and knowledge about the product technologies and efficiencies. The two dimensions must complement each other, not compete. The business area managers can also be country managers, providing a link across the matrix. Yet, the balance is difficult to maintain.

ABB tics the matrix together in many ways. English is the common language. ABB has a 21-page bible, which spells out the responsibilities of the matrix managers - an amazingly low degree of formalization. The bible speaks directly to the responsibilities of the business area manager and what is decentralization and accountability. ABB's Abacus information system helps provide the glue, lets managers how they are doing, and highlights potential problems. Even so, ABB's main problem remains communications; it must be worked throughout the organization and frequently face-to-face. ABB's matrix managers are a special group who can keep the foci on local and business area and maintain a balance across a broad range of issues. ABB's matrix illustrates that the information processing fundamentals apply whatever the size of the matrix organization.

The matrix configuration is an example of two of Galbraith strategies. The dual advantage of lateral relationships, along with specific product/market focus is achieved through the coexistence of the divisional and the functional form in the matrix configuration.

The limitations are numerous. Managers must be comfortable with uncertainty, willing to share authority, and yet be responsible for team results. These attitudes and skills can be rare among managers. Matrix configurations are costly and require lots of managers-one for each function and project. Finally, a matrix configuration can be disastrous - neither effective nor efficient. When the matrix team breaks down, conflict emerges and nothing goes well. In brief, a well-run matrix can be a marvel of effectiveness and efficiency; a poorly managed matrix can be a disaster and ruinous. Managerial attitude and skill are essential.

Ad Hoc Configuration

The ad hoc configuration is one extreme version of the matrix organization with only one department. If we take this one-step further, the ad hoc configuration does not have a department at all. The members of the organization do not have a place to go back to when the project is finished; it is temporary or ad hoc. See Figure 2.10.

Often such a temporary project is rather loosely structured. It may be a group of experts that all are at the same level with no formal boss. One example is a group of researchers that team up to carry out a specific research project and publish the results. Frequently, an associated staff may be experts as well. There are no levels at all. Everything happens at the same level and is only loosely coupled. There is no centralized decision-making and there are no written rules. The project can be self-organized, but with an overall organizer or director.

Figure 2.10. An Ad Hoc Configuration

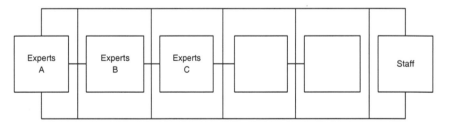

Coordination is achieved through many and often heated meetings. The essential glue is obtained through focus on the common goals that were the basis for the creation of this organization. The incentive system is very important in such an organization; there must be a reward for each individual or group. An ad hoc configuration is not a very stable organizational configuration. In Galbraith's terms, an ad hoc configuration may be a team or a taskforce.

Bureaucracies

The bureaucracy has been confused with a functional configuration; and, they can have many aspects in common. However, they are conceptually distinct.

A bureaucracy can be defined as an organization with the following characteristics:

- Division of labor
- Well-defined authority hierarchy.
- High formalization.
- Impersonal in nature.
- Employment based on merits.
- Career tracks for employees.
- Distinct separation of members' organizational and personal lives.

The above is often called a machine bureaucracy.

These are restatements of Weber's (Gerth and Mills, 1946) original bureaucracy characteristics:

- The principle of fixed and official jurisdictional areas, which are generally ordered by rules, is followed.
- The principle of office hierarchy and of levels of graded authority, a firmly ordered system of which higher levels supervise lower levels, is followed.
- The modern office is based upon written documents.
- Office management usually presupposes thorough and expert training.
- The management of the office follows general rules, which are less stable, are less exhaustive, and can be learned.

A primary characteristic, perhaps the primary one, is the adherence to rules.

A functional configuration can be a bureaucracy. A divisional configuration may have these characteristics. Some matrix organizations may be bureaucracies, although this is less likely.

The purpose of high formalization is in most cases (discussed later) to obtain standardization. Standardization can also be obtained through socialization or professionalization. If a bureaucracy is staffed by highly skilled professionals and if some of the standardization is obtained via the professionalization, giving the professionals some decision authority, it is called a *professional bureaucracy*. Medical clinics, law firms, and accounting partnerships are examples of professional bureaucracies.

We find machine and professional bureaucracies everywhere - in the private sector and particularly in government. That an organization should be of bureaucratic form is used to embellish the recommendation regarding a particular functional or divisional configuration.

New Forms and Organizational Description

Recently, there has emerged great interest in new organizational forms. These new forms are intriguing as they nominally deviate from the well-known functional, divisional, matrix, and bureaucratic configurations. Yet, we want to examine what is new, and what is modified from what we already know. We live in a fast-paced world with new information technologies and quick response customer environments. The pace of corporate life has increased and the pace of change has quickened. The rapid pace of change does not necessarily mean that existing fundamental concepts are no longer applicable. Information processing is now faster and less expensive; does this change the underlying basis for organization? We argue that new forms do require new research, but we also want to argue that what we have learned from years of research and experience remains relevant.

A fundamental question here is whether the new forms require a new vocabulary for describing the organization, or whether the traditional organizational descriptors included in this chapter still apply. That is, can we describe the new forms in terms of configuration, organizational complexity, formalization, coordination mechanisms, incentives, etc. or do we need a new set of descriptors. We suggest that the traditional terms still apply and further we can utilize the extant organizational research in our understanding of new forms.

New forms may be new values or a new combination of values for the organizational descriptors herein. An older form might be a bureaucratic configuration accompanied by high formalization, high complexity and procedural based incentives. A new flexible form may have an ad hoc configuration with low formalization, low complexity and result-based incentives. The flexible form does not then require new conceptual descriptors; the flexible form takes on new values for the well-known organizational descriptors. That is, a new vocabulary is not the issue; our research and understanding of organizations should address the new combination of organizational values that we have not researched and understood well.

Indeed, we do want to understand the flexible organization and the situations where it is appropriate to use.

The information processing view of organization then suggests that the new forms of organization involve the same management issues as older forms of organization. There is more similarity than differences between new and old forms. Information processing remains the means by which the organization is coordinated and activities fit together. The information processing perspective is therefore useful to analyze new organizational forms. According to this perspective, a new organizational design should balance the demand for information processing and the capacity for information processing. Thus, the views and propositions presented in this book can be used to evaluate new as well as old organizational forms, using the traditional organizational descriptors.

In Chapter 6, the issue of flexibility is discussed in the capsule on hyper competition. In Chapter 8 strategy and new organizational form are examined. We return to new forms and explore some conditions when a new form would be appropriate in Chapter 10.

Virtual Network

The term virtual organization has been used to describe a number of organizational variations. It has been used to describe a temporary network of independent companies linked by IT. It has also been used to describe companies, which produces virtual products or services (Davidov and Malone, 1993). However, another definition relates to an organization not having a clear physical locus. Many of the definitions of a virtual organization require the use of IT for coordination. Nevertheless, while IT systems may make it much easier to make a virtual organization successful, it is not required for an organization to be virtual. Partly following DeSantis et al (1999) we define a *virtual network* configuration as a configuration where a significant part of the units or the employees are spatially separated and operate in an asynchronous way. This may require special IT systems or other special means for coordination. The dimension of virtuality can be combined with the simple, functional, divisional, matrix and ad hoc configuration. It can be distributed geographic teams in large organizations, or small organizations working together. It may be teams where independent people and groups who are linked across boundaries, to work together for a common purpose. The virtual network may have multiple leaders, lots of voluntary links and interacting levels. However, it can also be a network, which is coordinated in a more formal way. The reasons for setting up a virtual network are multiple. It may provide a closer relationship with customers or suppliers. In this way, it may be more flexible and adaptable to the increasing uncertainty in the environment. It may also be possible for an organization to get access to resources that would otherwise not be possible. We will discuss these issues in detail in subsequent chapters.

International Configurations

Are there unique configurations for international organizations? In their research, Bartlett and Ghoshal (1988; 1989) have developed four configurations - global design, multi-national design, international design, and transnational design - and have described their organizational characteristics. The organizational characteristics are the configurations of assets and capabilities, role of overseas operations and the development and diffusion of knowledge (see Table 2.4).

The multi-national design is decentralized on a nationally self-sufficient basis to focus on the separate local market and to keep knowledge within the national subunit. This is a classic example of a divisional configuration, where each country unit is a division.

The global design characteristics are centralized on a global scale with implementation at the local level and knowledge at the headquarters. Although the correspondence is less than exact, the global design operates similar to a functional organization.

The international design is a hybrid design. Some of its activities are centralized and others are decentralized. The strategy is to leverage parent capabilities. Knowledge is transferred from the headquarters to the units.

The transnational design is a very complex design. The company's assets and capabilities are dispersed, interdependent, and specialized. National units are integrated worldwide. Knowledge is developed by and shared among the units worldwide. Bartlett and Ghoshal (1989) argue that the transnational design goes beyond a matrix configuration, yet the two have much in common.

Table 2.4. Organizational Characteristics of International Structures

Organizational Characteristics	Multinational	Global	International	Transnational
Configuration of assets and capabilities	Decentralized and nationally self-sufficient	Centralized and globally scaled	Sources of core competencies centralized, other decentralized	Dispersed, interdependent, and specialized
Role of overseas operations.	Sensing and exploiting local opportunities	Implementing parent-company strategies	Adapting and leveraging parent-company competencies	Differentiated contributions by national units to integrated worldwide operations.
Development and diffusion of knowledge	Knowledge developed and retained within each unit	Knowledge developed and retained at the center	Knowledge developed at the center and transferred to overseas units	Knowledge developed jointly and shared worldwide.

Source: Bartlett and Ghoshal (1989).

Philips, the Dutch electronics giant, has a classic multi-national design. Each country unit is quite independent of the other in its development, manufacture, and sales of products. The strength is the exploitation of local opportunities, but cooperation and coordination across borders have been lacking. Its capability to exploit multi-national markets has been questioned.

Matsushita, the Japanese consumer electronics manufacturer, has a global design. The Japanese headquarters runs the corporation worldwide. National manufacturing and sales units implement the headquarters' plans. Local adaptation is not facilitated.

L.M. Ericsson, the Swedish electronics and telecommunications manufacturer, has a transnational design. Its operations, research, development, manufacturing, and sales - are dispersed among several countries. Yet these activities are integrated to take advantage of opportunities worldwide. Large-scale personnel transfer is one means of obtaining the necessary communications needed for coordination among these several activities. Although there are many matrix characteristics, the transnational design involves additional coordinating and integrating mechanisms.

Phatak (1992) discusses four basic international configurations. In the pre-international division phase, the domestic organization is expanded with an expert department and may be further expanded with foreign subsidiaries. Such an organizational structure will be chosen when export activities are rather low and in a development phase.

In the international division configuration, all international activities are in one international division. If market diversity and product diversity are low and foreign activities are relatively small, then an international division configuration may be the optimal choice.

If the market diversity grows, a more global configuration may be necessary so that activities are not in one division but in more divisions based on geographical areas.

The global functional configuration is a functional organization where each function operates in many international markets and areas. For this configuration to be advantageous, the organization has to produce narrow standard product lines.

Finally, Phatak (1992) discusses the multi-dimensional global configuration, which is an international-oriented matrix structure based on functional areas and market and product areas.

Habib and Victor (1991) also use four basic international configurations similar to the configurations presented above: worldwide functional configuration, international division structure, geographic region structure, and international matrix structure as discussed in the capsule on ABB.

Basically, the international configurations are only variations on the more basic configurations discussed in this chapter, particularly when an information processing view is adopted. International or not, each configuration has to be designed in detail, and international issues have to be taken into account. In this book, we primarily address our basic configurations and only occasionally refer to international issues specifically.

In this section on configuration, we have defined a number of alternative organizational configurations. Organizational configuration, due to its graphical representation in charts, is perhaps the best-known characteristic of an organization. We now consider other equally important organizational characteristics.

ORGANIZATIONAL COMPLEXITY

The organizational configurations specify the basis for the division of the activities. The organizational complexity defines the breath, depth and dispersion of the configuration. Organizational complexity is the degree of horizontal, vertical, and spatial differentiation. The number of job titles and the proportion of employees with an advanced degree or many years of specialized training measure horizontal differentiation. Horizontal differentiation is greater when there are several small tasks and specialization by experience, education, and training. Vertical differentiation is the number of hierarchical levels between top management and the bottom of the hierarchy. It can be measured by the number of vertical levels in the deepest unit as well as the average number of levels for the organization as a whole. Spatial differentiation can be measured as the number of geographic locations, the average distance from headquarters, and the proportion of dispersed personnel. It is greater when there are many locations of facilities and personnel.

As the degree of organizational complexity increases, the difficulty of coordination issues and the requirements for information processing increase as well. For example, with a high degree of specialization, each specialty may have to be coordinated. With a low degree of specialization, the coordination is internalized either within a group or even within a person. The complication of coordination in the high specialization case comes from two sources. First, the amount of information that has to be exchanged is high. Second, each specialty may have difficulties understanding the signal sent from another specialty. Each group has its own jargon, values, and form of communication. In an international organization, coordination may be further complicated by language and culture.

The very nature of the airline business demands a large spatial differentiation. As SAS expanded abroad, spatial differentiation increased. Horizontal differentiation is also high for SAS; it has a large number of specialists, including both flight and ground personnel. The vertical differentiation has traditionally been high; however, recent organizational changes to flatten the organization have reduced the vertical differentiation. Overall, SAS has high complexity, where technical task specialization enhances operating efficiencies.

Horizontal Differentiation

Horizontal differentiation refers to specialization within an organization. The simple organization shown earlier in Figure 2.1 has no well-specified horizontal differentiation; Peter, the top manager, assigns work to George, Jane, and Jens at will. The prototypical functional configuration in Figure 2.2 has clearly specified departmental tasks and responsibilities. Mack Trucks in Figure 2.3 has a well-specified, moderate, horizontal differentiation. Horizontal differentiation is a general concept and not restricted to any configuration. Note that at the functional task level, there are more organizational units here, but it also could be low. GE has a high degree of horizontal differentiation with its many divisions and large number of functions. For the matrix configuration in Figure 2.8, there is a moderate degree of horizontal differentiation, as the project, product, and customer dimensions do not increase the horizontal differentiation. Ad hoc configurations can have high or low degrees of horizontal differentiation, depending on the collection of task specialization. Finally, horizontal differentiation and task specialization are fundamental aspects of bureaucracy-whether machine or professional.

A recommendation about level of horizontal differentiation has to be associated with the specific configuration that is considered. Additionally, the particular kind of specialization may also depend on the level of professionalization. We have mapped the number of job titles and the proportion of the employees that hold advanced degrees or have many years of specialized training into the measure of horizontal specialization. This view of horizontal differentiation as a practical measure of organizational complexity has a general agreement in the literature (Hall, 1991). Specialization may increase efficiency. People may be selected to do what they do best, and by doing it regularly, they may learn how to do it well. Additionally, if persons or departments do many different things, .then there are costs involved in moving from one task to another. These setup costs can be minimized by proper specialization. Specialization can be based on functions, products, customer groups, or geographical areas. It also may be based on production processes.

However, increased horizontal differentiation and task specialization come with a cost. Increased horizontal differentiation creates a need for coordination among the specialized units. Without specialization, each individual cuts and tops his own nails. But consider two nail makers - one who specializes in cutting nails and another who puts tops on them. The nail cutter and the nail topper must coordinate their activities so that the number of nails cut equals the number of nails topped. The two individuals can coordinate their activities very simply by talking with each other. However, 10,000 individuals with specialized tasks need to be coordinated, a problem that is not easily resolved. We now have a very large information-processing task. There are numerous information-processing strategies. A 10,000-person organization will have some organizational configuration with a hierarchy that provides the dominant information processing approach. Normally, information for coordination and control follows the hierarchy. Galbraith (1973; 1974) offers various strategies that either (1) increase information processing by becoming faster within the hierarchy or bypass the hierarchy with liaison and integrative mechanisms (2) decrease the need for information processing by rearranging the tasks into more nearly independent units or standardize the operation. Whatever scheme may be used to coordinate the specialized activities, the general relation is that greater horizontal differentiation increases the need for information processing to obtain the required coordination and control.

Vertical Differentiation

Vertical differentiation relates to the depth of the hierarchy. Since the configuration may not be symmetric, the measure has to take that into account. We follow Hall, Hass, and Johnson (1967) and use the number of levels that separate the chief executive from those employees working at the bottom of the organization combined with the average number of vertical levels (total number of levels per number of subunits) as the basis for our measure of vertical differentiation. The measure shows that the vertical differentiation may not be symmetrical. It may be deeper in one area than in another.

The simple configuration shown in Figure 2.1 has only two levels and thus a low degree of vertical differentiation. The functional configuration - that of both the prototype in Figure 2.2 and Mack Trucks in Figure 2.3 - has more vertical levels and thus higher vertical differentiation. However, we cannot measure precisely the degree of vertical differentiation because the configurations are not complete to the lowest level of the organization. Similarly, the divisional configurations - the prototypical and GE - have considerable vertical differentiation, but the charts are incomplete. Matrix organizations are usually asymmetrical where the functional side is the deeper than the project, product, or customer side, and thus the functional dimension yields the degree of vertical differentiation. An ad hoc configuration is usually not very deep with minimal number of levels. Again, bureaucracies, particularly machine bureaucracies, tend to have many vertical levels.

The horizontal and vertical differentiation measures are not unrelated. The higher the horizontal differentiation, the higher the vertical differentiation; the relationship depends on the span of control. The span of control defines the number of subordinates a manager directs. If the span of control is wide, many subordinates are directed. If it is narrow, few are directed. The optimal span of control depends on the complexity of the task that subordinates perform and the subordinate's skill level. This issue is dealt with in Chapter 3.

Delayering, Organizational Complexity and IT

Organizations are delayering; what does that mean? It means to take out a horizontal layer in the hierarchy, by eliminating an organization layer and its jobs. Delayering usually takes place in the middle of the organization and is likely to affect middle management. So an organization that had five levels will now have four levels, frequently with new titles and redefined functions. Delayering does not eliminate the activity or necessarily mean that the old level did no work; it means that this work will have to be done by individuals above and below in the hierarchy. Adjustments will have to be made from the bottom to the top of the hierarchy, not just the nearby levels. Renaming all the levels is a way to signal the extent of the change as well as new job descriptions and titles. So, delayering is a large organizational change which can affect everyone in the organization.

Delayering decreases the vertical differentiation of the organization. If one level is eliminated, then the vertical differentiation is lower and then, the organizational complexity has decreased. It is a less complex organization. The organizational complexity can be changed by delayering, but it can also be changed by decreasing the horizontal differentiation.

Horizontal differentiation is a measure of how finely the work is divided upon into jobs or how specialized the jobs in the organization are. Very specialized jobs indicate a high horizontal differentiation. When the organization enlarges the jobs and tasks, the horizontal differentiation is lowered. When two jobs are made into one or the work is less specified, then the horizontal differentiation is lower. This, too, decreases the organizational complexity.

Organizations with a high level of organizational complexity require greater information capacity to coordinate across the various tasks and jobs, and also up and down the longer hierarchy. The organizational design tradeoff is to balance the costs of the greater information processing with the benefits from the greater organizational complexity. With new information technology and lower relative information processing cost, the tradeoffs involve all three variables. We would expect organizations to become less complex and delayering is one way to decrease the organizational complexity as we realize the returns from new information technology.

Increased vertical differentiation also comes with a coordination and information processing cost. First, increased vertical differentiation increases the number of individuals in the organization, and they must be paid; these individuals process information-orders and information up and down the hierarchy. Second, increased vertical differentiation increases the number of individuals who handle a given piece of information. Consider again our two-person nail example. The organization can become a three-person organization by creating a boss or top manager, whose job is to coordinate the cutter and the topper. The boss does not cut or top but gathers information, makes plans and tells the cutter and topper what to do. These are valued activities and increase the total production of nails. With 10,000 individuals, there are likely to be a large number of vertical levels. Consider again the earlier examples of Mack Trucks and GE.

Recently, many organizations have addressed the issue of number of vertical levels. Generally, increased efficiency in information technology and systems has permitted the elimination of costly middle-management levels. The current trend to eliminate middle management and flatten the organization decreases the vertical differentiation of the organization.

Spatial Differentiation

The third measure related to the organizational complexity of the organization is the geographical dispersion of the activities in the organization. It is relatively easy to measure this concept. We use the number of geographical locations, their average distance from the main office, and the location of the dispersed personnel as our measure of spatial differentiation (Hall et al., 1967). The simple organization is likely to have one location. Mack Trucks has one manufacturing location where the functional heads are located. However, GE has locations all over the world. In general, international organizations have a high degree of spatial differentiation. Spatial differentiation is one characteristic of the virtual network.

It is quite clear that increased organizational complexity implies that additional efforts have to be put into coordination and control. This is particularly true for the horizontal and vertical differentiation. The effect of increased coordination and control due to spatial differentiation depends at least on the information system capabilities. In the past, this concept was particularly important because geographical distance slowed down communication - both with respect to amount and complexity. This may no longer be the case. With modern information systems, it is possible to communicate as easily with a person 5,000 miles away as with the person next door. In other cases, electronic communication may be inappropriate, and the location of personnel is very important.

FORMALIZATION

For many organizations, it is efficient to obtain a standardized behavior of the members of the organization. This standardization can lead to low cost, high product quality, and generally efficient operations.

Formalization is one way to obtain such standardized behavior and, as such, is a means to obtain coordination and control. Formalization represents the rules in an organization. Whether the rules have to be in writing has been discussed in the literature, and in most empirical studies the measurement of formalization has been related to written rules (Hall, 1991). The rules and procedures can be many and fine-tuned or be few and not so fine. The formalization often will vary depending on the particular part of the organization. The production department is more likely to have a high degree of formalization than the R&D department.

The empirical studies on formalization have usually measured formalization in terms of the quantity of written rules. Additionally, formalization involves measurement and degree of compliance.

We measure formalization as the degree to which there exists formally stated rules in writing. Additionally, we measure how much latitude employees are allowed from standards - the level of compliance. It is important to consider whether supervisors and managers make decisions under rules, procedures, and policies and whether these are in writing. Generally having more written rules means having higher formalization.

Formalization may be seen as the means to both increase the information processing capacity and decrease the demand for information processing. When the information processing demand originates from more sales, more customers, or more clients, standardized rules and general policies can keep information processing at low levels in the hierarchy. At the same time, one also may argue that the amount of information that has to be processed for each product, customer, or client is reduced: "Formalization of a decision-making language simply means that more information is transmitted with the same number of symbols" (Galbraith, 1974).

SAS is a highly formalized organization. There are numerous rules on how to accomplish tasks-change an airplane tire, inspect a jet engine, write a roundtrip ticket, and so on. There are hundreds of thousands of detailed instructions. Further, there are many unwritten rules - particularly for professional employees such as accountants.

In contrast, a simple organization has few, if any, written rules, showing very low formalization. Mack Trucks, with a functional configuration, has elaborate written rules and follows them to a large degree. GE, likewise, has a large number of stated rules, although formalization varies across divisions. Matrix configurations are less bound by rules and rely heavily on give and take to obtain coordination. Ad hoc configurations abhor rules. Finally, written rules are fundamental to a bureaucracy and provide a fundamental characteristic of a bureaucracy. For a machine bureaucracy, the rules are written; for a professional bureaucracy, they are brought to the organization by the professionals.

Formalization must not be confused with standardized behavior per se. Standardized behavior can be obtained by a number of different means. The rules may not be by the organization but by a professional association. Medical Doctors and CPA's do a number of things in a standard way because their professional organizations tell them to do so. Professionals also may do things in a standard way due to their training. Professionalization is the term for standardization obtained in this manner.

Social norms and group pressure may also lead to standardized behavior, and the organization may or may not be able to control these norms. The behavior of employees also can be modified by the use of incentives. Some incentive systems can lead to standardized behavior, and others may not. If the incentive is associated with process, it usually leads to standardized behavior. If it is linked to results, it may not.

CENTRALIZATION

Centralization is the degree to which formal authority to make discretionary choices is concentrated in an individual, unit, or level (usually high in the organization). Decentralization is low centralization. We measure centralization by how much direct involvement top managers have in gathering and interpreting the information they use in decision-making and the degree to which top management directly controls the execution of a decision. The above issues are important in determining who has authority to influence a decision aside from actually making the decision.

Centralization is related to who makes which decisions. This includes establishing the budget, exercising control over evaluations and rewards, and being involved in hiring and firing personnel. It also includes issues such as purchasing supplies and equipment and establishing of programs and projects. Finally, related to formalization is the issue of how exceptions are handled. No matter how many rules there may be, they cannot cover all possible situations.

Traditionally, SAS has been a highly centralized organization with most decisions made at the top. More recently, SAS has become more decentralized where midlevel managers are given much more discretion over service decisions and latitude to meet the customers' needs. Yet at the same time, strategic issues remain at the top.

A simple configuration is usually very centralized, and the top manager makes all the important decisions. Sometimes, a simple organization can be decentralized when the top manager lets others decide operational issues. Functional configurations require coordination across functional units; centralized decision-making is a frequent approach. Mack Trucks is rather centralized. A divisional configuration does not require operational coordination across divisions and lends itself to greater decentralization. However, top management is usually deeply involved in budgetary matters and frequently in strategic decisions, even for the divisions. Each division may be centralized or decentralized. Adhocracies can be either, but somehow the required coordination must be realized. Bureaucracies can be either; their desired predictable behavior and outcomes can be obtained by formalization and low centralization. Exception decisions are normally centralized.

From the above it is obvious that centralization is a means for coordination. The relationship between centralization and formalization has been discussed by a number of researchers (Zeffane, 1989). Both are means of coordination. There can be a rule telling what to do, or the top management can each time tell what to do. Therefore, in cases where both centralization and formalization could be high, formalization is most often lower than it otherwise would be. This issue is particularly dependent on the size of the organization and is discussed in a later chapter.

Centralization and decentralization are directly related to the information processing capacity in an organization. The notion of bounded rationality assumes that individuals have limited capacity to read, store, and interpret information. The actual amount of information an individual can process depends on skill and educational levels. As the information processing demand increases, more individuals have to be involved in decision-making, and decentralization increases. Empowerment of individuals usually means greater decentralization and decision-making at lower levels in the organization.

INCENTIVES

One of the important dimensions of the organizational design is the incentive system; that is, the way individuals and their activities in the organization are evaluated and compensated. A shared purpose is central to the definition of organization. That shared purpose can be supported by introducing an incentive system. In recent years, there has been a heated discussion of top-management compensation. Words like "shareholder value" have been common in the popular financial press. The resulting financial scandals in Enron and Worldcom have much to do with the incentive and compensation schemes in these companies.

Table 2.5. *Reward system design and implementation dimensions*

	Reward Form	
	Monetary Form	Non Monetary Form
Unit of analysis	Job	Person
Value comparison	Internal	External
Reward measures	Behavior	Results
Reward level	Individual	Business unit
Pay increase	Fixed	Variable
Administrative level	Centralized	Decentralized
Timing	Lead	Lag
Communication	Open	Closed

Source: Adapted from Heneman and Dixon(2001)

There is little doubt that incentives affect behavior; that is the intention. This notion is supported by research in psychology and economics, as well as in everyday observations. In relation to organizational design, we examine incentives from a strategic organizational design point of view. This means that we take into account the basic principles for designing specific incentive systems for the group or individual and the nature of their activities in an organization. We do not treat the issue of setting up a specific incentive contract for the individual. A basic issue in determining an incentive is whether it is based on behavior or results. Both, agency theory and transaction cost theory take this view (Kowtha, 1997). In these two theories, the incentives are dependent on skills of the individuals and uncertainty of the task or situation, among other elements. Heneman and Dixon (2001) list nine design and implementation dimensions in their alignment of the reward system with the organizational design (see Table 2.5).

The basic principle for the design of an incentive system is whether it is behavioral/procedural or results/output oriented. Behavioral/procedural incentives focus the individual on the efficiency of standards. The results/output incentives focus the individual on the effectiveness of meeting the goals of the organization. Further, an important issue is whether behavior/procedure or results/outcome can be monitored and evaluated. Is the monitoring possible on an individual or a group basis? Information is always costly and the measurement of activities for incentives - individual or group - is costly. Further, the alignment of the incentives, the measurements and the goals can be problematic, (Kerr, 1975). In our treatment of the incentive system for the strategic organizational diagnosis and design, we will focus on a typology consisting of four incentive basic principles as shown in Table 2.6.

Table 2.6. Basic Incentive Principles

	Individual	Group
Procedural	IP	GP
Results	IR	GR

The design of the actual incentive system is based on the particular situation and can be classified into one of the four types above.

COORDINATION AND CONTROL

Lawrence and Lorsch (1967) concluded that the more differentiation is required, the more integration is needed. Organizations are formed to achieve a set of goals. For efficiency, the work in the organization may be divided into a number of separate tasks. To obtain common goals, the activities must be coordinated.

So far four major means to obtaining coordination and control have been presented: formalization, centralization and incentives. They may be used in various combinations, and there are a number of ways to implement each of them. Coordination and control have two sides. One is to make sure that enough relevant information is available at the right time to be able to make the right decisions. The second is to make sure that the right decisions are made. Incentives can be related to both process and results, and can be associated within either groups or individuals. Incentives may be unrelated either to process or results to groups, or to individuals. Lawrence and Lorsch (1967) studied coordination intensity in the different industries and found that the means to obtain that level of integration was very different, even when the level of integration was the same. A fit between all the various elements in the organizational design has to be obtained.

There are numerous means to exchange information and to obtain coordination. This includes the development of *rules* and *procedures* and information exchange in the form of *meetings, task forces, integrators*, or other *liaison activities*. It also includes *direct supervision, planning, forecasting*, and *budgeting*. Finally, coordination using elaborate *IT systems* for direction, monitoring, supervising, communicating, and providing the right information at the right time are used to coordinate the activities in an organization. Here integrated systems like SAP, Navision and Movex are examples of systems that make sure that same information is available to the appropriate bodies in the organization. Chenhall (2003) surveys the findings on management control systems from contingency-based studies from the early 1980's to 2002. He lists more than twenty-five different kinds of control mechanisms in this context.

To summarize, coordination can be obtained by a number of means ranging from direct supervision to autonomous groups using both the hierarchy and formalization and including incentives. Additionally, various kinds of lateral procedures may be appropriate. It is also obvious that a proper information system is important. The information system can be characterized according to a number of dimensions. An obvious dimension is the amount of information that can be transferred and processed. Further, the media richness of the information system is of interest. Media richness relates to the type of information that can be processed and the type and speed of feedback. Face-to-face meetings have higher richness than a short written note. It is interesting to observe that the revolution in information technology increases both of these dimensions: the amount of information and the media richness.

Information systems are also categorized as sequential or concurrent. In some operations, it may be important that you use a sequential information system, but when LEGO plans to release the same spaceship design on the same day in 40,000 stores, then the system has to be concurrent. New fax systems and electronic mail are modern information systems that can be both sequential and concurrent. In the choice of coordination and information system, the particular combination of tools is important. The choice has to fit the demand for information processing capacity.

SUMMARY

In this chapter, we considered how to describe an organization for management and design purposes. We argued that an information processing description is relevant: how does the organization use information to coordinate its activities and accomplish its goals?

A number of organizational concepts were introduced and related to coordination. The organizational configuration is the way the organization divides its work into smaller tasks for management-simple, functional, divisional, matrix, ad hoc, or virtual. An organization may have a number of configurations operating at the same time in different parts of the organization. Complexity, formalization, and centralization were defined and illustrated. Additionally, means for coordination and control and incentives issues were introduced. Throughout we have referred to SAS, Mack Trucks, GE, and Michael Allen to illustrate the concepts.

CHAPTER 3

Leadership and Management Style

INTRODUCTION

The Institute of Applied Computer Science, in Odense, Denmark, is a small research-oriented organization located at Science Park. Its main activities are related to high technology applications of computer science in various organizations and projects. It has been involved in technology transfer projects for a number of European companies. The basic idea is that Applied Computer Science will transfer ideas from research laboratories that can be used in practice. It has particularly targeted companies with a relatively low research and development budget that do not have staff that can make the transfer. For these companies, a technology transfer project is a rare event. The activities are typically done in the framework of national research programs and European Union (EU) R&D programs such as ESPRIT, BRITE/EURAM, SPRINT, and COMETT. The Institute of Applied Computer Science is a small organization currently employing seventeen highly skilled employees. It was established in 1986 at the request of the local industry, which needed to better organize the transfer of knowledge from Odense University (now University of Southern Denmark) and a technical college. A group of local business people and representatives from Odense University and the technical college formed the company, raised a small amount of capital, and hired a teacher, Benny Mortensen, at the technical college to run the company. Benny Mortensen had good contacts both at the University and in local businesses and quickly linked the company to EU grants. His knowledge and enthusiasm have contributed significantly to the success of the company. In particular, he has established cooperation with ten European partners.

The company has developed expertise in project management and has managed both small and large EU projects. Benny Mortensen has created relationships with more than fifty Danish companies. The Institute of Applied Computer Science has worked for and with European companies such as British Aerospace, Agurta Helicopters, MARI computer systems, and Verilog, among others. Additionally, he has working contracts with research institutes and universities in Milano, Madrid, Kiel, Delft, Utrecht, London, Manchester, and Oxford. Lately, the company has worked for US companies like Boeing. It also received the local IT-award for entrepreneurial excellence. The organization is relatively loose, and the CEO spends many days on the road getting new contracts, but he knows what goes on in each project. He prides himself on actively seeking out new business opportunities that challenge the organization. He wants the organization to grow, but growth is not the primary goal. The CEO's enthusiasm is his own approach to motivation, and he prefers to lead by example, believing that others will follow with equal enthusiasm and hard work. He likes the technical details of each project but does not impose his views on the group. If projects fit his longer-range goals for Applied Computer Science, he does not interfere but gets involved because of personal professional interest and his commitment to the firm's high-quality reputation. The CEO's leadership style definitely helps shape the firm-both its strategy and organizational structure.

SAS: The Leadership and Management Style

Jan Carlzon was hired in the second year of deficit after seventeen years of continuous profit; something had to be done. Carlzon turned out to be the person to do it, and by the spring of 1981, he was made president of the entire SAS group (Carlzon, 1985).

Jan Carlzon had been president of the Swedish Linjeflyg, where he had changed deficit into profit in just two years by changing the concept of the organization. He had gained a reputation of being a man of vision with a unique ability to realize that vision. A new vision was needed at SAS in 1980.

The new strategy had to be followed up by a drastic change in leadership and management style. Top managers would no longer make decisions alone, but each person would be responsible and take part in the day-to-day running of the company. It was particularly emphasized that employees were to be informed and not instructed. As a first sign of this changed attitude toward personnel, Carlzon began informing employees about new policies as soon as the board of directors approved his strategy outline. All 20,000 employees received a red booklet called "Now We Start Fighting," which outlined the main concepts of the new strategy. Through this action, Carlzon showed that he considered employees to be a resource that should take active part in creating the changes necessary for SAS's future success. Carlzon himself explained his leadership style by using the following metaphor: "If you give two employees a block of stone each and ask them to carve a square out of it. If you tell one of them that his stone is going to become a part of a large castle whereas you only ask the second person to carve a square, the first person is by sure going to like his job better than the last" (Lennby, 1990). And this was Carlzon's idea - to make a clear strategy that everybody would understand, and to tell everybody how they could contribute to make the strategy become successful. Because the new strategy focused strongly on the front-line personnel's role in the customer's perception of service, it was important for those on the front line to give good service in their daily contact with the customer; Carlzon called it the "moment of truth". This meant that front-line personnel should be able to solve problems as they occurred without consulting their superiors. This decentralized decision-making made it necessary to delegate responsibility, but if employees were to assume responsibility, they needed education. As a result of educational efforts and the increased focus on front-line personnel, employees suddenly felt they were appreciated and were willing to lend a helping hand in the attempt to bring SAS back on its feet. Employees - colleagues or not - helped each other in order to deliver the best product to the customer without giving thought to who got credit for the work. At the presentation, Carlzon talked about the importance of practicing "Management by Love rather than Management by Fear," meaning that by showing employees respect and trust they would strain every nerve in order to show that they were worth the credit.

One of Carlzon's strengths was his ability to motivate employees by appealing to the idea of the organization's common goal. Carlzon managed by motivation rather than through control, and this was why employee motivation vanished when Carlzon started making cutbacks. In terms of our description later in this chapter, Carlzon had:

- A high preference for delegation.
- A general level of detail in his decision-making.
- A proactive decision style.
- A future orientation in his vision and a relatively long-term view.
- A high risk preference.

He also promised motivation through inspiration and not detailed control. His leadership style can be categorized as an entrepreneur.

Jan Carlzon's strategy and leadership style worked well for some years, but risky decisions again caused a financial crisis. As a result SAS wanted a new CEO.

Jan Stenberg became SAS president and CEO in April 1994. Stenberg's leadership style was very different to Carlzon's. Stenberg's main focus was to bring SAS back on its feet financially. This implied rationalizations of up to SEK 2.6 billion, and involved firing 3000 employees. These decisions were obviously taken in the top management group without much consultation of employees. Stenberg's personality was also very different to Carlzon's. Contrary to Carlzon who was always in high spirits when he was in the limelight, Stenberg did not like the media attention, and he only talked directly to the employees when absolutely necessary.

Stenberg managed to get SAS back on its feet and produce profits unlike ever before. Many of his decisions were taken within the top management group. He did not pay a great deal of attention to inform the employees of how decisions would be implemented. The general lack of formal information to employees led to employee frustration, resulting in strikes, especially among cabin crew and pilots, threatening to deteriorate SAS's punctuality and reliability and hence the company's competitiveness.

Although the appointment of Stenberg has increased centralization at SAS, the centralization has mainly been on strategic issues and not operational issues on which employees still have a relatively expanded decision-making authority. Stenberg has a relatively high risk preference, although not as high as Carlzon's. He is future oriented but with focus on short term results. Stenberg is mainly concerned with general directions in his decision-making, and although he motivates with more control than Carlzon, he still motivates by "fewer controls" rather than "many". His leadership style is categorized as one with some preference for delegation, high task orientation, proactive decision-making, long-term time horizon, neutral risk preference and mixed motivation and control. His leadership style seems to be a mixture of leader and producer.

Jørgen Lindegaard succeeded Jan Stenberg in 2001. The basic idea just before Jørgen Lindegaard was hired was to expand SAS within a growth strategy. But problems occurred immediately after Jørgen Lindgaard came to SAS. SAS was found guilty of violating the EU antitrust laws and had to pay a high fine. The vice-president for aviation was forced to resign. The board was accused of being aware of the irregularities and decided to resign. Then, September 11 happened and shortly after a SAS flight crashed in Milan with all passengers killed. The crash was due to errors in the airports air control. So the new CEO was face with a very critical situation right from the beginning. Further, low price airlines like Ryan Air and Virgin have cut into some of SAS profitable businesses. Jan

Lindegaard has been focusing on the overall planning, maintaining a high level of decentralization within operations. He has a long term view which can be seen from the way he has incorporated airlines like Spaniar and Braathes into the SAS group. He has also managed to have SAS better positioned on the financial market by creating one SAS stock (instead of three). He seems to manage by motivation with focus on the strategic issues. He is concerned with risk without being risk averse. He is very visible in the media, keeping a high public profile. Although, he has had to react to the crisis his leadership style tends to be proactive. In the typology of this chapter, Jørgen Lindegaard can be categorized as a leader.

1. Which organizational structure (configuration, centralization, complexity, formalization, coordination and control, and incentives) would you think each of the three CEOs would prefer?
2. What would be the structural component that each of the three CEOs would change first, if it was possible?

LEADERSHIP AND ORGANIZATION

Does the leadership style determine the organizational structure, or does the reverse hold true? From the literature it is not clear that any directional cause-effect relationship exists. Correlations are more easily demonstrated. In our view, the cause-and-effect relationship is not as important as the fit; cause and effect may work in both directions. For a given management group, management will try to tailor the structure so that it fits the needs of the management. Child (1972) argues that management has considerable discretion in its choice of the organizational structure - more discretion than contingency theory implies. First, the contingency factors of size, technology, environment, and strategy are incomplete determinants of organizational structure. Thus, management has a choice of structure within constraints given by other contingencies. Second, at a higher level, management can choose the contingencies themselves over time. Management can change the organization's size, its technology, and the environment it operates in and adapt its strategy. It can adopt, and probably should adopt, a new organizational structure when needed, however, taking into account the cost of making the changes.

Robbins (1990) argues that management and its power relations account for a significant part of the variation in the choice of structure. Lewin and Stephens (1994) have argued that "social-psychological attitudes of chief executive officers and general managers are critical contingencies in organizational design and strategy." Management will choose the structure within the limits given by the other contingency factors. But in some situations no feasible solution exists; it is not always possible for the management to find a structure that both fits their needs and creates a fit with the size, technology, environment, and strategy of the organization. In many cases, practice has shown that a change in the structure causes a change in the management group. We have seen many examples of organizations where a change from a functional configuration to a divisional configuration accompanied the introduction of a new CEO, as was the case for SAS. In the functional configuration, the CEO is involved in more detailed, short-term decision-making; in the divisional configuration, the CEO is mainly involved in long-term, strategic resource allocation decisions. "Obviously ... incorporating managerial philosophies into current structural contingency theories is of major importance" (Miles and Creed, 1995).

Similar concerns relate to the fit between the structure and employees other than the CEO. There are managers at all levels, and their views and style generally affect the operation of the organization. Individual managers play an important role, as does the way they interact. In general, the organization is a social information-processing unit, and the way the individuals process and exchange information is an important factor for the functioning of the organization. Interaction among people depends to a great degree on organizational configuration and properties. It can therefore be expected that people in the organization with decision-making power will try to influence the choice of organizational structure, as discussed above. Here we want to consider the effect of the leader on the choice of the organization, where the leader wants an organizational design which fits his/her own preferences and is also efficient and effective.

For the organization to function well, there should be a fit between what the organization wants to do; how it wants to do it; and the people who have to do it. It may be desirable to have a very complex and decentralized organization with a low degree of formalization. However, if employees do not have the necessary skills or will not assume new responsibilities, such a structure may not work. Also, individuals with a need to assume responsibilities may not function well in a very centralized structure with a high degree of formalization. The chosen structure should accommodate the type of people who work in the organization, or the organization should hire and train people so they can function well. The information processing capacity of the individuals in the organizations affects the information processing capacity of the organization. The use of information is an important issue in this context (Arrow, 1974). Professionals may develop their own terminology that enables them to communicate efficiently among themselves, but makes it difficult for them to communicate with others. The degree of horizontal differentiation may require or prohibit such developments. For international organizations multiple language skills are important. In general, organizational culture or climate plays an important role in the way individuals and organizations process information. From this point of view, personnel planning and education are investments in information processing capacities. There should be a fit between the demand for information processing and individual capacity (Galbraith, 1973; Levitt et al., 1999). This kind of fit is an important aspect of organizational behavior (Schermerhorn et al., 1991). In Chapter 4 the relation between the climate in the organization and the organizational structure is treated and related to the way people process information. In this chapter we focus on the fit between leadership and management style, and organizational structure.

LITERATURE REVIEW ON LEADERSHIP

Leaders and managers have different skills, values, and personalities. The literature on leadership is extensive. Hunt (1991) claims that there are more than 10,000 empirical studies on leadership. Different organizational structures require different skills and attitudes. Therefore, structure affects the desired type of leader or manager, and the requisite skills and education of the employees. A number of authors have developed leadership behavior typologies. See Hunt (1991) for a survey.

We are more concerned with the counter relation: does the leadership or management style affect the choice and fit of the organizational structure? In the design phase, we argue that leadership style is an important contingency for the choice of structure. This question is particularly important for small and medium-size organizations that are owner run; however, it is an important contingency for all organizations.

In this chapter, we develop the framework for leadership style for the organizational design recommendations for the OrgCon.

A classic way to categorize leadership styles is by applying McGregor's (1969) Theory X and Theory Y. He suggests a specific view of the skills and attitudes of the members of the organization. Theory X presumes that people dislike work, shirk responsibility, and seek formal direction whenever possible. They must be coerced, controlled, or threatened with punishment to achieve desired goals. In contrast, Theory Y posits that people like to work and will exercise self-direction and self-control. According to Theory Y, people accept and seek responsibility. Many Theory Y people can make good decisions and have a high degree of creativity. A Theory X or Theory Y leader may develop a leadership style that fits his or her view of the members of the organization. From an information processing point of view, Theory Y individuals can process more information and more complex information than Theory X individuals. Theory Y information is more forward looking and general; Theory X information is more control oriented and detailed.

Kotter (1988) and Hunt (1991) make a similar distinction between a leader and a manager. A leader develops future visions, communicates with those whose cooperation is needed, and uses motivation and inspiration rather than controlling and problem solving. A manager does planning and budgeting, monitors results in detail, and develops planning, staffing, and delegation structures. A manager is more internally oriented than a more externally oriented leader. A leader can process more complex information; it is unclear which leader type - leader or manager - processes the greatest amount of information.

Mintzberg (1980) developed three major groups of leadership roles: (1) interpersonal roles which include those of figurehead, leader, and liaison; (2) the informational category which includes the roles of monitor, disseminator, and spokesperson; and (3) the decisional category that consists of the roles of entrepreneur, disturbance handler, resource allocation, and negotiator. Mintzberg argues that every leader or manager plays each of these roles to some extent over time. The importance and intensity of each role vary with the level in the organization and the type of organization.

Luthans, Hodgetts and Rosenkrantz (1988) have developed a typology with four groups: (1) routine communication consisting of exchanging information and handling paperwork; (2) traditional management including activities such as planning, decision-making, and controlling; (3) networking including the activities interacting with outsiders, socializing, and politicking; and (4) human resource management consisting of motivating, disciplining and punishing, managing conflict, staffing, and training and development. Again, it is hypothesized that all leaders are engaged in all of these activities to a certain degree. Particular leadership positions require an optimal balance between the activities and their implementation.

Mintzberg's and Luthans, Hodgett and Rosenkrantz's categories are related to different types of information and how they are handled. The type of organization affects the type of decisions that the manager or leader has to make and how he or she has to process information.

Similarly, Yukl's (1981) typology is related to the information network, information processing, and decision-making. He also developed a typology of four groups, each with a number of subgroups: (1) making decisions: consulting and delegating, planning and organizing, problem solving; (2) giving or seeking information: monitoring, clarifying, informing; (3) building relationships: supporting, networking, managing conflict and team building; (4) influencing people: motivating, recognizing and rewarding. The four groups relate both to skills and behavior. A particular manager or leader will play various roles, depending on the particular organization. Managers can, to some degree, be taught how to play these various roles. However, the personalities and values of the particular manager or leader will also determine how this manager or leader will play each role.

Autocratic behavior and democratic behavior have been viewed as two extreme positions (Likert, 1967). This is also related to the leader-participation model proposed by Vroom and Yetton (1973), where the information and who makes the decisions are the most important factors. Many of these issues are depicted in the competing-values leadership approach developed by Quinn and Rohrbaugh (1983), Cameron and Quinn (1999) and Denison, Hooijberg and Quinn (1995). It is summarized in Figure 3.1.

Figure 3.1. Competing Values

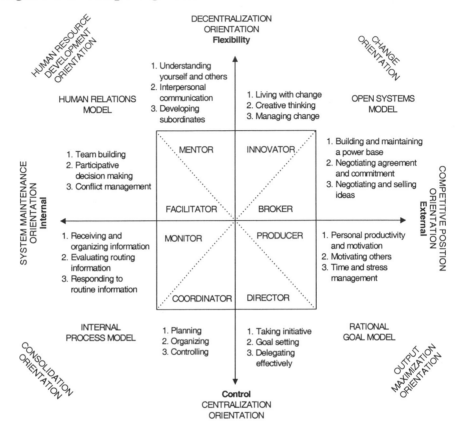

Source: Hunt (1991, p. 163).

Complementary values are next to each other, contrasting values are opposite. For example, one may want the organization to be adaptable and flexible but also stable and controlled. In the mind of the leader, these states are assumed to be opposites, and these opposites cannot exist at the same time. Quinn and his associates argue that managers narrowly pursue one viewpoint and need to break out and pursue other viewpoints, particularly opposite ones. Individuals who are able to balance these values are referred to as master managers or Janusian, after the Roman god. The competing values approach also captures many of the leadership styles like charismatic leadership, supportive management, etc.

A review of the leadership or management typologies reveals that many of the dimensions are similar to the dimensions used to describe the organizational structure. The delegation issue relates to decentralization. Planning and control relate to formalization. In general, the dimensions in Figure 3.1 relate to all the informational issues that are important for organizational design, including reward and motivational issues.

The important issue is how effectively the various leadership styles and their information processing capacity affect organizational structure. Fiedler (1977) proposed a contingency model of leadership behavior and its effect on the task structure. He found that if the task structure is structured, then the leader position should take a strong power position, and if the task structure is unstructured, then the leader position should be weaker with respect to power. From our information-processing point of view, if the task structure is unstructured, then the demand for information processing is high and decentralization is appropriate; a highly structured task is compatible with centralization.

Miller and Toulouse (1986b), Miller and Dröge (1986a), and Miller, Kets de Vries, and Toulouse (1982) investigated the relationship between leadership style or personality and strategy and organizational structure. They consider three personality dimensions: flexibility, the locus of control, and the need for achievement. First, flexibility refers to the adaptability of a person's thinking and social behavior. Leaders that are flexible are informal, adventurous, confident, assertive, and egoistic. Inflexible leaders are deliberate, cautious, industrious, mannerly, guarded, methodological, and rigid. Second, locus of control is measured on an internal-external scale. An internal manager believes that the consequences of his behavior stem from his own efforts, while an external manager sees the events of his life as beyond his or her control. Third, managers or leaders with a high need for achievement prefer to work on a problem rather than leave it to others; they have a strong preference for structural work situations with quick and concrete feedback. Such managers strive to meet standards of excellence.

Miller and Toulouse (1986b) encountered two qualifying conditions (Table 3.1). The first one is that the relationship between leadership style and personality is strongest when the organization is small. The second is that the leadership style has more effect on the strategy and structure when the organization faces a dynamic, unpredictable, and changing environment than when it operates in a stable and simple environment. Their empirical study of ninety-seven Canadian firms generally supports their hypotheses.

Table 3.1. Expected Relationships Between CEO Personality and Organization[a]

	Flexible Personality	Internal Locus of Control	High Need for Achievement
Strategy	Niche-focused	Innovative	Broad aggressive marketing
Decision making	Intuitive Short-time horizon Reactive Risk taking	Informal Long-term Proactive Risk neutral	Analytical Long-term planning Proactive Risk aversion
Structure	Informal Unspecialized Much delega- tion of authority Few controls Few liaison devices	Informal Mixed Much delegation of authority Mixed Mixed	Formal Specialized Little delegation of authority Many controls Many liaison devices
Performance	Successful in small firms and dynamic environments	Successful in any size but especially so in dynamic environments	Successful in large firms and stable environments

Facilitating Conditions: We expect that the relationships between CEO personality and organizational variables will be higher in (a) small organizations, and (b) dynamic environments.

[a] The attributes listed are for high scores on the variables in question. For example, the higher the score on CEO flexibility, the more strategies will be focused, decision-making intuitive, and so on.

Source: Miller and Toulouse (1986b).

Miller and Dröge (1986a) investigated the relationship between need for achievement and structural dimensions of centralization, formalization, and complexity using the same sample of firms as Miller and Toulouse (1986b). In particular, they investigated the correlation related to the age and size of the organization. They found that leadership style is more important in small, young organizations than in large, old organizations. Miller (1991) extended the analysis to include the tenure of the CEO. He found that the longer the tenure, the less likely it was that the CEO and the organization would adapt to changes and uncertainties in the environment. He also found that such a mismatch would decrease organizational performance.

The various empirical studies by Miller and his colleagues use different typologies of leadership style and personality. The effect and relationship are not entirely clear (Miller and Toulouse, 1986b). They generally found that flexibility is an important measure, but they also found that the relationship between flexibility, delegation, and formalization becomes insignificant when controlled for need for achievement and locus of control. They argue that this is due to the significant negative relationship between flexibility and the need to achieve. The need to achieve is a more important predictor of appropriate structural dimensions, but Miller also uses the locus of control in earlier studies (Miller et al., 1982). Referring to Table 3.1, the flexible personality is at one end of a scale, and high need for achievement is on the other, with internal loci of control in between. This is also consistent with the significant negative relationship between flexibility and the need to achieve that Miller and Toulouse found. The decision-making attitudes of a flexible leader show that he or she is less involved in decision-making than a leader with a high need for achievement.

The importance of the relationship between leadership style and organization and strategy is quite clear. The problematic issue is to define a typology that is appropriate for establishing this relationship.

DIMENSIONS AND TYPOLOGIES

Our review of the various typologies on leadership behavior shows that leadership style can be described by how the leader makes decisions, handles information, relates to risk, and motivates and controls subordinates. These dimensions determine whether the leader is a Theory X or Theory Y leader, an autocratic or democratic leader, or a flexible or need-for-achievement leader. The different typologies look very much the same on most of the above-mentioned dimensions. As discussed in the previous section, it is involvement in information processing and decision-making that matters.

Each leadership or management style can thus be assessed on the following information-processing and decision-making dimensions, which are the dominant dimensions that we found in the models reviewed above:

- Preference for delegation.
- Level of detail in decision-making.
- Reactive or proactive decision-making.
- Decision-making time horizon.
- Risk preference, and.
- Motivation and control.

In Table 3.2 we have summarized how the various typologies score on the dimensions that are important with respect to organizational structure. Beginning on the left, McGregor's Theory X and Theory Y model is contrasted, (McGregor, 1969). A Theory X manager prefers not to delegate and is formal, detailed, and reactive in decision-making. He or she has a short-time horizon, is risk averse, is not motivating, and has many controls. A Theory Y manager prefers to delegate, and is general, proactive, and prefers to be a longer-term decision-maker. He or she has high-risk preference, prefers to motivate, and has few controls. Similarly, Zalesnik (1977) and Kotter's (1988) managers and leaders model demonstrates the contrasting styles. Likert's (1967) autocratic and democratic model is similarly contrasting. Finally, Miller, Kets de Vries, and Toulouse's.three categories of high need for achievement, internal locus of control, and flexible personality follow the same pattern. (Miller et al., 1982; Miller et al., 1986b)

Consider the similarity among Theory X-managers, autocratic, and high need for achievement. Then consider Theory Y-leaders, democratic, and flexible personality. On the six dimensions that are particularly related to organizational design issues, the similarity is striking.

Table 3.2a. Management Preference According to Different Typologies

	Theory X	Theory Y	Managers	Leaders
Preference for delegation	Low	High	Low	High
Level of detail in decision-making	Formal detailed	General directions	Very detailed	Low
Reactive or proactive decision-making	Reactive	Proactive	Reactive	Proactive
Decision-making time horizon	Short term	Long term	Short, related to plans	Future visions
Risk preference	Risk Averse	High	Risk averse, planning oriented	Risk taking related to future visions
Motivation and control	Little motivation, many controls	Motivation by fewer controls	Monitoring details	High motivation and inspiration

The typologies of McGregor, Zalesnik, and Likert fit together well on the six dimensions. The three categories investigated by Miller fit less well. The correspondence matches on the dimensions of delegation, level of detail, risk taking, and motivation and control. They differ, however, on the proactive and reactive and time-horizon dimensions.

From an intuitive point of view the four dimensions on which they agree are closely related to the amount and complexity of information processing. The time-horizon perspective is more difficult. Depending on the situation, a long-term or short-time horizon may demand complex information processing, but the quantity may be lower. In our model, we put a lower weight on this dimension than on the four where all agree. There is also disagreement in the reactive and proactive dimension. In most cases a reactive style reflects an inappropriate leadership behavior. In particular, one might discount the reactive decision-making perspective for the flexible personality as suggested in Table 3.2b. Additionally this dimension might be given a low weight and used more to warn against inappropriate leadership behavior.

Table 3.2b. Management Preference According to Different Typologies

	Autocratic	Democratic	High need for achievement	Internal locus of control	Flexible personality
Preference for delegation	Low	High	Low	Some	High
Level of detail in decision-making	High	Low	Analytic and high level	High task oriented	Relatively low
Reactive or proactive decision-making			Proactive	Proactive	Reactive
Decision-making time horizon	Usually short	Usually long	Long-term planning	Long-term	Short-time planning
Risk preference	Risk averse	High	Averse	Neutral to averse	High
Motivation and control	Controls	Motivation	Many controls	Mixed motivation and control	Few controls

The contrasting categories of leaders and managers above are ideal types and are end-points of a continuum. They are only two of the possible 2^6 or 64 possible styles. For instance, it is possible that a leader/manager would have a style which has a low preference for delegation, general directions, proactive, long term orientation, high risk preference and motivation by control. This is one of the 62 remaining possibilities which is neither a leader nor a manager as described above. Similarly, the three leadership categories proposed by Miller and his associates: high need for achievement, internal locus of control and flexible personality, do not match with the leader/manager categories; they fall in the one or more of the remaining 62 categories. We would be surprised if all individuals were only in the leader or manager categories; or equally, we would be surprised if individuals fell in all 64 categories. Individuals likely fall in more than two categories, but less than 64.

MEASURING LEADERSHIP STYLE

In the literature there are a number of leadership frameworks which focus on the decision-making activities and information utilization by the leader. This could also be related to the roles leaders and managers play. Mintzberg (1980) developed three major groups of leadership roles: (1) interpersonal roles; (2) the informational category; and (3) the decisional category. Mintzberg argues that every leader or manager plays each of these roles to some extent over time. The importance and intensity of each role vary with the level in the organization and the type of organization.

Luthans, Hodgetts and Rosenkrantz (1988) have developed a typology with four groups: (1) routine communication; (2) traditional management including decision-making, and controlling; (3) networking; and (4) human resource management. Particular leadership positions require an optimal balance between the activities and their implementation.

Similarly, Yukl's (1981) typology is related to the information network, information processing, and decision-making. He also developed a typology of four groups: (1) making decisions and problem solving; (2) giving or seeking information; (3) building relationships; and, (4) influencing people.

Mintzberg's and Luthans, Hodgett and Rosenkrantz's and Yukl's categories, as well as Vroom and Yetton's (1973), are related to different types of information usage. The type of organization affects the type of decisions that the leader or manager has to make and how he or she has to process information. These authors categorize leadership as a complex, but small set of decision-making and information activities.

In their classic book on the behavioral theory of the firm, Cyert and March (1963) focus on decision-making and how the organization uses less-than-perfect information to make decisions and implement those decisions. Their decision-making style and uncertainty avoidance concepts capture and summarize well the six decision-making dimensions that we found in the literature. The six dimensions can thus be summarized by two aggregate measures of leadership style.

- *Preference for delegation.*
- *Preference for uncertainty avoidance.*
  - o   level of detail in decision-making
  - o   reactive or proactive decision-making
  - o   decision-making time horizon
  - o   risk preference
  - o   motivation and control.

The first dimension is the preference for delegation and the second dimension is the remaining five elements.

Our first dimension captures the desire to delegate, which is a related description of decision-making style. The preference for delegation is a widely understood and used notion about the behavior of leader/managers. Yukl (1981) includes delegation as related to planning and problem solving as his first dimension. The role of delegation is implicit in Mintzberg's (1980), and Luthans Hodgetts and Rosenkratz's (1988) dimensions of leadership. The preference for delegation reflects a decision-making style, which is the first element of the Theory X,Y, Kotter's (1988) leadership/manager typology, Likert's (1967) democratic/automatic typology and Miller et al's (1982) three categories. Cyert and March (1963, p. 121) proposed problematic search, where search is motivated by the problem, simple-minded and biased by the experience of the decision-makers. Problematic search is a heuristic, which utilizes the leader's time efficiently by keeping the problem solving relatively simple and familiar to what he/she understands with confidence. It is efficient to focus directly on the problem as it is likely that a reasonable solution will be found and found quickly. To use one's past experience as a guide is also efficient as it takes less time to find a solution by association than a de novo approach. A leader is likely to delegate when he/she finds it efficient in terms of his or her own time availability and further when he/she thinks the organization will make decisions congruent with his/her preferences and experience; delegation is also a heuristic for the leader. The leader wants some degree of predictability when delegating. Of course, some leaders may prefer to make decisions on their own for a variety of reasons: self-confidence vis a vis others, value of own experience for the situation, predictability and a sense of control. A leader will delegate decisions when he/she is confident in the organization and finds delegation efficient in his/her own use of time.

For the second dimension, we chose the term "uncertainty avoidance" following Cyert and March (1963, p. 119). They describe uncertainty avoidance to include:

- Avoid correctly anticipating events in the distant future by using short-run reaction to short-run feedback, i.e., solve pressing problems rather than develop long run strategies.
- Avoid anticipating the environment by negotiating with it.

The match of the five elements in our dimension of uncertainty avoidance is congruent with the Cyert and March (1963) concept. A manager can implement Cyert and March's short-run reaction to short-run feedback and avoid the uncertainty of long run anticipation and commitments by a fine level of detail, reactive style, short term decision-making, risk avoidance and control.

Referring to Table 3.2, McGregor's Theory X-manager, Kotter's manager, Likert's autocratic manager are high on uncertainty avoidance as they are detail oriented, reactive to events, short run, risk averse and control oriented. Theory Y-leaders and democratic leaders are low in uncertainty avoidance as they are less detail oriented, less reactive, more long run, less risk averse and motivate through inspiration rather than controls. Miller's categories of high need for achievement, internal locus of control and flexible personality are all a mix of high and low uncertainty avoidance. A flexible personality is reactive, short term, yet not risk averse and not control oriented. A high need for achievement manager is risk averse with many controls, yet proactive. The internal locus of control manager falls at the midpoint. All three are neither high nor low in uncertainty avoidance, but for different reasons. Overall, uncertainty avoidance captures well a dimension in leadership style.

The two dimensions of decision-making style and uncertainty avoidance are shown in Figure 3.2. The various leadership and management styles reviewed earlier can now be placed in the decision-making style and uncertainty avoidance dimensions.

Figure 3.2. The Two-dimensional Leadership Model

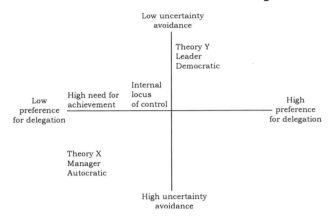

We observe that the leadership descriptions fall into two elements of the four possible categories, except for Miller's three categories. These are ideal types of leader and manager. A leader has high preference for delegation as a decision-making style and has a low level of uncertainty avoidance; this is a low level of microinvolvement[1]. A manager has a low preference of delegation and a high degree of uncertainty avoidance; this is a high level of microinvolvement. There are two other categories which we call entrepreneurs and producers. An entrepreneur has a low preference for delegation and a low preference for uncertainty avoidance. A producer has a high preference for delegation and a high preference for uncertainty avoidance. Thus we have four categories of management and leadership styles:

- Leader
- Producer
- Entrepreneur
- Manager.

Døjbak et al (2003) find partial support in an empirical investigation of 245 Danish SME's using a factor analysis that individual firm

---

[1] In earlier editions of this book, we used the concept of microinvolvement to describe leadership style. Here, we use the two dimensions of preference for delegation and uncertainty avoidance. The relation is that high microin volvement is the lower left quadrant of Figure 3.2 and low microinvolvement is in the upper right quadrant.

CEO's fall along two dimensions and into four categories as developed above.

LEADERSHIP STYLE AS A CONTINGENCY

The leadership style can explain a good portion of the variance not explained by the contingency factors: size, technology, environment, and strategy (Child, 1972). The prescriptive normative issue is, however, less clear (Robbins, 1990). The same is true for the studies by Miller or his associates. Robbins (1990) and Lewin and Stephens (1994) suggest that management's preference for control has a great effect on the choice of the organizational structure. Using the arguments presented in Chapter 1, the descriptive studies can be used as a basis for a normative model if we assume that only the efficient and effective organizations generally survive. CEOs view themselves as organizational architects (Howard, 1992). They want to shape and organize their organizations. In fact, it is the responsibility of the CEO to determine that the organization is appropriately organized to serve its purpose. However, it is not surprising that CEOs try to create an organization that fits their skills and views.

The fit between the leader and the organization suggests that the information processing capacities and capabilities meet the leader's preferences and also are efficient and effective for both the leader and the organization. Galbraith (1973) indicated that the information processing capacity of the individual and the organization must be sufficient to cope with the uncertainty of the task. First, the leader prefers an organization which fits his/her own preferences for delegation and uncertainty avoidance; leadership style creates a contingency for the organization. The organization must also fit with the climate, external environment, strategy, technology and size. The organizational design must include all these factors for efficiency and effectiveness.

Using the preference for delegation and preference for uncertainty avoidance which are related to the leader's information processing capacities, we have developed a series of propositions. The propositions relate to formalization, centralization, complexity, incentives and coordination mechanisms as well as to configuration.

To illustrate that the fit between the leader and the organization is a dynamic and ongoing process and a challenge, we examine the lifecycle of an organization briefly here and in more depth in Chapter 10. The general lifecycle approach to leadership and organization (Cameron and Whetton, 1981) suggests a congruency between the leadership style and the configuration throughout the life of the organization. In the lifecycle view, the organization goes through the entrepreneurial stage, the collectivity stage, the formalization stage, and the elaboration stage via growth and decline. In the entrepreneurial stage the organization is informal and non-bureaucratic. The manager has to be creative and technically oriented. When the organization moves to the collective stage due to growth, a few formal systems begin to appear. The manager has to delegate some decision-making authority. However, the manager is still involved in many of the details. The installations and use of rules, procedures and control are important when the organization reaches the formalization stage. Communication becomes more formal and less frequent. Specialization becomes important. Top management is less involved in the day-to-day operation but more concerned about long-term strategic decisions. More decisions are delegated, and coordination and control are keys to success. In the elaboration stage, the organization matures. The organization has to fight the bureaucratization to be more innovative, or the organization will slide into a decline. The organizational structure has to be very elaborate (Daft, 1992). The evolution shows how the organization adapts to the changing demands for information processing capacity. The organizational design is dynamic and has to be adapted to the organizational situation. Additionally, the requirements of the management are not the same in the different stages. This may cause different kinds of misfits. In many cases management either has to change preferences or leave the organization.

The leader group (which may be one individual) affects the best choice of the organization structure. This effect can be modeled in a two stage model describing the leader group's preference for decision-making in four leadership styles and their effect on the structure. This is shown in Figure 3.3.

Figure 3.3. The Two-stage Model on Leadership

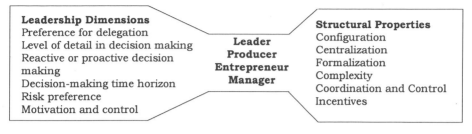

Describing a Leader

We describe a leader in terms of decision-making preference for delegation and preference for uncertainty avoidance which composed of five dimensions: level of detail in decision-making, reactive or proactive decision-making, decision-making time horizon, risk preference, and motivation and control.

Referring to Table 3.2 above, a high preference for delegation is consistent with Theory Y leaders, a democratic approach and a flexible personality - the individual we categorize as a leader.

The five dimensions of uncertainty avoidance for a leader include a more aggregate approach to detail. A Theory Y leader with a democratic approach is less involved in details as is a democratic leader. On the second dimension, a proactive leader can engage in more general problems and create opportunities facing the organization. Creative thinking, a vision for the future, or similar concepts - all are consistent with a proactive orientation. Third, leaders tend to have longer term horizon, even if they are excellent at dealing with the present and short term problems. Next, leaders take risks - big risks where the downside results are more likely to occur and may imply significant high losses. Finally, leaders lead by inspiration, not detailed control of activities. Miller and Toulouse's (1986b) empirical work supports the relations that leaders motivate by inspiration and not by direct control.

An individual with a high preference for delegation, a long-term horizon, aggregate information, general decision-making, a proactive approach, risk taking, and motivation by inspiration is likely to be a leader.

Leader Propositions

If an individual has a high preference for delegation of decision authority, then the individual is likely to be a leader.

If an individual has a preference for very aggregate information when he makes decisions, then the individual is likely to be a leader.

If an individual prefers to be proactive in his decision-making process, then the individual is likely to be a leader.

If an individual has a preference for long-term horizon in his decision-making process, then the individual is likely to be a leader.

If an individual has a low level of risk aversion, then the individual is likely to be a leader.

If an individual prefers to motivate the employees, then the individual is likely to be a leader.

A Leader's Effect on Structure

A leader as described above is an individual who delegates and does not avoid uncertainty. Here, we develop the effect of the leader style for the organization.

Generally, a leader wants a configuration where he/she can focus on large issues and the future, but does not require the leader to be involved in the operations or details of the organization. The leader wants a general level of predictability, but can tolerate ranges of outcomes. The leader sets policy and direction, but wants an organization that can function on its own. This is efficient for the leader as his/her attention or information processing capacity is more focused on policy than details. Delegation is a heuristic for the leader to use his/her time efficiently to accomplish his/her goals. The ad hoc organization needs general direction and goal setting, but individual specialists can make decisions within the general framework. This works well for a small organization. For larger organizations, the divisional organization requires policy and direction for the divisions, but decisions can be delegated to the divisions themselves. The matrix configuration requires goals and general direction; it functions well when the matrix is delegated to make decisions within the matrix itself. These organizations are efficient for the leader as each configuration can operate on its own.

A leader can assume a degree of uncertainty avoidance and wants an organization that embraces uncertainty, but with a purpose. He/she has a preference for decentralization or decision-making throughout the organization. This is the direct result of the leader group's decision-making style. This turns out to be very important, and a misfit here will usually hamper organizational performance (Burton et al, 2003). If the organization is designed with a high degree of centralized decision-making, and the leader group wants to delegate, then he or she may let decisions go without resolution and create delays. The results are confusion and frustration for all. Miller and Toulouse (1986b) and Miller and Dröge (1986a) find very strong support for this relation.

A leader prefers a low formalization where there are few rules and standards for the operations of the organization. These rules impede a leader's flexibility to do new things and do them differently. The leader wants to change things quickly.

A leader prefers an organization where jobs and functions are flexible and can be changed to fit the new, emerging and existing challenges. The coordination and control of the activities is not a central concern. Yet, the information processing demands are great within the organization. The high uncertainty requires lots of information and communications to cope (Galbraith, 1977). Generally, a leader prefers an organization where the demand for information processing does not exceed his/her capacity. But having the right information in the right place at the right time is a big risk for this organization. Relational information processing systems including email and telephones work well here (Hunter, 1998). The organizational complexity requires a media rich information system. This is in line with Robbins (1990), who suggests that an advanced information system can enhance the coordination and control even in very complex organizations.

The incentives for this organization should be results based, i.e., individuals need to focus on getting the job done well, and the exact means of accomplishments are less important, but within general guidelines and policy. Individuals should be rewarded for getting the job done - not following detailed procedures.

In order to coordinate activities, rich information is important. Coordination emerges from the individuals in the organization than being imposed. Here the leader takes on the uncertainty of what can happen, but leads through policy and motivation of the employees. The leader inspires others to take on the goals, challenges and opportunities of the organization.

Leader Propositions on Structure

If an individual is a leader, then the organization configuration should be ad hoc for a small organization and divisional or matrix for a large organization.

If an individual is a leader, then the centralization should be low.

If an individual is a leader, then the formalization should be low.

If an individual is a leader, then the complexity should be medium.

If an individual is a leader, then the incentives for the organization should be results based.

If an individual is a leader, then the coordination and control should be through general oversight and review, loose coordination, meetings, liaison, and rich information.

Figure 3.4. Leader: Description and effect on Structure

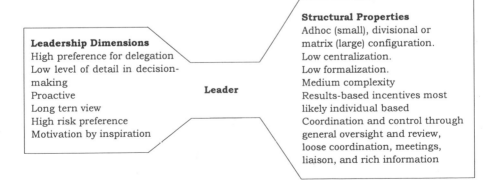

Leadership Dimensions
High preference for delegation
Low level of detail in decision-making
Proactive
Long tern view
High risk preference
Motivation by inspiration

Leader

Structural Properties
Adhoc (small), divisional or matrix (large) configuration.
Low centralization.
Low formalization.
Medium complexity
Results-based incentives most likely individual based
Coordination and control through general oversight and review, loose coordination, meetings, liaison, and rich information

Describing a Producer

We describe a producer in terms of decision-making preference for delegation and preference for uncertainty avoidance which is composed of five dimensions: level of detail in decision-making, reactive or proactive decision-making, decision-making time horizon, risk preference, and, motivation and control.

Referring to Table 3.2 above, a high preference for delegation is consistent with Theory Y leader, a democratic approach and an individual with a low locus of control; the individual we categorize as a producer.

The five dimensions of uncertainty avoidance for a producer include individuals who want to be involved in the details of decision-making and what is going on in the organization. If an individual wants to understand, but not make decisions on line-by-line items, he is likely to be categorized as a producer, as supported by studies by Miller and Toulouse (1986b). They also fit the competing leadership model shown in Figure 3.1. Second, the producer reacts to these events as they occur. A reactive manager is often faced with emerging detailed crises and operational problems. Third, the producer takes a short-term orientation to avoid the uncertainty of longer term commitments. Fourth, a person who is very risk averse does not want to let anything be determined by chance and wants to know what is happening. The risk-aversion dimensions appear in most typologies either indirectly or directly. This fits the view of being a Theory X manager and an autocratic person. Fifth, the distinction between motivation and control is very important and is summarized in the last row of Table 3.2. Miller and Toulouse's (1986b) empirical work supports these relations.

We categorize an individual with a high preference for delegation, a short-time horizon, detailed information, a reactive approach, risk aversion, and a desire for control as a producer.

Producer Propositions

If an individual has a high preference for delegation of decision authority, then the individual is likely to be a producer.

If an individual has a preference for very detailed information when he makes decisions, then the individual is likely to be a producer.

If an individual prefers to be reactive in his decision-making process, then the individual is likely to be a producer.

If an individual has a preference for short-term horizon in his decision-making process, then the individual is likely to be a producer.

If an individual is risk averse, then the individual is likely to be a producer.

If an individual prefers to control the employees, then the individual is likely to be a producer.

A Producer's Effect on Structure

The producer is described above as an individual who delegates and avoids uncertainty. Here, we develop the effect of the producer's style for the organization.

Generally, a producer wants a configuration where the producer can delegate, but know details and maintain control. The divisional configuration permits the producer to delegate strategy within policy limits and most all operational decisions to the divisions. Similarly, the matrix organization requires delegation of decision-making. The functional organization is not a good fit as it requires more direct involvement in operational detail than the producer desires. Here again, the matrix and divisional configurations are efficient information processing approaches for the producer although, it might be argued that the producer is too involved in details, but it serves the need to avoid uncertainty.

A producer has a preference for decentralization. This is the direct result of the manager's decision-making style and preference to delegate. Greater centralization requires that the producer make decisions that he/she would prefer others make.

A producer prefers a medium to high formalization where there are rules and standards for the operations of the organization. These rules and their application give the producer the desired predictability and helps avoid the uncertainty that may result in an organization with fewer rules.

A producer prefers an organization where jobs and functions are well defined and the producer can predict what each subunit will do. The information demands are great, as the producer focuses on details of operations to avoid uncertainty; he/she has a great need to know, even if the producer prefers to delegate.

The incentives for this organization should be results based, i.e., individuals need to focus on getting the job done well. The producer is interested in results for the organization and less concerned with the means provided general guidelines are not violated. The results-based incentives provide the producer with a motivational device that provides both inspiration and control, i.e., focused work in the issues of concern to the producer.

In order to coordinate and control and avoid uncertainty, the producer uses explicit and detailed goals, frequently developed by others. These goals are frequently converted into resource implications and budgets which become a major operations control and a means to avoid financial uncertainty.

Producer Propositions on Structure

If an individual is a producer, then the organization configuration should be a divisional or matrix, but not a functional.

If an individual is a producer, then the centralization should be low.

If an individual is a producer, then the formalization should be medium or high.

If an individual is a producer, then the complexity should be medium or high.

If an individual is a producer, then the incentives for the organization should be results based.

If an individual is a producer, then the coordination and control should be through general oversight and review, resource allocation and budgets, detailed explicit goals, and meetings.

Figure 3.5. Producer: Description and effect on Structure

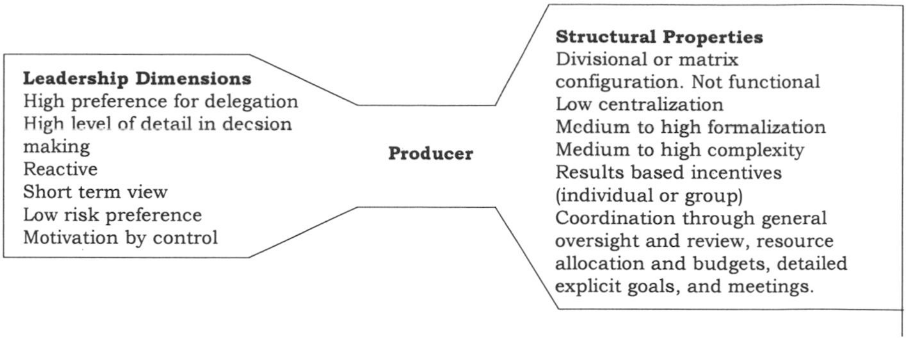

Leadership Dimensions
High preference for delegation
High level of detail in decsion
making
Reactive
Short term view
Low risk preference
Motivation by control

Producer

Structural Properties
Divisional or matrix
configuration. Not functional
Low centralization
Medium to high formalization
Medium to high complexity
Results based incentives
(individual or group)
Coordination through general
oversight and review, resource
allocation and budgets, detailed
explicit goals, and meetings.

Describing an Entrepreneur

A low preference for delegation is consistent with Theory X leaders, an autocratic approach and a high need for achievement; the individual we categorize as an entrepreneur.

The five dimensions of uncertainty avoidance for an entrepreneur are like those of a leader which include a more aggregate approach to detail. On the second dimension, an entrepreneur engages in more general problems and created opportunities facing the organization. Creative thinking, a vision for the future, or similar concepts - all are consistent with a proactive orientation. Third, entrepreneurs tend to have longer term horizon, even if they are excellent with dealing with the present and short term problems. Next, entrepreneurs take risks - big risks where the downside results can be significant both as likely to occur and with significant loss. Finally, entrepreneurs lead by inspiration rather than through detailed control of activities (Miller et al., 1986b).

An individual with a low preference for delegation, a long-term horizon, aggregate information, general decision-making, a proactive approach, risk taking, and motivation by inspiration is likely to be an entrepreneur. The information processing demands for an entrepreneur are very high for making decisions and keep a focus on the longer term in the face of uncertainty.

Entrepreneur Propositions

If an individual has a low preference for delegation of decision authority, then the individual is likely to be an entrepreneur.

If an individual has a preference for very aggregate information when he makes decisions, then the individual is likely to be an entrepreneur.

If an individual prefers to be proactive in his decision-making process, then the individual is likely to be an entrepreneur.

If an individual has a preference for long-term horizon in his decision-making process, then the individual is likely to be an entrepreneur.

If an individual has a low level of risk aversion, then the individual is likely to be an entrepreneur.

If an individual prefers to motivate the employees, then the individual is likely to be an entrepreneur.

At the Institute of Applied Computer Science, Benny Mortensen was very involved in making decisions. When the new organization was struggling to survive, he took risks and had a short-time horizon. He delegated a little but in general was very involved in the activities. His decision-making was neither reactive nor proactive and he used a mixture of motivation and control. Generally, he was closer to a need-for-an-achievement manager than a flexible leadership style. Benny Mortensen has a relatively high preference to be involved. For this situation he had an appropriate leadership style. However, only a few changes in one or two of the dimensions would jeopardize the appropriateness.

An Entrepreneur's Effect on Structure

An entrepreneur is described above as an individual who prefers not to delegate and does not avoid uncertainty. Generally, an entrepreneur wants a configuration where he/she is in control and uses his/her personal authority. The entrepreneur determines the strategy and frequently the operating decisions as well. For the small organization, the simple configuration where the entrepreneur is in charge and makes most decisions is a good fit. For the larger organization, the functional organization where the departments and specialties are coordinated and controlled by the entrepreneur is a good fit. The ad hoc organization might work if the entrepreneur can let others do their own specialized activities without direct interference. The machine bureaucracy is not a good fit; the entrepreneur does not have sufficient freedom to drive the organization in the direction that he/she desires.

An entrepreneur can assume a high degree of uncertainty and wants an organization that embraces uncertainty, but with a purpose. He/she has a preference for centralization around himself/herself. This is the direct result of the entrepreneur's decision-making style.

An entrepreneur prefers a low formalization where there are few rules and standards for the operations of the organization. These rules may impede one's flexibility to do new things and do them differently. The entrepreneur wants to change things quickly.

An entrepreneur prefers an organization where jobs and functions are flexible and can be changed to fit the new, emerging and existing challenges. The information processing demands are great within the organization. The high uncertainty requires lots of information and communications to cope (Galbraith, 1977). Generally, an entrepreneur wants an organization where the demand for information processing does not exceed his/her capacity. But having the right information in the right place at the right time is a big risk for this organization. The organizational complexity requires a media rich information system. The information revolves around the entrepreneur who must deal with lots of complex information.

The incentives for this organization should be results based, i.e., individuals need to focus on getting the job done well, and the exact means of accomplishments are less important, but within general evolving directions as given by entrepreneur. Individuals should be rewarded for getting the evolving job done - usually there are no detailed procedures to follow and if there are, the procedures are likely to be dysfunctional.

In order to coordinate activities, rich information is important. Coordination revolves around the entrepreneur who wants to direct operations. The entrepreneur takes on the uncertainty of the results.

The leader inspires others to take on the goals, challenges and opportunities of the organization.

Entrepreneur Propositions on Structure

If an individual is an entrepreneur, then the organizational configuration should be simple for a small organization and functional for a large organization, but not a machine bureaucracy. It might also be an ad hoc configuration.

If an individual is an entrepreneur, then the centralization should be high.

If an individual is an entrepreneur, then the formalization should be low.

If an individual is an entrepreneur, then the complexity should be low or medium.

If an individual is an entrepreneur, then the incentives for the organization should be results based.

If an individual is an entrepreneur, then the coordination and control should be through direct intervention, and meetings.

Figure 3.6. Entrepreneur: Description and effect on Structure

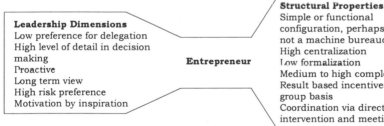

Leadership Dimensions
Low preference for delegation
High level of detail in decision making
Proactive
Long term view
High risk preference
Motivation by inspiration

Entrepreneur

Structural Properties
Simple or functional configuration, perhaps adhoc - not a machine bureaucracy.
High centralization
Low formalization
Medium to high complexity
Result based incentives on a group basis
Coordination via direct intervention and meetings

Describing a Manager

A low preference for delegation is consistent with Theory X managers, an autocratic approach and an individual with a close locus of control; the individual we categorize as a manager.

The five dimensions of uncertainty avoidance for a manager include individuals who do not delegate decision-making authority and want to be involved in details of what is going on in the organization. A manager is a more involved decision-maker who prefers to gather his own information and use detailed information. If the decision-maker wants to make decisions on line-by-line items, he is likely to be categorized as a manager, as supported by studies by Miller and Toulouse (1986b). They also fit the competing leadership model shown in Figure 3.1. Second, the individual reacts to these events as they occur. A reactive manager is often faced with emerging detailed crises and operational problems. Third, the manager takes a short-term orientation to avoid the uncertainty of longer term commitments. Fourth, a person who is very risk averse does not want to let anything be determined by chance. He/she wants to be in control. The risk-aversion dimensions appear in most typologies either indirectly or directly. This fits the view of being a Theory X manager and an autocratic person. Fifth, the distinction between motivation and control is very important and is summarized in the last row of Table 3.2. Miller and Toulouse's (1986b) empirical work supports these relations. The information processing demands on a manager are very high. The focus on short-term decision-making and risk avoidance may lead to a neglect of longer term strategic challenges. The manager uses his/her time efficiently within the bounds of operating the organization.

We categorize an individual with a low preference for delegation, a short-time horizon, detailed information, detailed decision-making, a reactive approach, risk aversion, and a desire for control as a manager.

Manager Propositions

If an individual has a low preference for delegation of decision authority, then the individual is likely to be a manager.

If an individual has a preference for very detailed information when he makes decisions, then the individual is likely to be a manager.

If an individual prefers to be reactive in his decision-making process, then the individual is likely to be a manager.

If an individual has a preference for short-term horizon in his decision-making process, then the individual is likely to be a manager.

If an individual is risk averse, then the individual is likely to be a manager.

If an individual prefers to control the employees, then the individual is likely to be a manager.

A Manager's Effect on Structure

The manager is described above as an individual who does not delegate and avoids uncertainty. Generally, a manager wants a configuration where the manager is in control and can utilize the authority of the position. The functional configuration where the manager is in charge of operations and the coordination issues is a good fit. The manager can exert his/her authority and be directly involved in the organization's activities. A machine bureaucracy is also a good fit. Here, the manager maintains control through the application of the rules and standard operating procedures. The functional configuration and the machine bureaucracy can operate in a similar manner; they are quite compatible. A manager is not a good fit where there is a more shared decision-making delegation and where uncertainty avoidance is lower. Ad hoc, matrix, professional burcaucracy and virtual networks - all require a more shared decision-making style where information is dispersed among many individuals in the organization and actions must be taken without direct intervention by the manager.

A manager has a preference for high centralization or decision-making at the top. This is the direct result of the manager's decision-making style; the manager wants a more centralized organization. This turns out to be very important, and a misfit here will usually hamper organizational performance (Burton et al, 2003). If the organization is designed to allow a high degree of decentralized decision-making, and the leader has a high preference for detailed decision-making, then he or she will tend to get involved despite the fact that he/she is not supposed to. The results are confusion and frustration for all. Miller and Toulouse (1986b) and Miller and Dröge (1986a) find very strong support for this relation.

A manager prefers a high formalization where there are rules and standards for the operations of the organization. As discussed above, the machine bureaucracy is a good fit for a manager. These rules and their application give the manager the desired control and help avoid the uncertainty that may result in an organization with less predictability of activity.

A manager prefers an organization where jobs and functions are well defined and the manager knows what each subunit and individual does. Yet, with a high level of differentiation, it is more difficult for the manager to coordinate and control the activities of the organization. The information processing demands are great. The result comes from an increase in the horizontal differentiation as well as from an increase in the vertical differentiation. A high horizontal differentiation requires the coordination of specialists, and a high vertical differentiation implies that many middle-level managers coordinate. Generally, a manager prefers an organization where the demand for information processing does not exceed his/her capacity. His/her capacity may, however, be enhanced by an information processing system or a decision support system. The manager prefers an organization where the complexity is managed and coordinated with rules and direct intervention - both require high levels of information.

This is in line with Robbins (1990), who suggests that an advanced information system can enhance the coordination and control even in very complex organizations. It is supported empirically by Miller and Dröge (1986a), who in their orthogonal, exploratory factor analysis found that "the complexity factor simply did not materialize" in their study of Canadian firms. But, the propositions are not supported by the empirical studies by Miller and Toulouse (Miller et al., 1986b), who found a significant negative correlation between a flexible personality and specialization-one component of differentiation. Additionally, they found a significant positive correlation between specialization and need for achievement. The control argument leads to a low degree of complexity. The counterargument relates to the horizontal differentiation where a specialization could lead to a higher performance, which would fit the attributes of a high-need-to-achieve leader.

The incentives for such an organization should be procedural, i.e., individuals have jobs, job description and rules to perform their responsibilities. Individuals should be rewarded for doing their jobs and doing them well, which means consistent with the rules and norms of the organization. An individual, who deviates from these norms whatever the motivation, may compromise the coordination demands of the organization and also create uncertainty which the manager wants to avoid.

In order to coordinate functional specialties, management has to control the decisions and activities of the organization. This organization demands lots of information to meet the needs of the manager.

Manager Propositions on Structure

If an individual is a manager, then the organization configuration should be functional and/or a machine bureaucracy, but not an ad hoc, a matrix, a professional bureau or a virtual network.

If an individual is a manager, then the centralization should be high.

If an individual is a manager, then the formalization should be high.

If an individual is a manager, then the complexity should be high.

If an individual is a manager, then the incentives for the organization should be procedure based.

If an individual is a manager, then the coordination and control should be through standards, rules, monitoring, and auditing.

Figure 3.7. Manager: Description and Effect on Structure

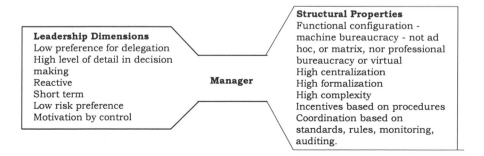

Leadership Dimensions
Low preference for delegation
High level of detail in decision making
Reactive
Short term
Low risk preference
Motivation by control

Manager

Structural Properties
Functional configuration -
machine bureaucracy - not ad
hoc, or matrix, nor professional
bureaucracy or virtual
High centralization
High formalization
High complexity
Incentives based on procedures
Coordination based on
standards, rules, monitoring,
auditing.

Interaction Effects of Size and Leadership Style

Miller and Toulouse (1986b) in their study found two qualifiers of the effect of management style on structure: size and the dynamics of the environment. They suggest that the relationship between leadership style and organizational structure is strongest when the organization is small and the environment is dynamic. If the organization is small, then management has a greater power and, therefore, may influence the structure more than if the organization is large. This is also often confounded with the fact that in small organizations management is both owner and founder of the organization. In a small organization the demand for information processing is normally smaller than in large organizations. Therefore, centralization can be higher without creating a misfit between the demand and capacity for information processing.

In a small organization, when there is a conflict about a recommendation between management's preference and the environment, then management's preference wins. We have found that to be generally true in our validation of the expert system. The certainty factors associated with the rules derived from the propositions in this chapter are generally larger than the certainty factors for rules derived from any other propositions, by a magnitude of two to three. This means that there should at least be two contingency factors in conflict with management's preference when management's preference does not win.

Miller and Toulouse (1986b) also found that leadership style is more important in small young organizations than in large old ones..

Propositions on the Interaction Effects of Leadership style and Management

The relationship between the management group's leadership style and organizational structure is stronger in small organizations than in large ones (see Chapter 5 for a discussion of size).

The relationship between the management's leadership style and organizational structure is stronger in more dynamic environments than in stable environment (see Chapter 6 for a discussion of environment).

The relationship between the management's leadership style and organizational structure is stronger in young organizations than in old ones.

MANAGING THE LEADER

If there is a misfit between the leadership style and the organization, which should be changed? There are many issues to consider and many possible changes: the organization can be changed to meet the preferences of the leader as we have discussed above, the leader can change his/her behavior to meet the needs of the organization, or the leader himself/herself can leave the organization. Let us focus on the leader who wants to meet the needs of the organization.

Individual skills as well as attitudes can be adapted. A fit between the management, employees and organizational structure can be obtained by hiring the right people, training them, and having organizational experience. Organizational changes may require changes in personnel. Changes also may be obtained through training and education. Most large organizations have extensive educational activities to ensure that employees have the right skills to perform well. Human resource management activities must fit the developments in the organizational structure and its task requirements. Numerous techniques have been developed for personal development (Milkovich and Boudreau, 1988).

The organization's incentive system influences individual behavior. Incentives based on behavior focus on activities for which the incentives are related. For example, if the incentive package for the leader is based on growth, managerial style may drift towards a preference for growth. Management's preferences may be changed by the incentive system. Individuals can learn to operate within a given organization. That is, an individual can learn to delegate or not; he or she can learn to become comfortable with risk avoidance or not. These are not trivial adjustments and require some time to learn. We tend to think it is easier to find a new leader than wait for the current one to adjust, or leaders may look to change the organization to meet his/her preferences rather than considering their own preferences

In SAS, as discussed in Chapter 2, major restructuring happened in 1975 when a crisis built up and one of the board's solutions was to hire a new CEO. A crisis is often the antecedent to a major re-structuring (Miller and Friesen P., 1980; Tushman and Romanelli, 1985). The question then arises about the attributes that the CEO should have. The propositions in this chapter give some answers to this question by using the following procedure. Apply the contingency factors excluding management style to find the best organizational structure. Then use the propositions to find the management style that would best support the recommended structure. The logic of "If X, then Y" allows you to conclude that "If Y is false, then X cannot be true." This allows us to exclude management styles that do not fit the recommended structure.

SUMMARY

In this chapter, we considered the fit between leadership style and organizational design. Consistent with the neo-information processing perspective, leaders have preferences for the level of detail they require in decision-making, for a reactive or proactive decision-making style, for risk preference, for a time horizon, for delegation, and for style of motivation. These leadership dimensions define a style that should fit the organization's complexity, formalization, and centralization as well as its organizational configuration.

CHAPTER 4
Organizational Climate

INTRODUCTION

"Navy retirees recruited by tech firms fit discipline, flexibility to new tasks." This was one of the headlines in the San Jose Mercury News, Sunday, February 15, 1998. The story tells how Silicon Valley corporations have started hiring retired and former military personnel. The story focuses on the similarities and differences in working in the military with its chain-of-command hierarchy and being in a seemingly less structured environment. "Military personnel are veterans of the teamwork that is regarded as an essential part of valley life. But the military requires clarity at every level about the task at hand to be critical for effective performance. What these newcomers to high tech often see are missed opportunities for nurturing a common sense of mission." "There's more freedom, more latitude to make things happen in the high tech companies," says Jack Gale a former Navy commander, but like in the Navy it is all about performance. The story not only stresses the problems and transitions the former military personnel had to go through to fit to their new positions but also what new ways of looking at things they bring in.

The important message is that people, their way of looking at things, their perceptions and values are indeed important. Applied Komatsu manager Mayekawa says companies hiring military personnel try to gauge their ability to operate in a civilian environment by asking how they would handle a hypothetical situation with a customer. "If they are too structured, the issue of culture comes into play," he says.

The Organizational Climate at SAS

Throughout its many years of steady growth, SAS had developed into a proud and respected organization - well known for its high quality and good service. SAS's reputation for high quality was widely attributed to the management's constant focus on new technologies, new aircrafts and new engines.

SAS was reputed for the loyalty of its employees. In return, SAS treasured its employees by offering most of them life-long employment relationships, and a wage level generally higher than most competitors.

With the appointment of Carlzon in 1980, many changes were initiated. To create a higher team spirit as well as heighten the motivation of employees, Carlzon made a big effort to align the corporate climate with the new strategy.

In this way, the management formulated common ideas and values that SAS should stand for in the future. The basic message was that SAS now had to create a service climate, meaning an increased focus on customers' wishes and demands.

Efforts were made to heighten the co-operation across functional areas such as cabin crew, cleaning-staff and technical staff. An amelioration of the SAS service could not be achieved if the various functional groups could not depend on each others to fill out their jobs to their utmost. This was necessary as all areas were highly dependent on each other.

Front line employees, who were most radically affected by the changes, were particularly enthusiastic. A team spirit large enough to evoke the admiration of most competitors was created.

Unions and middle managers were harder to convince, however. The changes brought along an obliteration of occupational demarcations and a reduction in the number of staff, which the unions obviously did not favor. Middle managers were not too enthusiastic either, as they had been deprived of much of their earlier authority.

Nevertheless, SAS created solid financial growth, and it was hard to convince anybody that Carlzon's strategy was not a success. Indeed, Carlzon was loved and admired by most employees as well as by most management consultants, and even competitors. The success of SAS was very much accredited to his personality.

Over time however Carlzon's entrepreneurial ideas got out of hand. In his effort to enlarge the company, SAS made investments in areas that were not related to its core products. The management slowly started to lose sight of its vision. As the earnings went down, so did the employees' admiration for Carlzon.

Employees were still excited about their new working environment but started criticizing Carlzon for his efforts to change the company via reductions and cutbacks. Further Carlzon seemed to have broken one of SAS's non written laws by not recruiting insiders for the company's management positions. Most employees believed that part of the reason for SAS's failing success was that the new managers from the outside did not have the same expertise in deciding what was good for SAS, as many of the senior employees would have had. Certain barriers were thus created between the management group and the employees.

The company suffered from the slowly increasing mistrust in Carlzon and his management group, and it was difficult for Carlzon to make any rationalizations without losing even more of the employees' faith, something he could not afford.

Stenberg, who was appointed in April 1994, did not conceal the fact that his major goal was to reduce the costs of SAS, and that this would necessitate major cutbacks and rationalizations.

Although many of Stenberg's rationalizations have led to employee frustrations, nobody can deny the fact that Stenberg has managed to get SAS back on its feet financially. Stenberg is respected for this, but in a less impassioned way than Carlzon was.

During the first period with Stenberg as CEO SAS probably was a less symbiotic company than it used to be. The company was been haunted by personnel strikes, particularly among the cabin crew, who were dissatisfied with their new working conditions as well as the many firings of their colleagues. The focus on costs was therefore in some periods at the expense of customers' wishes, threatening to deteriorate the company's competitiveness. Additionally, employees were discontent with not being let in on the management's plans.

Jan Stenberg followed a different personnel policy from Carlzon's, and generally did not use as much open dialogue to communicate his plans and visions to the employees. His overall sense of purpose was nevertheless probably very clear to most of the employees; Jan Stenberg wanted results! At the end of Stenberg's period the good financial results slightly loosened up the tight situation with less conflict and more trust.

Jørgen Lindegaard was expected to spearhead SAS expansion plans. In an interview he stated that one of his major tasks were to incorporate thousands of new employees in SAS. That vision only lasted a very short time. The major problems that SAS and other airlines faced at the end of 2001 forced Jørgen Lindegaard to announce major cutbacks. The sequence of unexpected problems including the SARS epidemic in 2003 implied that Jørgen Lindegaard had to introduce more cost savings and reorganizations. Further, he had to take away some very costly agreements with respect to the airline crews. Jørgen Lindegaard has been very open about the problems, and he made some strategic choices that allowed him to move activities to affiliated airlines if the SAS airline personnel did not accept his proposals. Apparently, he has managed to do this without destroying the trust and respect of the employees. The conflicts have been relatively minimal given the serious situation. The employees seem to have a somewhat ambivalent view on the readiness for change. They do not like it but acknowledge that change is necessary.

SAS has been going through a number of changes in the organizational climate. First, it seems that there was a high level of trust, a little conflict, equitable rewards, but a high resistance to change. This would lead towards a group climate. Carlzon with his entrepreneurial spirit worked hard on changing the resistance to change, creating a developmental climate.

When problems started to emerge the level of trust fell, conflicts ensued, and Carlzon and the management group started losing credibility. Additionally, a resistance to change reappeared. This drove the climate towards an internal process climate.

Stenberg had the same ideas as Carlzon: to change the resistance to change. Coming from an internal process climate he moved towards a rational goal climate. At the end of his period the climate moved towards a development climate as trust and respect increased.

Jørgen Lindegaard is facing a climate between a developmental and group climate. Using the descriptions by Hooijberg and Petrock, SAS has gone from being a friendly place to work, passing by an entrepreneurial phase, to a tough demanding place to work where the leaders are hard drivers, producers, and results-oriented and back to a situation that still is tough but also becoming more friendly. (Hooijberg and Petrock, 1993).

1. How did the climate change in the different periods described above?
2. What was the effect on structural possibilities?

Tosi (1992) argues that the scope of contingency theory should be broadened to include culture, organizational culture, and individual and group variables. So, how do the organizational climate and culture affect performance and are there issues of fit related to the appropriate choice of the organizational structure - its configuration, properties, and performance? In the literature the discussion has caused a good deal of controversy. First, the meaning of each - culture and climate - is unclear. Second, the difference between organizational climate and organizational culture is often blurred. Third, the relationship to performance is somewhat ambiguous.

In this chapter we will therefore argue that a climate measure is a complement factor to obtain an efficient and effective organization. The organization's climate, which captures how individuals feel about the organization, has to be in alignment with the organizational structure. An individual's feelings affect his/her ability to process information and thus, is an important measure in the determination of the information processing capacity of the organization. Therefore, climate affects the choice of structural properties. Additionally, there are important fit/misfit issues between the organization's climate and the other contingency factors in the multi-dimensional contingency model. These will be dealt with in Chapter 9.

We first review the concepts of climate and culture. We then examine climate and develop an approach to measuring climate for an organization. The competing values approach is a means to summarize and categorize climate. We further develop a number of relations between the four climate categories and the organizational design, i.e., the implications of the organization's climate for design. Finally, we discuss how the organizational climate may be changed and affected by the other variables in the multiple-contingency model.

CLIMATE AND CULTURE

Organizational culture and climate are often used in the literature to describe similar issues. Denison (1996) asks in his survey paper: "what is the difference between organizational culture and organizational climate?" He concludes that "on the surface, the distinction between organizational climate and organizational culture may appear quite clear: climate refers to a situation and its link to thoughts, feelings, and behavior of organizational members", while "culture, in contrast, refers to an evolved context (within which a situation may be embedded)." He is then led to the conclusion that climate and culture research "should be viewed as differences in interpretation rather than differences in the phenomenon." The discussion continues as can be seen in "The Handbook of Organizational Culture and Climate" (Ashkanasy et al., 2000).

Although organizational climate and culture are often used interchangeably, they have different roots. Let us begin with some definitions (Webster's Ninth New Collegiate Dictionary):

Climate: The prevailing influence or environmental conditions characterizing a group or period, atmosphere, the prevailing set of conditions.

Culture: An integrated pattern of human knowledge, belief and behavior that depend upon man's capacity for learning and transmitting knowledge to succeeding generations, the customary beliefs, social forms and material traits of a racial, religious or social group.

These definitions are a good point of departure. Culture is a pattern of knowledge, belief and behavior that emerge, including social forms. Social forms and knowledge in general include the organizational structure, for example, formalization and decision-making.

Organizational climate has been defined as the "relatively enduring quality of the internal environment of an organization that a) is experienced by its members, b) influences their behavior, and c) can be described in terms of the values of a particular set of characteristics (or attitudes) of the organization" (Tagiuri and Litwin, 1968). The climate is the "ether" within which an organization exists. "In an overall organizational model, climate can be seen as an intervening variable in the process between input and output and one that has a modifying effect on this process. Climate effects organizational and psychological processes, and thus acquires an influence over the results of organizational operations" (Ekvall, 1987).

The organizational culture is the organization itself...the form, beliefs, norms, social patterns, the way things are done, the symbols, rituals, etc. Schein (1992) defined culture as: "A pattern of shared basic assumptions that the group learned as it solved its problems of external adaptation and internal integration, that has worked well enough to be considered valid and, therefore, to be taught to new members as the correct way to perceive, think, and feel in relation to those problems." Culture is thus derived and observed in the emergent behavior of the group or the social form or organization. It is learned and transmitted from one generation to the next. The organizational culture is bound up with the form and properties of the organization itself. Patterns of behavior may emerge from the organizational form and the organizational properties themselves. Management creates and changes the form and properties primarily due to the influence the "pattern of behavior" has in the organization. It seems that the culture is integrated into the organizational design and the organizational design levers are also organizational culture levers.

The culture emerges from the form and acquires properties as the organizational form is implemented. Perhaps, the most explicit new organization element discussed as parts of the culture are the rituals, rites and symbols of management and these should not be over-looked. They are part of the organizational structure as they can be seen as means of coordinating the organizational activities.

From a design view, one would like to argue that some cultures are more efficient or more effective than others. Many popular books prescribe how to create an organizational culture that will enhance performance. Siehl and Martin (1990), however, argue that the evidence is very weak in the normal positive science sense: "...culture-financial performance proposition consists of variants of a contingency argument. Differences in the empirical results may also be attributed to the fact that studies have been carried out in different settings and in different countries. This raises the important question of how the national culture may affect the organizational climate (Hofstede, 2001). The effect of national culture is treated in Chapter 6 as one dimension of the environment.

The most common variant is a claim that firms with cultures congruent with their business strategies are better performers than firms that lack this congruency. When a firm's culture clashes with its strategy, confusion and conflicts of interest increase, strategies are resisted, and the firm's financial performance ultimately may be impaired. In contrast, when the culture and strategy are synchronized, such difficulties are said to be reduced, with a concomitant beneficial effect on financial performance. Evidence supporting this has, however, been confined, for the most part to short, almost anecdotal, descriptions of case studies of single organizations" (Siehl and Martin, 1990). They also state "when researchers look at organizations "as if" they were cultures, they can enrich our understanding of organizational life in new and unexpected ways. Already, interest in cultures has opened and reopened neglected areas of inquiry for example, by exploring the organizational relevance of symbolism, structuration, semiotics, deconstruction, and ideology." "These studies bring us far beyond the traditional variables of organizational theory, such as structure, firm size, technology, job satisfaction, motivation and leadership" (Siehl et al., 1990).

In the discussion of organizational culture and climate so far, little was said about the actual relationship between dimensions of culture, climate, structural design and performance. Some studies relate directly the various dimensions to performance while others try to develop culture and climate typologies (Poole and McPhee, 1983) from which a relationship to performance can be established.

One of the successful typologies that is heavily researched is the competing values approach, which we introduced in Chapter 3, Figure 3.1 (Quinn and Rohrbaugh, 1983). The competing values approach uses four categories to describe four cultures of the organization:

- The group culture (a friendly place to work where people share much of themselves).
- The developmental culture (dynamic, entrepreneurial and creative place to work).
- The internal process culture (formalized and structured place to work).
- The rational goal culture (results-oriented where leaders are hard drivers, producers and competitors).

Each culture describes an ideal model and may not be found in its pure form in any organization. Rather, each real world organization will display some aspects of each cultural model with greater weight for one model than another, i.e., an organization is more like a group culture than a rational goal culture, but there are aspects of both. Additionally, different parts of the organization may be more like one type while other parts of the organization may be more like another type.

Figure 4.1. The Competing Values Model

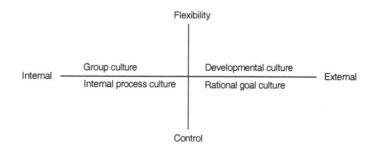

The competing values framework captures two basic dimensions of the organization (control-flexibility and internal-external). On the control-flexibility axis, the group and developmental cultures are more flexible and the internal process and rational cultures are more control oriented. On the orthogonal axis, the group and the internal process are more internally oriented where the developmental and rational cultures are more externally oriented.

The competing values framework is a very versatile typology for capturing the complexity of a variety of management issues. The applications include strategy (Bluedorn and Lundgren, 1993), human resource policies (Giek and Lees, 1993; Yeung et al., 1991), organizational change (Hooijberg et al., 1993), management information systems (Cooper and Quinn, 1993) as well as culture (Cameron and Freeman S.J., 1991; Denison, 1990), and climate (Zammuto and Krakower, 1991). The initial development by Quinn and Rohrbaugh (1983) examined organizational effectiveness criteria. The competing values approach has been robust across many applications and reliable as a measurement instrument. Thus, the competing values approach enables us to sort out the relationship between culture and climate.

Poole (1985) states that climate seems to be a feature of, rather than a substitute for, culture. The same view was adopted by Zammuto and Krakower (1991) who related climate to the four categories in the competing values approach. They use organizational characteristics (centralization, formalization, long-term planning), climate measures (trust, conflict, morale, equity of rewards, resistance to change, leader credibility, and scapegoating) and strategy dimensions (reactive/pro-active orientation) to categorize the culture into group culture, developmental culture, internal process culture, and rational culture. They concluded that cultural type is related to differences in organizational climate. As we shall discuss later their study allows us to develop four types of climates that correspond to the four types of culture thus allowing us to obtain relationships between climate dimensions and organizational characteristics.

The many empirical studies using the competing values approach give us a rich set of knowledge to fill in the organizational psychological climate on the top left-hand side in Figure 1.2 in Chapter 1.

Quinn and Spreitzer (1991) analyzed in two studies two different instruments to describe the competing values categories. Both instruments seem to capture the categories well. The implication, with some caution, is that it is possible to combine results from different studies using different instruments to describe the competing values typology. The categorizations seem to be rather robust. Using this reasoning the competing values approach provides us with relationships between strategy, environment, technology, leadership, organizational form, and psychological climate. The competing values approach specifically allows us to sort out the various dimensions of organizational climate and culture. From an information processing point of view, we are interested in including those dimensions that affect the information processing capacity of the members of the organization.

In brief, the general support that culture is an important factor for efficient organizations is mostly case and anecdotal evidence, not the normal positive science studies. However, interpreted more broadly, culture may be considered as looking at organizations as particular sets of situational and structural patterns which are consistent - provide a total fit. This is called the "configurational perspective" by Delery and Dotty (1996). In the context of culture, the competing values approach provides such ideal patterns including climate as one of the dimensions.

Climate and culture are thus important to consider in the design of the organization. The relationship between climate and performance is linked via behavior. However, culture is commonly defined and used to include both dimensions - how the individuals in the organization process information as well as more structural dimensions. In this way culture is both the means and the ends in the design process. Particular culture typologies thus implicitly assume that a proper fit has been obtained. We are searching for a fit with the way individuals in the organization process information and the way the organization should be structured.

Introducing climate and culture opens and enriches the managerial framework for a deeper and more comprehensive understanding of design choices. Culture is part of any organization, whether designed or not. However, organizational culture mixes organizational properties with the behavior of the individuals. The organizational climate is the part of the culture that is more directly related to the general behavior of the individuals in the organization and thus a candidate to be included in the multi-contingency model.

The correct fit between behavior of the members of the organization and the organizational structure is the key to improved performance. Again, the individuals' capacity to process information is an important element in assessing the information processing capacity of the organization. From a design point of view it is thus important to develop a set of climate dimensions that captures the individuals' perception of the work environment and its effect on the individuals' performance and thus on the performance of the organization.

LITERATURE REVIEW ON CLIMATE

Climate was defined in the dictionary as the prevailing influence on environmental conditions characterizing a group or period, atmosphere, the prevailing set of conditions.

Even within the more narrow set of literature on climate there has been confusion about the relevance and definition of the concept and the relationship with organizational structure and organizational culture (Denison, 1996; James and Jones, 1974; Schneider, 1990). This review does not pretend to survey all the research within the climate area, but will try to outline different views and research directions that will allow us to integrate a measure of climate into the multi-dimensional contingency model.

One reason for the confusion in the literature can be found in the use of climate to represent seemingly different concepts. First, climate can be seen as organizational climate or psychological climate. Ekvall (1987) states that the organizational climate arises in the confrontation between individuals and the organizational situation. According to James and Jones (1974) organizational climate can also be viewed in two different ways: "a multiple-measurement-organizational attribute approach" or "a perceptual measurements-organizational attribute approach".

Is Ethics part of Climate?

The conduct of business involves conflicts of interests, often with an intrinsic moral or ethical dimension. In recent years, value conflicts have increasingly been associated with organizational incentive systems. Performance-related bonus, profit sharing plans, employee stock-ownership plans, option-based compensation, and Golden Handshake programs have inherent potentials for generating value conflicts between personal and organizational interests, for example, the Enron, Xerox, and ABB alleged scandals.

The climate of an organization has major impacts on the behavior of the members of an organization; and since managers can have significant influence on the development of the value system of an organization (Sims and Keon, 1997), it is necessary to understand what affects the psychological and ethical climate of an organization.

Theoretically, (Schneider and Reichers, 1983; Schneider and Snyder, 1975) it has been claimed that a relationship exists between ethical climate and traditional psychological climate. The ethical climate is a macro-level organizational characteristic consisting of a shared set of norms, values, and practices of the organizational members regarding appropriate behavior in the workplace (Cullen et al., 1993; Victor and Cullen, 1987; Victor and Cullen, 1988; Victor and Cullen, 1990). The construct measures the prevailing values of an organization in a combination of ethical standards rooted in moral philosophy and cognitive moral judgment structures (Kohlberg, 1971) and dimensions of loci of ethical concern borrowed from sociological theories (Gouldner, 1957; Merton, 1957) on groups and reference groups. The cognitive moral development theory is based on a definite sequence of stages of development of moral reasoning, which is related to managerial growth and the ethical quality of professional decision-making, whereas the sociological theories give guidance to the orientation of the organization in line with the Gemeinschaft/Gesellschaft tradition.

Empirically, Lemmergaard (2003) has tested the relationship between ethical and psychological climate, and found that ethical climate is a distinct but supplementary dimension to more traditional climate dimensions. Consequently, this study argues that the traditional climate must be extended to include dimensions of ethics, values, and morality in order to capture the complete picture of organizational climate(s).

Although the ethical climate construct is not directly useful in an everyday organizational context, it gives valuable insights into the new paradigm of values-based management, which builds on shared organizational values and beliefs. The values-based management paradigm is in its infancy and characterized by a continuous discussion among researchers and practitioners about which theoretical foundation and consequently which technical tools are most appropriately applied under this paradigm. Since values are the decisive motive that drives action, knowledge about values and value conflicts need to be a focus area for the organization of the future.

Both these approaches are confounded with organizational structure and processes and the general organization situation in the same way as was discussed with respect to organizational culture. The organizational climate is measured using variables like individual autonomy, the degree of structure imposed as the positions, reward orientation, consideration, warmth, and support. This is also the case in the treatment of organizational climate dimensions presented in Litwin and Stringer (1968) where organizational climate is measured along the following dimensions: structure, responsibility, warmth, support, reward, conflict, standards, identity, and risk.

It is obvious from the above that measures and dimensions of organizational climate and organizational culture can, as discussed earlier, be confused.

In contrast the definition of the psychological climate seems stricter in the sense it refers to perceptions held by the individuals about the work situation. James and Jones (1974) summarize the psychological climate to be a set of summary or global perceptions held by individuals about their organizational environment. The psychological climate is a summary evaluation of actual events based upon the interaction between actual events and the perception of those events. The psychological climate has been measured using dimensions such as disengagement, hindrance, esprit, intimacy, aloofness, production emphasis, trust and consideration.

Koys and DeCotiis (1991) define the psychological climate as "an experimental-based, multi-dimensional, and enduring perceptional phenomenon, which is widely shared by the members of a given organizational unit." They continue to state that the psychological climate is the description - and not the evaluation - of experience. As such, the psychological climate is different from, e.g., job satisfaction. In their survey Koys and DeCotiis report more than 80 different dimensions found in the literature, which has been labeled a climate dimension.

They set out to find a theoretical-meaningful and analytical-practical universe of all possible climate dimensions. They established three rules for a dimension to be included in the universe:

- Has to be a measure of perception.
- Has to be a measure describing (not evaluating).
- Must not be an aspect of organizational or task structure.

These rules attempt to sort out the confusion and also distinguish the measures from the organizational climate and culture measures. The rules make sure that the psychological climate measure is not confounded with the organization's structural properties.

Applying these rules to the more than 80 dimensions used in climate measurements and combining dimensions that were actually the same despite the fact they had different names reduced the number of dimensions to 45. Koys and Decotiis were then able to categorize these 45 dimensions into the 8 summary dimensions.

- *Autonomy*: The perception of self-determination with respect to work procedures, goals, and priorities.
- *Cohesion*: The perception of togetherness of sharing within the organization setting, including the willingness of members to provide material aid.
- *Trust*: The perception of freedom to communicate openly with members at higher organizational levels about sensitive or personal issues with the expectation that the integrity of such communications will not be violated.
- *Pressure*: The perception of time demands with respect to task completion and performance standards.
- *Support*: The perception of the tolerance of member behavior by superiors, including the willingness to let members learn from their mistakes without fear of reprisal.
- *Recognition*: The perception that member contributions to the organization are acknowledged.
- *Fairness*: The perception that organizational practices are equitable and nonarbitrary or noncapricious.
- *Innovation*. The perception that change and creativity are encouraged, including risk-taking into new areas or areas where the member has little or no prior experience.

Koys and DeCotiis (1991) tested the validity and reliability of their summary scales using a separate sample of managerial and professional employees and found that their dimensions were both valid and reliable. The results suggest that trust and support may be combined bringing the dimension of psychological climate down to seven.

The dimensions of climate developed by Koys and DeCotiis fit the notion by Rousseau (1988) where she says that "climate is a content-free concept, denoting in a sense generic perceptions of the context in which an individual behaves and responds"

From these dimensions it is seen that the concept of psychological climate does not interfere with the right-hand side of the multi-dimensional contingency model (Figure 1.2), and thus they can more readily be candidates for measuring climate in a contingency model for organizational design. "In an overall organizational model, climate can be seen as an intervening variable in the process between input and output, and one that has a modifying effect on this process. Climate affects organizational and psychological processes, and thus acquires an influence over the results of organizational operations" Ekvall (1987).

The conclusion of this survey on climate in an organizational context is that seven dimensions seem to exist that measure the individuals' perception of the organization's psychological climate. As such they can be a basis for a definition of climate in a multi-contingency model for organizational design.

MEASURING AND CATEGORIZING CLIMATE

The psychological climate refers to the beliefs and attitudes held by individuals about their organization. The climate is an enduring quality of an organization that (1) is experienced by employees, and (2) influences their behavior. It should be thought of and measured "at the organizational level of analysis" (Glick, 1985). Climate is an organizational characteristic - not a characteristic of each individual in an organization. However, we do look to individuals as the source of information on the climate. It is their perception about the organization that we measure.

As mentioned earlier, Zammuto and Krakower (1991) investigated the relationship between the competing values categories: group culture, developmental culture, internal process culture, and rational culture and organizational characteristics (centralization, formalization, long-term planning), climate measures (trust, conflict, morale, equity of rewards, resistance to change, leader credibility, and scapegoating) and strategy dimensions (reactive/pro-active orientation). For their study, Zammuto and Krakower thus define climate as part of culture. Using the Koys and DeCotis rules on the Zammuto and Krakower model, organizational climate can be measured using the seven climate variables that they included in their study; and further, the climate can be categorized into four types using the competing values framework. We will name these particular profiles:

- The Group Climate.

- The Developmental Climate.
- The Rational Goal Climate.
- The Internal Process Climate.

Zammuto and Krackover (1991) defined their climate measures as follows:

- *Trust:* An organization has a high level of trust when the individuals are open, sharing and truthful, where individuals place their confidence. An organization has a low level of trust when the individuals are closed, guarded, unsharing, untruthful, and creates an atmosphere of anxiety and insecurity.
- *Conflict:* An organization has a high level of conflict when there is a high opposition of forces, goals and beliefs, which are experienced in friction and disagreement among the individuals. An organization has a low level of conflict when there is harmony in goals, beliefs, which yields a spirit of cooperation among the individuals.
- *Morale:* An organization has a high level of employee morale when the individuals are confident and enthusiastic about the organization - an Esprit de Corps. An organization has a low level of employee morale when the individuals lack confidence and enthusiasm about the organization and individuals lack a sense of purpose and confidence about the future.
- *Rewards:* An organization is equitable in its rewards when individuals accept rewards as fair and just without bias or favoritism. An organization is inequitable in its rewards when individuals see favoritism, bias, and non work related criteria as the basis for rewards.
- *Resistance to change:* An organization has a high resistance to change when individuals believe the inertia is high and presume and desire that "we will do things tomorrow as we did them today." An organization has a low resistance to change when individuals embrace change as the normal circumstance and relish that "tomorrow will be different."
- *Leader credibility:* The leader credibility is high when individuals have belief in its leadership; there is a sense of respect, inspiration and acceptance of decisions and actions.

The leader credibility is low when the individuals lack respect and do not accept the legitimacy of authority.

- *Scapegoating*: An organization has a high level of scapegoating when individuals believe that the responsibility for actions will be shifted to others - top management, staff, employees, or outsiders. An organization has a low level of scapegoating when individuals believe that the responsible individuals assume the responsibility for the failure of actions.

It is interesting that Zammuto and Krackover (1991) and Koys and DeCotiis (1991) both define climate using 7 relatively similar dimensions.

Autonomy	—	Credibility
Cohesion	—	Conflict
Trust/support	—	Trust
Pressure	—	Scapegoating
Recognition	—	Morale
Fairness	—	Equitable rewards
Innovation	—	Resistance to change

The dimensions do not fit completely together one-by-one, but they are very similar and the totality of the seven dimensions are indeed very similar. Since the empirical research using the competing values approach provides us with a number of relationships between climate and organizational properties as well as with other contingency factors we will use the Zammuto and Krackover dimensions.

In a study of 246 Danish firms, Burton et al (2003) found that the seven dimensions by Zammuto and Krackover (1991) could be used to categorize the organizational climate. Using a cluster analysis they found that the seven dimensions could describe four climate types very similar to the types found in the study by Zammuto and Krackover (1991). Table 4.3 shows that the two studies resulted in very similar results.

Comparing the four climate types, one notices that the group and developmental climates have very similar scores, but they differ on their resistance to change. Similarly, the internal process and rational goal climates are very similar. The main difference is again their resistance to change. Using a factor analysis Burton et al (2003) also suggest that trust, morale, rewards, equitability, and leader credibility may be considered one factor which they called "tension".

Table 4.3.   Comparing the results by (Zammuto et al., 1991) (Burton et al., 2003)[a]

	Group		Develop-mental		Rational goal		Internal process	
Trust	H	*H*	M	*H*	L	*L*	L	*L*
Conflict	L	*L*	L	*L*	H	*H*	H	*MH*
Morale	MH	*H*	MH	*H*	M	*L*	L	*MH*
Rewards equitability	H	*H*	M	*H*	L	*L*	L	*L*
Resistance to Change	M	*H*	L	*L*	M	*L*	H	*H*
Leader credibility	H	*H*	H	*MH*	ML	*L*	L	*L*
Scapegoating	L	*L*	ML	*L*	H	*H*	H	*MH*

[a]The results by Burton et al (2003) are in italics. The scores are high (H), medium/high (MH), medium (M), medium/low (ML), and Low (L).

Hooijberg and Petrock (1993) characterize the four corresponding climate types from the point of view of the competing values framework: *"The group climate could be described as a friendly place to work where people share a lot of themselves. It is like an extended family. The leaders, or head of the organization, are considered to be mentors and, perhaps even parent figures. The organization is held together by loyalty or tradition. Commitment is high. The organization emphasizes the long-term benefits of human resource development with high cohesion and morale being important. Success is defined in terms of sensitivity to customers and concern for people. The organization places a premium on teamwork, participation, and consensus.*

The developmental climate could be described as a dynamic, entrepreneurial and creative place to work. People stick their necks out and take risks. The leaders are considered to be innovators and risk tak-

ers. *The glue that holds organizations together is commitment to experimentation and innovation. The emphasis is on being on the leading edge. Readiness for change and meeting new challenges are important. The organization's long-term emphasis is on growth and acquiring new resources. Success means having unique and new products or services and being a product or service leader is important. The organization encourages individual initiative and freedom.*

The rational goal climate could be described as a results oriented organization. The leaders are hard drivers, producers, and competitors. They are tough and demanding. The glue that holds the organization together is the emphasis on winning. The long-term concern is on competitive actions and achievement of measurable goals and targets. Success is defined in terms of market share and penetration. Competitive pricing and market leadership are important. The organizational style is hard driving competitiveness.

The internal process climate is a formalized and structured place to work. Procedures govern what people do. The leaders pride themselves on being coordinators and organizers. Maintaining a smooth running organization is important. The long-term concerns are stability, predictability, and efficiency. Formal rules and policies hold the organization together."

These descriptions fit very well with the scores in Table 4.3.

CLIMATE AS A CONTINGENCY

The discussion in the previous sections of this chapter provides us with a limited set of dimensions that fit nicely into the multi-contingency model. The three rules developed by Koys and DeCotiis (1991) sort out the confusion about using dimensions from both the right-hand-side and left-hand-side of the model in Figure 1.3. This allows us to develop a two-stage model relating the climate dimensions to structural properties via the climate typology based on the competing values approach. The model is depicted in Figure 4.2.

This model first maps the seven climate dimensions into four climate categories from which propositions between climate and structural form are developed. These seven dimensions, or "questions should focus on the specific organizational units with recognized boundaries, not an ambiguous 'work environment'" (Glick, 1985).

The four climate types each represents four ways to process information and thus which organizational structure that will be most effective and/or efficient.

Figure 4.2. The Two-stage Model on Climate

Climate Measures		Structural Properties
Trust		Configuration
Conflict	**Group Climate**	Centralization
Morale	**Rational Goal Climate**	Formalization
Rewards equitability	**Internal Process Climate**	Complexity
Change resistance	**Developmental Climate**	Coordination and Control
Leader credibility		Media Richness
Scapegoating		Incentives

Describing a Group Climate

We describe the organization's climate in terms of the seven characteristics about the organization: trust, level of conflict, employee morale, rewards, the resistance to change, the leader credibility and the level of scapegoating (Burton et al., 2003; Zammuto et al., 1991).

A group climate has a high degree of trust; it is open and promotes sharing of information among its members. Further, a group climate usually has a low degree of conflict. If conflict exists, it is constructive and tends to strengthen the group, rather than destroy the group, i.e., there can be disagreement for the group purpose itself. This is usually coupled with a high or moderately high degree of employee morale. Individuals feel that they belong and are part of the group. The equitable distribution of rewards is also consistent with the group climate. These rewards need not be equally distributed, but there must be a sense of fairness where the basis for the distribution is understood and accepted by the individuals in the organization. Group climates are normally resistant to change. In a group climate, leaders have a high degree of credibility with the group and can act on behalf of the organization. There is a high degree of trust and little scapegoating. Failure can be recognized and accepted without the transfer of blame to those who are not responsible. Overall, the group climate has a consistent pattern of beliefs and attitudes about desirable behavior. A group climate has an open and a free flow of information among the individuals in the organization. Trust, low conflict and equity of rewards encourage individuals. Further, high morale and little scapegoating are advantageous. Information becomes a kind of public good which is shared and widely available. Information is more likely to be "broadcast" than "channeled." "Need to know" is replaced by "everybody knows." There are few secrets. The group climate can handle complex sets of information.

The seven propositions below describe a group climate:

Group Climate Propositions

If the level of trust in the organization is high, then the climate is likely to be a group climate.

If the conflict in the organization is low, then the climate is likely to be a group climate.

If the employees' morale is high or medium, then the climate is likely to be a group climate.

If the organizational rewards are given with equity, then the climate is likely to be a group climate.

If the resistance to change is high, then the climate is likely to be a group climate.

If the leader credibility is high, then the climate is likely to be a group climate.

If the level of scapegoating is low, then the climate is likely to be a group climate.

The greater the number of these propositions that are true, then the greater the likelihood that the organization's climate is a group climate. The implications of a group climate for the fit of the structure are developed next.

The Group Climate Effects on Structure

A group climate has implications for the structure of the organization. The group climate sets an atmosphere which is compatible or fits a specific set of structural properties. An adhocracy fits with a group climate. The fluidity of the adhocracy is consistent with a high level of trust, employee morale and commitment. Sometimes, an adhocracy has conflict, but such controversy can be directed to the organizational purpose that fits a group climate. The group is a very stable climate with a high resistance to change. A group climate could also fit with other configurations, e.g. a matrix, but a bureaucracy seems unlikely to work well. With the high degree of trust the group climate provides the basis for a virtual network configuration (Handy, 1995).

A group climate requires a low to medium organizational complexity with a low vertical differentiation, i.e., there are not many layers in the organization. There is less emphasis on the division of work or detailed task definition. The group itself takes responsibility in more aggregate form. Task boundaries are not necessarily sharp and well defined for the individual. This is also consistent with a low formalization where there are few written rules and again the organization is more fluid in its operations (Hunt, 1991; Quinn et al., 1983; Zammuto et al., 1991). Normally, a low formalization is coupled with high centralization, but for a group climate there is evidence that the group climate requires low centralization (Hunt, 1991; Zammuto et al., 1991). In the group climate, decisions can be made appropriately throughout the organization. The span of control can be wide. There is no requirement for hands on control. However, activities must be coordinated and controlled by using meetings and many liaison relations. These mechanisms must then serve the function that otherwise would be obtained through either centralization or formalization (Hunt, 1991). The group climate can process complex and large amounts of information relative to its size. Coordination via meetings and liaison is appropriate. The incentives must be results oriented since there are few procedures. Further, these incentives should reward the group as matters of equity and reinforcement of the group itself.

The above discussion can be summarized in the following propositions:

Group Climate Effect on Structure

If the climate is a group, then the configuration should be an adhocracy or a matrix possibly with elements of a virtual configuration.

If the climate is a group, then the complexity should be low to medium with low vertical differentiation.

If the climate is group, then the formalization should be low.

If the climate is group, then the centralization should be low.

If the climate is group, then the span of control should be wide.

If the climate is group, then the coordination and control should be via integrators and group meetings.

If the climate is group, then the media richness should be high with a large amount of information.

If the climate is group, then the incentives should be result based with a group orientation.

In brief, the group climate provides the necessary information; structural mechanisms can be reduced; and the organization should employ configurations and structural mechanisms which complement and enhance the abundance of information already available in the group climate. A group climate can primarily enhance efficiency but may support effectiveness as well. The statements on the group climate are summarized in Figure 4.3.

Figure 4.3. Group Climate: Description and Effect on Structure

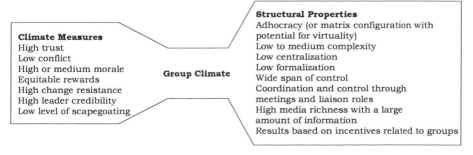

Climate Measures
High trust
Low conflict
High or medium morale
Equitable rewards
High change resistance
High leader credibility
Low level of scapegoating

Group Climate

Structural Properties
Adhocracy (or matrix configuration with potential for virtuality)
Low to medium complexity
Low centralization
Low formalization
Wide span of control
Coordination and control through meetings and liaison roles
High media richness with a large amount of information
Results based on incentives related to groups

Describing the Developmental Climate

Some of the characteristics for the developmental climate are similar to those of the group climate. For both, the trust is high, conflict is low, and the morale is high, with relatively equitable rewards. The significant difference is the resistance to change, which was high in a group climate, but is low in a developmental climate. However, the developmental climate scores are not as extreme as for the group climate. The evidence (Burton et al., 2003; Zammuto et al., 1991) is that the trust for a developmental climate may not be so high as for a group climate, although the difference seems minimal. For the developmental climate, there is a greater focus on the growth of the organization itself. This is the basis for the low resistance to change. Similarly, the rewards can be more individual with less attention to the internal equity as perceived by the group. Individual contribution for the organization is more important. This is a more external orientation where success is in part realized more outside the organization.

There are also small differences with respect to leader credibility and the level of scapegoating. The developmental climate has different information characteristics to the group climate. The group climate will focus relatively more on internal information while the developmental climate focuses more on external environmental information. Environmental information is likely to have more value for development and growth. Additionally, compromise is important (Quinn and Kimberley, 1984).

There are also seven propositions describing the developmental climate:

Developmental Climate Propositions

If the level of trust in the organization is high to medium, then the climate is likely to be a developmental climate.

If the conflict in the organization is low, then the climate is likely to be a developmental climate.

If the employees' morale is high or medium, then the climate is likely to be a developmental climate.

If the organizational rewards are given with high to moderate equity, then the climate is likely to be a developmental climate.

If the resistance to change is low, then the climate is likely to be a developmental climate.

If the leader credibility is high to medium, then the climate is likely to be a developmental climate.

If the level of scapegoating is medium, then the climate is likely to be a developmental climate.

Developmental Climate Effect on Structure

The matrix configuration is suggested as an appropriate structure for the developmental climate. As for the group climate, an adhocracy is also suggested. The medium to high trust will support a virtual network structure. Both climates require fluidity and "give and take" to be successful. It is this similarity that we want to emphasize. Organizational complexity, formalization and centralization have the same recommendations for the two climates: complexity- medium or perhaps low, formalization-low, and centralization-low-medium (Hunt, 1991; Quinn et al., 1983; Zammuto et al., 1991). The supporting arguments follow the same logic and empirical evidence. The medium complexity includes a low vertical differentiation and a medium span of control. Here the external focus suggests that a tall organization might be less adaptable than required for the external and growth orientation of the developmental climate. The low formalization and low to medium centralization may be recommended for different reasons. The change requirements of the environment and the need for growth require flexibility that is not compatible with greater formalization and centralization. The coordination and control mechanisms are meetings with negotiations (Hunt, 1991). More detailed planning is likely to be required here than for the group climate where there is a more implicit understanding about what to do. Similarly, high media richness and a large amount of information are required; here there should be a richer kind of information for a more outside focus. Finally, the incentives should be results based, but more individually oriented with a developmental climate. There is less focus on the group and there are likely to be external success factors that can be interpreted for the individual.

Here too, the information processing characteristics of the developmental climate affects the configuration and structure. As with the group climate, the developmental climate is more open and yields an abundance of information; the organization structure should be complementary. A more open structure is desired which recognizes the already abundant information. The matrix configuration, low vertical differentiation, a medium span of control, low formalization, low-medium centralization, integrators and meetings, all are consistent with an abundance of information. Here, this information should include extensive external content. A developmental climate will primarily support effectiveness.

The relationship between the developmental climate and organizational properties are stated in eight propositions:

Developmental Climate Effect on Structure

If the climate is developmental, then the configuration should be a matrix. It could also be an adhocracy. The developmental climate supports a virtual network structure.

If the climate is developmental, then the complexity should be medium with low vertical differentiation.

If the climate is developmental, then the formalization should be low.

If the climate is developmental, then the centralization should be low or medium.

If the climate is developmental, then the span of control should be medium.

If the climate is developmental, then the coordination and control should be via planning, integrators and meetings.

If the climate is developmental, then the media richness should be high with a large amount of information.

If the climate is developmental, then the incentives should be results based with an individual orientation.

Figure 4.4 summarizes the developmental climate descriptions and effects.

Figure 4.4. Developmental Climate: Description and Effect on Structure

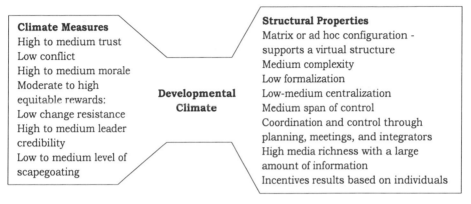

Climate Measures
High to medium trust
Low conflict
High to medium morale
Moderate to high
equitable rewards:
Low change resistance
High to medium leader
credibility
Low to medium level of
scapegoating

Developmental Climate

Structural Properties
Matrix or ad hoc configuration -
supports a virtual structure
Medium complexity
Low formalization
Low-medium centralization
Medium span of control
Coordination and control through
planning, meetings, and integrators
High media richness with a large
amount of information
Incentives results based on individuals

Describing the Internal Process Climate

The internal process climate is quite different from the two climates described above. The internal process climate is characterized by low trust. There is not a sharing and open atmosphere among the individuals as each is more inward and guarded. Conflict is high in the organization and disagreement over both means and ends are prevalent. The employees' moral is medium to low although there is somewhat conflicting evidence. Intuitively, it fits that morale would be low under an atmosphere of distrust and conflict.

Rewards are perceived to be given inequitably. Here too, it seems reasonable that rewards would be viewed with low equity in this atmosphere. There is a high resistance to change. Perhaps this is less intuitive as it might be argued that a change, any change, would be welcome. But the evidence suggests that individuals prefer to keep what they have and not engage in activities that could lead to a different situation, although it could be better. Perhaps it is the leader credibility, which is low and the low level of trust that helps explain this reluctance. There is little faith in the leader and consequently not much hope that the situation would improve. The high level of scapegoating seems consistent with this story about the internal process organization.

The internal process climate does not possess the capacity to process a lot of information. The organization structure must supply the requisite information processing capacity. There is not a norm of sharing and openness. Information tends to be private and within the role. Information is passed on within prescriptions and according to procedures. Information is closely associated with the job or task, or "a need to know." The spontaneous information links are largely missing, or not utilized.

The discussion above can be summarized in the following propositions:

Internal Process Climate Propositions

If the level of trust in the organization is low, then the climate is likely to be an internal process climate.

If the conflict in the organization is high, then the climate is likely to be an internal process climate.

If the employees' morale is medium to low, then the climate is likely to be an internal process climate.

If the organizational rewards are given inequitably, then the climate is likely to be an internal process climate.

If the resistance to change is high, then the climate is likely to be an internal process climate.

If the leader credibility is low, then the climate is likely to be an internal process climate.

If the level of scapegoating is high, then the climate is likely to be an internal process climate.

Internal Process Climate Effects on Structure

The internal process climate has implications for the structure. Generally, the structure should be tighter with an emphasis on the mechanism of the organization itself. In many ways, the organization must be the complement to the lack of "togetherness" among the individuals. That is, if working conditions promote an internal process climate, the ideal organizational structure is a bureaucracy.

This is consistent with a functional organization where the coordination is realized through the organizational mechanisms themselves. Virtuality will be almost impossible with an internal process climate. This is consistent with a high organizational complexity where there are many specialized roles and a tall organization. Formalization should be high with lots of written rules, which govern actions and behavior (Hunt, 1991; Zammuto et al., 1991). With a low level of trust and a high level of conflict many written rules are needed to ensure that the information is processed in the right way (Quinn et al., 1984).

Centralization should also be high where decisions are made at the top (Hunt, 1991; Zammuto et al., 1991). Formalization and centralization can both be means to coordinate activities, but high centralization and high formalization may be more than is required. The internal process climate is quite demanding for mechanisms to maintain order and control in the traditional sense. Although the span of control can be wide, this notion is reemphasized with a call for rules and procedures for coordination and control (Hunt, 1991). The media richness can be low with a relatively moderate amount of information required (Hunt, 1991). Given the climate and the atmosphere and the organization itself, the incentives should be procedur based for the individual. An individual should have a well-defined job and responsibility with rather narrow limits and the reward issue is whether the individual does the task appropriately or not. The internal process climate requires well-structured and well documented information processing to be efficient. The internal processing climate does not enhance effectiveness.

The propositions below summarize the structural requirements for an internal process climate as also shown in Figure 4.5.

Internal Process Climate and Effect on Structure

If the climate is internal process, then the configuration should be a bureaucracy or functional. It cannot be a virtual network.

If the climate is internal process, then the complexity should be high with both high vertical and horizontal differentiation.

If the climate is internal process, then the formalization should be high.

If the climate is internal process, then the centralization should be medium or high.

If the climate is internal process, then the span of control should be wide.

If the climate is internal process, then the coordination and control should be by rules and procedures.

If the climate is internal process, then the media richness should be low with a moderate amount of information.

If the climate is internal process, then the incentives should be procedures based with an individual orientation.

Figure 4.5. Internal Process Climate: Description and Effect on Structure

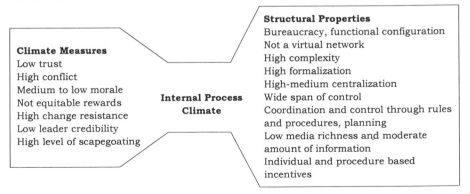

Climate Measures
Low trust
High conflict
Medium to low morale
Not equitable rewards
High change resistance
Low leader credibility
High level of scapegoating

Internal Process Climate

Structural Properties
Bureaucracy, functional configuration
Not a virtual network
High complexity
High formalization
High-medium centralization
Wide span of control
Coordination and control through rules and procedures, planning
Low media richness and moderate amount of information
Individual and procedure based incentives

Describing the Rational Goal Climate

The rational climate is closer to the internal process climate than to the group and developmental climates, although they are different. The main difference is the score on the resistance to change The rational goal climate is structured with an emphasis on planning, productivity, and efficiency (Quinn et al., 1984).

Information processing in the rational goal climate is similar to internal process climate but with a greater emphasis on environmental/external information. The low level of trust, high conflict, etc., lead to a private view of information; sharing and exchange of information does not occur spontaneously. The rational goal climate is a very competitive environment to work in. It is not to be expected that the employees will be loyal to the organization in the sense that high turnover can be expected. With the low resistance to change, reorganization at the personnel level can be expected with a very tough competition for the prestigious jobs. This can be summarized in the following propositions:

Rational Goal Climate Propositions

If the level of trust in the organization is low, then the climate is likely to be a rational goal climate.

If the conflict in the organization is high, then the climate is likely to be a rational goal climate.

If the employees' morale is low-medium, then the climate is likely to be a rational goal climate.

If the organizational rewards are inequitable, then the climate is likely to be a rational goal climate.

If the resistance to change is low to medium, then the climate is likely to be a rational goal climate.

If the leader credibility is low, then the climate is likely to be a rational goal climate.

If the level of scapegoating is high, then the climate is likely to be a rational goal climate.

Rational Goal Climate Effects on Structure

The rational goal climate implications for the structure are similar to those of the internal process climate. This is not surprising as the climate descriptions themselves were in the same spirit. Nonetheless, there is a variation in emphasis and tone between the internal process climate and the rational goal climate. The rational goal climate calls for a less "tight" structure and has a more external focus to its orientation. To begin, a divisional configuration is recommended. The more external focus is compatible with the divisional structure. It will be difficult to operate with a high degree of virtuality. Complexity should be high with high specialization of tasks. However, formalization and centralization should be medium whereas they should be high for the internal process climate (Hunt, 1991; Quinn et al., 1983; Zammuto et al., 1991). This is a less "tight" mechanism of organization and is compatible with the greater external focus here. The span of control should be medium. Coordination and control should use more process in planning and meetings than for the internal process climate. Media richness should be high with a medium richness of the information. Finally, the incentives should be more results-oriented for both the individual (Hunt, 1991) and the group. This, too, is very compatible with the more external competitive orientation of the rational goal climate and fits very well with the divisional configuration recommended here.

Here again, the rational goal climate with its limited information processing requires a higher level of information processing within the structure of the organization. The divisional configuration mitigates some need for information. The organization is structured to gather, communicate, and interpret the requisite information to manage. High complexity divides the tasks, which must be coordinated and controlled with medium formalization, medium to high centralization, and an emphasis on planning and meetings. In brief, the rational goal climate requires that the organization itself procure the requisite information. The rational goal could enhance both efficiency and effectiveness.

The propositions below summarize the structural recommendations for the rational goal climate:

Rational Goal Climate Effect on Structure

If the climate is rational goal, then the configuration should be divisional.

If the climate is rational goal, then the complexity should be high.
If the climate is rational goal, then the formalization should be medium.

If the climate is rational goal, then the centralization should be medium.

If the climate is rational goal, then the span of control should be medium.

If the climate is rational goal, then the coordination and control should be via planning and meetings.

If the climate is rational goal, then the media richness should be medium with a large amount of information.

If the climate is rational goal, then the incentives should be results based with a group and individual orientation.

Figure 4.6 summarizes the propositions on the rational goal climate.

Figure 4.6. Rational Goal Climate: Description and Effect on Structure

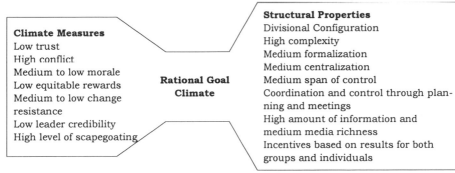

MANAGING THE CLIMATE

We argued in Chapter 1 that fit is important. To obtain fit both sides in the model in Figure 1.3 should be aligned. This means that to obtain a contingency fit with climate and structure; you may change the structural properties or you may try to adjust the climate.

To change the perception of the climate in an organization quickly may not be that easy. Generally the climate is rather stable (Campbell et al., 1970). It may take a long time to build trust, morale, and a perception that rewards are given in an equitable way, but it may be very easy to destroy.

Management may try to build a special psychological climate, by instituting a number of rituals and by showing over time that they are open and can be trusted. A special climate may also be instituted by hiring people with special views and values. Theory X-people will normally show less trust and be more resistance to change than do Theory Y-people (see Chapter 3). Cameron and Quinn (1999) suggest that climate can be changed and they provide a set of guidelines for changing culture and climate within the competing values framework.

The choices of strategy, environment, and the organizational design itself provide feedback to the people employed in the organization and may affect their perception of the working climate and thus affect the way information is processed. For example, the anti Microsoft movement both in government and on the Internet is said to affect the perception of the Microsoft employees that Microsoft is a "great" place to work. This may affect the work morale at Microsoft. Again a proper alignment is called for. We will return to the issue of mutual alignments in Chapter 9 in the context of managing misfits.

SUMMARY

The organizational climate is an important factor in determining the design of the organization. Climate is one of several factors, including the leadership style, size, environment, technology, and strategy which influence an appropriate choice of the organizational design. In Figure 1.3, the organization's climate then is one of the left-hand factors in determining the organization's structure on the right-hand side. For example, the climate helps set the appropriate level of centralization - a high level of trust suggests a less formalized and more decentralized organization.

The organization's climate and its culture are intertwined in every day managerial usage and in some of the literature. Yet, they are quite distinct concepts. Climate is a prevailing condition or atmosphere in an organization, where the culture is pattern of behavior including commonly held beliefs and social forms. The climate is more enduring and given, where the organizational structure is embedded in the culture. Climate is more given; culture can be learned and transferred from one individual to another. In terms of design, climate is a contingency on the left-hand side of our multi-dimensional model in Figure 1.2. Culture is emergent in the choice of the right-hand side design elements.

The organization's climate is a measure of the organization - not a measure of the individuals in the organization. Nonetheless, we look to the individuals for our information about the climate. Climate is then measured in terms of trust, conflict, morale, and equity of rewards, resistance to change, leader credibility and the level of scapegoating. Individuals "know" what the climate is and they are our source of information. These measures are reliable and valid, i.e., individuals can judge these issues and respond in a consistent and meaningful way.

The competing values approach can then be utilized to summarize and categorize the climate into four types: group, rational goal, internal process and developmental. The seven measures above can be mapped into the four climate categories as suggested in Figure 4.2. For Example, a high level of trust is an element of a group climate, which in turn suggests a more decentralized organization.

Each of the four climate types is consistent with a different organizational structure. The group climate not only indicates greater decentralization, but also less formalization and greater use of meetings for coordination. A less trusting climate should be less decentralized and more formalized in its structure, with a greater emphasis on fixed assignments to achieve coordination.

The support for these recommendations comes in part from the literature, but also from an understanding of the behavior of the organization and how it uses information. A high trust organization tends to have a more open and free flow of information and thus, coordination can be obtained with a good deal of localized decision-making or decentralization. In the less trusting organization, information is less shared and more guarded and localized and thus, coordination is realized within more formalized structures and more centralized decision-making.

The climate-structure relations are more complex and multi-dimensional going beyond trust and decentralization. Climate is not only trust, but includes six other dimensions; the organizational design incorporates not only centralization, formalization and means for coordination, but the configuration, rewards, etc. as presented in Figure 4.2. Further, climate is complemented by other contingencies in determining an appropriate design as given in Figure 1.2. The multi-dimensional contingency model is more complete with the incorporation of climate as a factor in the organization's design.

In this chapter, we have incorporated climate as a contingency factor in the multi-dimensional contingency model of organizational design. Climate gives us more important information to help determine an appropriate design which is effective and efficient.

CHAPTER 5
Size and Skill Capabilities

INTRODUCTION

The Institute of Applied Computer Science (introduced in Chapter 3) was negotiating a contract with a European Union research agency. This contract would be the single largest contract in the history of the company. The contract required Applied Computer Science to employ about forty people and to be the coordinating unit for a number of research teams located in other European countries. Benny Mortensen was not worried about the scientific and technological aspects of the new project but expressed concerns about running a company that would be twice its current size. He realized that he no longer would be as heavily involved in every project and that more traveling and time spent on recruiting would leave him less time to have detailed knowledge of how projects were doing. To a large extent, the success of the company depended on his ability to utilize resources efficiently, thus cutting cost and time, by using knowledge from one project to improve another project.

The growth of the Institute of Applied Computer Science generated some management concerns. The manager was concerned that quality and cost control could not be maintained as he became less involved in particular projects. His management style and personality were important for the success of Applied Computer Science. Its growth may require a structural change, but Benny Mortensen probably will try to develop a structure that fits not only the company's particular size and growth pattern but also his own style and needs. Furthermore, the working conditions for employees of Applied Computer Science will change, too. It is important that there is a fit between working conditions and the individuals employed by the organization.

It is obvious that the increase in size changes the managerial and organizational problems of this company. As the size increases, the

CEO may have to increase the company's level of decentralization. Centralized decision-making has been the main source of coordination of organizational activities. Now Benny Mortensen needs to think about other ways to ensure coordination. Project cost and quality were also supervised by the CEO. During the startup many administrative procedures were carried out by Benny Mortensen. The Institute of Applied Computer Science does not have people in administrative functions, but with growth, administrative positions may have to be created. This will allow Benny Mortensen to use his information-processing capacity on coordination and decision-making. With increased decentralization and specialization, formalized rules or incentive mechanisms may also be changed.

The small company–with its low formalization and high centralization–had a competitive advantage by being flexible and fast in its decision-making. Introducing a high degree of formalization may be appropriate with respect to cost and quality but may hamper flexibility. As the company grows, Benny Mortensen should reassess the organizational structure.

In recent years, size has entered the discussion in two ways. First, in many industries such as pharmaceuticals and banking, a merger wave has been seen creating new organizations much larger than before. Second, and somewhat in contrast, many have preached "small is beautiful" with the results of downsizing and outsourcing.

The Size of SAS

When SAS was founded in 1954, it could fly to sixty-seven cities (Buraas, 1972). By 1990, SAS had eighty-five destinations and had alliances with 291 destinations from Copenhagen, 181 destinations from Oslo, and 254 destinations from Stockholm with no more than one stopover (SAS, 1990, p. 24). SAS had 20.000 employees in 1990.

A new era in SAS's history emerged with the appointment of Jan Stenberg as SAS President and CEO in April 1994. The underlying idea for Stenberg's new strategy program was a renewed focus on SAS's core business: the airline. The goal was to create a cost-effective company. To reach this goal, an extensive rationalization program was initialized, aiming at a total cost reduction of SEK 2.9 billion. Shortly after his appointment, 2,930 employees were fired, primarily pilots and cabin crew. Later, the number of people in SAS has increased primarily by the acquisition of affiliated airlines. In 2003, SAS has 35000 employees. In the newly agreed cost reduction plan, the number of people will be reduced by 4000 people. Most of the people in SAS are highly skilled with many years of training and education. Despite the recent cutbacks, SAS remains a large corporation.

For SAS, the large number of employees creates a need for a large capacity to process information.

1. Which structure(s) would fit the large size of SAS?
2. Is the large size of SAS creating inertia such that other contingency factors have reduced importance?

Generally, size is an important contingency for organizational design. The size paradigm has been around for some time (Spencer, 1998). Yet, there have been a number of controversies about size and its effect on the organization.

In this chapter we will discuss the effect of size of the organization on the organizational structure from an information processing point of view. This will involve not only the number of people but also their ability to make decisions and handle information.

LITERATURE REVIEW ON SIZE[1]

Organizational size has been of interest to social scientists for a long time. Over ninety years ago it was asserted that the effect of size "is a character of social bodies, as of living bodies, that while they increase in size they increase in structure" (Spencer, 1998). Sociologists and organizational theorists have investigated the relation between size and structure - that is, size as an imperative factor explaining the organizational structure (Blau, 1970; Blau and Schoenherr, 1971; Hickson et al., 1979; Kimberly, 1976; Meyer, 1972; Pugh et al., 1969; Slater, 1985). In these and other works, size is operationalized in a variety of ways: number of employees, number of products or services, total sales, number of divisions, and so forth. One path, in particular, seems to be well traveled - to investigate the effect of administrative intensity or headquarters' burden. Child (1973) states "more studies have probably been carried out on the proportion of employees occupying administrative or supportive roles than on any other single aspect of organization structures".

---

[1] This section is updated from Burton, Minton, and Obel (1991).

An information processing model of the firm provides yet another perspective, suggesting that the size of the firm is limited due to the costliness of coordinating the activities within an administrative structure (Williamson, 1975). This is in general agreement with Starbuck, who suggests that, beyond some cost-optimal point, "managerial problems become inordinately complex as size increases producing progressively higher production costs" (Starbuck, 1965). Even here, size is not well defined. The absence of a common view on how to measure size (or, more specifically, on the use of differing measures or dimensions of the broadly defined "size" construct) has made findings from size and structure hypothesis tests somewhat equivocal.

Of the many debates in the literature, one is particularly illustrative. In testing prior work, Donaldson states that organizational diversity (of products, particularly) is more highly associated with structure (divisionalization, in particular) than is size - whether measured in sales, assets, or number of employees (Donaldson, 1982). Grinyer, responding to Donaldson, suggests that "more complex divisional structures tend to be more bureaucratic," using "decentralized operating decisions" (Grinyer, 1982; Grinyer and Ardekani, 1980). Furthermore, Grinyer suggests that while Donaldson's hypotheses may hold for product-based divisionalization, they do not hold for divisionalization based on other attributes, such as geographic dispersion or some minimum size as a threshold for divisionalization property. In short, Grinyer suggests, "it would be a mistake ... to take too simplistic a view on this issue. Growth, size, diversification, and divisionalization are all strongly interconnected" (pp. 342–343). Child, commenting on the Donaldson-Grinyer debate, suggests that size expressed as a threshold below which divisionalization is inappropriate and above which appropriate is also an oversimplification (Child J., 1982). Clearly, a lack of specificity with regard to the size construct makes equivocality a basic problem (if not a hallmark) of such research.

The Measurement of Size

Inquiry into the concept of size has been constrained by problems involving the operationalization of its measurement. Kimberly states that such studies have been conducted in a theoretical wasteland since hypotheses have typically been addressed apart from available theories of complex organization (Kimberly, 1976). Slater finds that most organizational theorists "think of organizational size mainly in personnel terms," that is, by counting organizational members (Slater, 1985).

Kimberly, on the other hand, suggests that other aspects or measures of organizational size (that is, other than number of members) may be useful and that attention should be paid to the physical, fiscal, input, and output dimensions as well (Kimberly, 1976). Kimberly concluded his 1976 work with three points. First, the ambiguous status of size (in the organizational literature) is due to it being too global a measure to permit a clear specification of its organizational role; size has numerous aspects, the theoretical and empirical aspects of which must be made more specific. Second, various aspects of size may fulfill differing causal or indicative roles in various types of organizations. Finally, effective organizational conceptualization depends on conducting inquiries with a dynamic (rather than a static) perspective.

The various measures of organizational size are not interchangeable due to dependence on differing conceptual referents. However, many researchers fall back on total member counts since such counts are likely to be highly correlated to a number of differently conceived nonpersonnel measures (Anderson and Warkov, 1961; Hawley et al., 1965; Pugh et al., 1969). In other words, personnel count is often considered a surrogate for other measures of organizational size. There are, of course, other appropriate measures of size.

Ask any CEO, "How large is your company?" The usual response is, "our revenue is 100 million." Revenues or sales are very often the size measure of interest. In a market economy, revenue is a natural way to think about size. But there are many others: assets, profits, countries, and the number of employees. All of these measures of size are reasonable and are the correct measure, but for different purposes.

Size as Imperative

While size itself has generated much interest, size as an organizational determinant has also been of widespread interest. The global use of size as an organizational measure is the hallmark of theorists who support a "size imperative" - that is, organizational size as a determinant of differentiation and organizational structure (Blau, 1970; Blau et al., 1971; Hickson et al., 1979; Meyer, 1972). By *imperative*, we mean a state or quality that "necessarily affect(s) organizations in a certain way" (Robey, 1982). Size, in this research, was seen to be positively related to increased differentiation, specialization, and formalization. These assertions are subject to some criticism (Aldrich, 1972; Hall et al., 1967; Mayhew et al., 1972). Results from these critical works are mixed, at best, and adherence to the size imperative has been debated extensively (Donaldson, 1982).

AN INFORMATION-PROCESSING PERSPECTIVE ON SIZE

Size as an imperative for organizational structure has not been a major concern in information processing models of organization, but one central question does regard the limitation of firm size. Williamson summarizes the proposition that "the argument ... comes down to this: The distinctive powers of internal organization are impaired and transactional diseconomies are incurred as firm size and the degree of vertical integration are progressively extended, *organizational form held constant*" (Williamson, 1975). In other words, increasing size (however measured) and high levels of centralized decision making are posited to have a detrimental effect on the quality of a firm's operations and, by extension, its outcomes. Earlier, Caplow emphasized the problems of communication that result from increases in the number of employees; as employees increase linearly, then the number of communication networks increases exponentially (Caplow, 1957).

   The information-processing requirements of an organization follow "who talks to whom about what." More simply, we can count who can talk to whom by examining the number of possible communication links in an organization.

Let us begin with a two-person organization and then increase the number of people.

Figur 5.1. Information Links

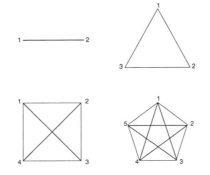

For a two-person organization, there is one link between the two. For three people, there are three possible links. For four, there are six. For five, the number grows to 10, and the pattern continues. It is clear that the number of possible links (1,3,6,10...) is going up faster than the number of individuals (1,2,3,4...). The general formula for the number of possible links is: $\binom{N}{2}$. For N Equal to 10, there are 45 possible links. For N equal to 100, there are 4950 links.

For organizations with a very large number of people, the number of links becomes almost unimaginable and it is impossible for everyone to talk with everyone else. Network organization, without some reduction of the communication possibilities, will quickly become too complex. We therefore invent mechanisms so that an individual can talk with only a few individuals in an organized way to accomplish the overall organizational task.

The simplest way to decrease the number of links is to put someone in charge and have everyone talk to that one person, but not to each other, i.e., create a hierarchy. The simple organizational configuration, presented in Chapter 2, is such an organization. Even so, it works only for small organizations as the "boss" becomes overwhelmed with the information; we call it "information overload." This approach can be replicated and then we have hierarchies with more than one level, i.e., there are bosses of bosses. A hierarchy can grow and it can have a large number of people. Hierarchies are efficient in reducing the need for information processing in an organization. Hierarchies limit the number of information links in an organization. But if there is a need for communication between two individuals who are far removed in the organization, a hierarchy can be an impediment to efficiency and effectiveness as the organizational tasks are not done or at least, not done well. Hence, we also seek a few selected non-hierarchical liaison mechanisms to link individuals across the hierarchy: matrix organization, committees, task forces, and ad hoc structures, to name a few. But we introduce only a few of the many possible links. The issue is to recognize and introduce the few, which are needed to process information for organizational efficiency and effectiveness.

New information technology changes the way people in an organization can communicate. A simple e-mail system allows one to broadcast to many people across hierarchical boundaries. Voice mail is another device that lets people communicate without the constraint of being present at the same time. Modern information technology forces one to revisit the notion of size. Brynjolfsson et al. (1994) suggest that there is a positive correlation between information technology investments and smaller firm sizes. Information technology may enhance coordination and monitoring capabilities leading to downsizing and outsourcing (Brynjolfsson et al., 1994). Brynjolfsson et al. (1994) also found that an increase of information technology reduces the vertical differentiation in firms (Brynjolfsson, 1994). If size is measured as the number of people, their ability to process information also has to be taken into account. In the context of incentives, skill level has been suggested as one characteristic that can be a proxy for information processing (Demski and Feltham, 1978; Kowtha, 1997).

Large firms are, however, seen as proportionally costlier to manage than small firms. Size here is a general measure of the firm's activities that could be related to the number of products and processes. It may also be positively correlated with the numbers of employees - but here the underlying notion of size, has to do with transactional requirements. This does not suggest any "one best way"; information-processing model of organization occur in a number of forms. In Galbraith's view, the central managerial problem is to reduce uncertainty (Galbraith, 1974). Two primary strategies are to reduce the need for information or to increase the capacity to process the required information. Size does not enter the model explicitly. Yet it is clear that more people, more products, and more processes increase the complexity of organizational decision-making. This increases uncertainty, which is defined as the difference between the information required and that available to manage the organization. Using common dimensions of size, it seems reasonable that size increases uncertainty, which leads to an increased need for coordination, which then influences the choice of the best information processing structure. This is an inferred size imperative–that is, that size can influence structure through size's effect on information processing demands as well as the information-processing capacity of the organization.

The individual skill level has an effect on the way individual process information and makes decisions. For example, it seems that people with high skill level will accept more uncertainty in a compensation framework than people with a lower skill level (Kowtha, 1997).

MEASURING SIZE AND SKILL CAPABILITY FOR DESIGN PURPOSES

The controversy about the effect of size on organizational structure is not diminished by our desire to measure the size of the organization for design purposes. Previous studies have established some positive and negative correlations between various size measures and design variables in organizational structure. But even those that agree on the correlations have very different views about what a small, medium, or large organization is. Robbins defines "a large organization as one having approximately two thousand or more employees" (Robbins, 1990). Carillo and Kopelman define small organizations as having fewer than ten employees, medium-size organizations as having eleven to nineteen employees, and the large organization as having twenty employees or more (Carillo and Kopelman, 1991). Smith, Guthrie, and Chen labeled organizations with less than thirty employees small, those with thirty to 300 employees medium, and those with more than 300 employees large (Smith et al., 1989). Miller, Glick, Wang, and Huber (Miller et al., 1991) use 475 employees as a threshold between small and large organizations. Some authors do not categorize organizations as small, medium, and large but map number of employees directly on to, for example, a vertical differentiation measure. In this case a functional relationship between number of employees and number of levels in the organization is established. The mapping can be linear or nonlinear. For example, Marsh and Mannari use a log transformation (Marsh and Mannari, 1989). In our validation process we found that the results of our expert system were particularly sensitive to the definition of size (Baligh et al., 1996). The reason is that size in itself has an effect but also that size turns out to be a moderator of the effect of other contingencies.

From our information processing perspective, we find that a size measurement is related to the number of people in the organization. However, the information processing capacity depends on the type of people in the organization. One problem with comparing various empirical results is that some studies deal with a very narrow homogenous set of organizations, while others have a much more heterogeneous set of organizations and do meta analyses (Carillo et al., 1991; Miller, 1987).

In our search for an appropriate measure of size that meets our criteria as being both practical and simple, we have used a measure that is the number of employees adjusted for their level of professionalization. This measures the information-processing demand on the organization design. There are a number of reasons for this choice. First, organizations differ and employ different types of people. We find the task they perform is less important, but that the way they handle information and make decisions is important. For a given level of size, the greater the degree of professionalization, the higher the possible level of decentralization, organizational complexity and formalized coordination mechanisms (Scott, 1998, p. 259-264). This is also related to the span of control. Despite the fact that the evidence is somewhat conflicting, it seems that the more complex the tasks that a person can do, the smaller the span of control of his superior (Scott, 1998). Higher levels of professionalization thus seem to increase formalization between units, but at the same time formalization within a unit may decrease. Therefore, if the span of control is small, a smaller number of employees will lead to a higher level of vertical differentiation, and we account for this in our compound measure. This argument may be moderated by the use of information technology as with some types of information technology comes is an increasing demand for professionalization.

Skill capability or level of professionalization is measured on a five-point scale. The proportion of employees that hold advanced degrces or have many years of specialized training is measured on a scale from 1 to 5 with 0 to 10 percent equal to 1, 11 to 20 percent equal to 2, 21 to 50 percent equal to 3, 51 to 75 percent equal to 4 and 76 to 100 percent equal to 5. This scale is similar to the scales used by Hage and Miller, Glick, Wang, and Huber (Hage, 1965; Miller et al., 1991)

The adjusted-size measure is calculated as the number of employees times the professionalization score. The adjusted-size measure is then mapped into small, medium, and large according to the scale shown in Table 5.1. The relationship between number of employees, professionalization, and size is shown in Table 5.2.

Table 5.1. Size Categorization

Adjusted size Measure		Size Category
	< 100	Small
101	– 500	Medium with less impact
501	– 1000	Medium
1001	– 2000	Large with less impact
2001	<	Large

Table 5.2. Thresholds for Categorizing the Adjusted-Size Measure, Number of People Employed and Professionalization

	Professionalization				
Size categorization	1	2	3	4	5
Small	<100	<50	<33	<25	<20
Medium with less impact	101–500	51–250	34–166	26–125	21–100
Medium	505–1000	251–500	167–333	126–250	101–200
Large with less impact	1001–2000	501–1000	334–666	251–500	201–400
Large	2001<	1001<	667<	501<	401<

An organization with fewer than twenty people will always be categorized as small. One with twenty to 100 employees will be categorized as small or medium with less impact. This means that if an organization has between twenty-one and 100 employees with a high degree of professionalization, it is a relatively small organization but without the restrictions on the organizational structure that come from being very small. However, once it passes 100 employees it is very likely that the organization cannot be seen as a small organization but has some properties of a larger organization. This is supported by empirical research that shows that companies with forty to 100 employees face transition problems from being a very small to being a larger organization (Boje et al., 1990). The same transition issues become important when moving from a medium to a large organization.

Returning to our introductory example, Benny Mortensen is rightly concerned that an increase in number of employees–most of who are professionals - is a cause for concern; the institute's size does change from small to a medium-size organization.

The size issue is not completely resolved by our adjusted measure, but we have found that it corresponds well with the literature and has worked well with respect to our validation process. The adjusted size fits the measure mentioned by Robbins (1990); in his treatment of size and discussion of technology, the size measure can be related to manufacturing companies with employees with a low level of professionalization. This is Woodward's mass-production organization (Woodward, 1965). At the other end of the scale, the organizations studied by Carillo and Kopelman were metropolitan branches of a financial service company, a type of organization supposedly staffed with employees with a high degree of professionalization (Carillo et al., 1991).

We have found that the size measure has worked well independently of the type of organization: service or production, or labor intensive or less labor intensive. It has worked well in different industries with different kinds of technologies.

SIZE AS A CONTINGENCY

Size affects the structure of the organization in at least two ways–in the effect of size that directly can be attributed to the adjusted-size measure, and in the effect of size on other contingencies. Only the first will be discussed in this chapter. The second kind of effect was presented in Chapters 3 and 4 and also is discussed in subsequent chapters.

As illustrated in the first section of this chapter, the change in size can have a significant effect on appropriate organizational structure. Simple information arguments based on bounded rationality assumptions lead to the conclusion that an increase in size would lead to a higher degree of organizational complexity and a higher degree of decentralization. A control-based argument suggests a need for higher degree of formalization. However, there are qualifications. In owner-controlled organizations the increase in size has a much lower effect than in other types of organizations. Pondy argued that a goal for owner-run organizations is to maintain control, and structural changes that yield lesser control are not likely to take place (Pondy, 1969). Geeraerts, studying 126 Dutch companies, showed that the correlation between centralization, formalization, and complexity was much lower in owner-run companies than in professional-run companies (Geeraerts, 1984).

The effect of an increase in size seems also to be higher for small organizations than for large organizations, leading to a nonlinear relationship between size and structure. We incorporate this nonlinearity by an adjusted-certainty factor. Size not only affects complexity, formalization, and centralization, but also has an effect on configuration. Small organizations tend to be most efficient, if they have a simple or ad hoc configuration. Additionally, for some configurations like the divisional, the organization should not be small. With this general background, we consider the size propositions.

Size Effects on Complexity

Generally an increase in size increases the complexity of the organization. As small organizations grow, so does their need for specialized staffs. When the organization is small, marketing, sales, and accounting functions may be done by the manager, perhaps with some help. When the organization grows, so do the supporting activities leading to marketing departments, accounting, and personnel departments. At the line dimension, growth may introduce functional specialization–such as in production. Both of these arguments lead to higher vertical and horizontal differentiation and in general to higher organizational complexity. As the size increases, the economics of specialization can be realized. Specialization should increase particularly when the increase in size is due to an increase in professionalization. As mentioned above if the span of control is constant then an increase in the degree of horizontal specialization also increases the degree of vertical differentiation. Thus, if the size of the organization is large the degree of complexity should be high and if the size is small the degree of complexity should be small. These statements find general empirical support as Miller (1987) finds twenty-seven empirical studies included in his meta-analysis, which support these propositions. However the relationship between size and complexity seems to be less strong in public organizations than in private organizations (Blau, 1970). This may be due to more rigid rules within a public organization with respect to possibilities for specialization, e.g. formation of departments.

Propositions on Size Effects on Complexity

If the organization is large and not public, then complexity should be high.

If the organization is large and public, then complexity should be medium.

If the organization is of medium size, then complexity should be medium.

If the organization is small, then complexity should be low.

Size Effects on Centralization

Perhaps the most intuitive proposition of all is that a large organization should be decentralized. A straightforward argument follows from an information processing view; namely, there is much information to process in a large organization and the decentralization of decision-making is desirable, even necessary to avoid top management overload and delay. The increase in information processing may come either from an increase in the number of employees or from an increase in professionalization. Privately owned organizations run most efficiently if the owner is actively involved in decision making. Goals and missions are then clear, and activities are coordinated. The relationship is empirically supported by Geeraerts (1984). Thus there is a tendency from a coordination point of view that the relationship between size and centralization is less predominant in private organizations. In many public organizations the goals are less clear and often are multidimensional. This allows for a higher decentralization where each department follows its own goals.

Small organizations should more likely be centralized, which is true in both public and private organizations. In small organizations centralization is a simple and less costly means for coordination than any other means.

Throughout, the argument is that larger organizations should be more decentralized; privately owned organizations are less extreme than public, but both follow the same pattern. However, Miller (1987) finds that the relationship between size and centralization is not clear and significant. Yet from a normative view, it is a very reasonable proposition that follows from our information-processing point of view and the notion of bounded rationality. As size increases, so does the demand for information processing and with the individual's limited information processing capacity, delegation can be one reasonable response. The argument is also related to the previous propositions on complexity. Size increases complexity and thus by itself creates greater information-processing demands. With the formation of specialized units, it may be virtually impossible for management to be on top of every detail. If management wants to make all decisions, these are delayed and not coordinated, and problems arise. Then he or she has to delegate and decentralization increases.

Size Effects on Centralization Propositions

If the organization is large and not private, then centralization should be low.

If the organization is a large private organization, then centralization may have to be low but it also could be medium.

If the organization is of medium size and private, then centralization may have to be high but it also could be medium.

If the organization is of medium size and not private, then centralization should be medium but it also may be high.

If the organization is small, then it is most likely that it should have high centralization but it also could be medium.

Size Effects on Formalization

As size increases, decentralization increases, and management loses control. An increase in formalization will enhance control and ensure coordination. Additionally, as size increases, it is more likely that the same activity has to be carried out more than once. It may therefore be cost efficient to find the best way to carry out the activity and then write a rule. Formalization thus decreases the amount of information to be processed, overcoming the increased need due to the increase in size.

Formalization should be low for small organizations. Here a cost argument explains the logic. Small organizations can easily be coordinated by a centralized decision-making. The cost of introducing formalization does not pay off. Thus the relationship between size and formalization seems to be straightforward and is also supported by Miller's meta-analysis (Miller, 1987).

However, it is not enough to increase formalization when needed; the formalization has to be the right type. If the rule is wrong, it is better to be without a rule.

The formalization effect may be moderated with a high degree of professionalization. Professional organizations are less formalized at the task level but remain equally formalized between organizational units. Generally, formalization increases standardization, which can also be obtained via professionalization. Professionals are trained to behave in a standardized way; certified nurses and accountants, for example, are trained to care for patients and prepare balance sheets, respectively. From an information-processing point of view, professionals with skills can process more information than less skilled employees.

Propositions on Size Effects on Formalization

If the organization is small, then formalization should be low.

If the organization is of medium size, then formalization should be medium.

If the organization is large, then formalization should be high.

If formalization is high and professionalization is high, then formalization between units should be high and formalization within units should be medium.

If formalization is medium and professionalization is high, then formalization between units may be medium or high, and formalization within units may be medium or low.

Size Effects on Configuration

It generally is agreed that if the organization is small, a simple structure is appropriate. In Mintzberg's discussion of his five configurations, the only configuration that fits the small organization is the simple one (Mintzberg, 1979). From an information processing point of view, the simple organization is the least costly configuration that can handle the information processing requirements of the small organization. Empirically, this is also supported by the study by Geeraerts (1984).

An organization needs a minimum size to have a more complex organizational configuration (Daft, 1992; Mintzberg, 1979; Robbins, 1990). The size has to cause a minimum information-processing demand to make the more costly complex configurations appropriate. This is consistent with ideas about the evolution of an organization. The organization starts small, then grows by exploiting its markets, and finally, continues to grow via diversification. In that process the configuration changes from a simple configuration to a functional configuration and ends with a multidivisional configuration in the last stage.

The simple configuration with centralized decision-making and loose organization is an appropriate organizational design, when the organization is small, young, and in its formative stage. It can respond quickly and foster innovation usually centered with the founder. When the organization obtains success and starts to grow, it usually moves into some kind of mass production. Formalization and specialization are necessary to obtain economies of scale and consistent level of quality. When local markets are exploited, then the organization starts to diversify to other markets. This forms the basis for divisionalizing the sales department. The next phase is reached when diversification is based on products. When that happens, the conditions for multidivisional configurations are satisfied. Throughout this process the organization grows in size, and organizational changes are made to meet the demand for information processing capacity.

Propositions on Size Effects on Configuration

If the organization is small, then it should be either an ad hoc or a simple configuration.

If the organization is large or of medium size, then the organization can be either a matrix, functional, divisional, machine bureaucracy, or a professional bureaucracy

---

Size Effects on Incentives and Coordination and Control

The size of the organization may not in itself have an effect on the incentive structure. However, it may indirectly influence the way incentives are set up. We found above that size has an effect on formalization and decentralization. A higher degree of formalization helps regulate behavior, potentially supporting a procedural-based incentive system.

On the contrary, if size is the reason for a high degree of decentralization then that may require an incentive structure based on results. Whichever incentive structure that will be the best will be related to obtaining organizational design fit. It is also related to whether efficiency or effectiveness is the primary focus. Thus size as such may indirectly affect the incentive structure. However, the particular size measure for an organization comes from the number of people and the average skill level. From an incentive point of view the skill level is important. Kowtha (1997), in a study of the effect of task uncertainty and the skill level of the employees, found strong support for the effect of skill level. A high skill level supports a results based incentive structure while a low skill level supports a procedural based incentive structure.

Incentives are one way to obtain coordination. If the size is small, direct supervision may be a possibility. If the size is large then more rigid coordination mechanisms may be needed. The particular type of coordination system will depend on other contingency factors as well organizational design fit considerations. Thus size is moderator for other factors determining the means for coordination and control.

Propositions on Size and Skill Level Effects on Incentives and Coordination and Control

If the skill level of the employees is low then the incentive structure should be procedural based.

If the skill level of the employees is high then the incentive structure should be results based.

If size is small then coordination and control could be obtained by direct supervision

If size is large then rigid and formalized means for coordination and control may be needed.

MANAGING SIZE AND SKILL

The main proposition in Chapter 1 is that a fit between contingency factors and organizational structure is necessary to obtain optimal performance. If there is a misfit, either the contingency factor value can be changed or the organizational structure can be changed.

How can size be changed? Most organizations have either an implicit or explicit goal that the organization should grow. The goal is most often stated in terms of profit or other financial terms but also can include operations. Generally, as was pointed out above, there is a high correlation between financial size measures and the measure of size. Organizations work hard to change their size.

The change in size can happen in an evolutionary way or can happen in big jumps. As the Institute of Applied Computer Science example shows, a big contract can suddenly change a company's size. Size also can increase by mergers and acquisitions. In Europe, there is a big merger wave. Different kinds of mergers lead to an increase in size but also to a number of other changes that affect organizational structure. This is not a new story. The above discussion was focusing on changes in the number of employees. However, the size measure may also change due to changes in the skill level of the employees. Training and experience are means to increase the skill level within human resource management (HRM). Formal education is another. Hiring people with a higher skill level is yet another way that the size measure can change. The changes in the average skill may be easier to obtain in small organizations than in large organizations.

So far we have discussed only increases in size, but size also may decrease. Due to retrenchments in various industries, many organizations have shrunk suddenly. It happens for small organizations, and in some cases they die during the process. It happens for large corporations as well. IBM, UNISYS, and SEARS are examples of organizations that have decreased their size.

Additionally, introduction of new information technology may reduce or change the number of type of employees in an organization. Many companies have tried to eliminate complete layers of middle management due to a more automated and efficient information and production technology. Modern technology has also made certain types of employees obsolete - mostly individuals with low or no education–while increased demands for new types of employees– mostly individuals with a high level of education or training.

The effect on the organization by an increase in size is not necessarily the opposite of the effect of a decrease (Weitzel and Jonsson, 1989), and the management process is quite different. However, there are many controversies about these differences (McKinley, 1993), and it is an area that is not much researched.

It may be that the process of downsizing an organization is not the reverse of growth from the point of view of organizational design– but should it be? Most research has focused on descriptive theory and has not had a normative point of view. From a normative point of view, there may be no difference between the most effective and efficient organizations; but from an organizational change point of view, the difference may be big.

SUMMARY

There are a number of size measures. The number of employees is the best measure of size for designing the appropriate organization. In order to consider the information processing requirements of the organization, this measure should be modified for the professionalization level of the organization - that is, greater professionalization yields effectively more employees than nonprofessional.

The size imperative indicates that a large organization should have high complexity, high formalization, and low centralization. A small organization should have low complexity, low formalization, and high or medium centralization. The size implication for the organizational configuration is less definite. The size of the organization is an important determinant of the appropriate organizational design.

CHAPTER 6

The Environment

INTRODUCTION

On Monday, January 13, 1992, the Danish business newspaper Børsen published a story about Samsonite's new distribution system for its European market. Samsonite produces luggage in Belgium for the European market. In each European country it had sold its products through a national company that had exclusive rights to import Samsonite's products. The national company then sold Samsonite products to stores. In Denmark, the firm Bon Goût had held the contract with Samsonite for twenty-three years, but Samsonite canceled the contract to sell directly to stores from its headquarters in Belgium. It developed an on-line order system that enabled it to sell to all countries in the European Union from Belgium, and allowed it to take advantage of changes within the new EU single market. Samsonite would also in a more direct way operate in different countries with different national cultures.

For both Samsonite and Bon Goût, the environment was changing. Deregulation within the EU led Samsonite to change the way it sold its products. For Bon Goût a well-established relationship with a supplier changed to a hostile environment. Not only did Samsonite cancel the exclusive-rights contract, but it also stopped supplying to Bon Goût and worked hard to transfer individual stores to a new on-line purchasing system. Bon Goût had to choose a strategy and organizational structure that could cope with this new situation.

The regulatory changes in the EU are one set of deregulations. Deregulation is a fact for many industries throughout Europe and North America. Traditionally, airlines, railroads, telecommunications, and banking, among other industries, have been closely regulated. The deregulation movement transcends national boundaries and political ideologies. The nature of the regulatory environment has changed and that change calls for adopting new business strategies and new organizational structures and capitalizing on new opportunities.

A basic thesis is that the change in regulation requires new business strategies, which in turn require different organizational structures than were efficacious in a regulated environment. More specifically, deregulation requires customer-oriented strategies and an organizational structure that can respond efficiently to customer preferences in a timely manner. Such innovative requisites seem self-evident, yet there are numerous pitfalls as well as windows of opportunity (Burton and Obel, 1986).

Deregulation implies changes in the environment. More generally, a primary characteristic of the environment is the degree of uncertainty, which may arise in numerous ways. For Bon Goût, uncertainty changed and increased. It used to have a stable relationship with its supplier and its customers. Now the supply side changed, but it also faced a new competitor. Additionally, although the market for Samsonite luggage has been fairly domestic, the new competitor had a more international profile. Given the degree of uncertainty, one can then indicate an appropriate internal organizational structure, if the firm is to be effective and efficient and, in the long run, viable.

Contingency theory, open systems theory, and strategy and structure arguments, among others, follow the common theme that there must be a fit between the environment and the organization (Chandler, 1962; Thompson, 1967; Lawrence and Lorsch, 1967). Each theory or model has its own features, and the implications of uncertainty suggest different specific adjustments. Understanding and fitting with the environment require an information processing capacity commensurate with the uncertainty in the environment (Galbraith, 1973). Environment is very important for the organization: despite numerous controversies, this much is agreed. Whether it is the environmental imperative of contingency theory or the environmental determinism of population ecology, the organization's environment must be considered (Lawrence et al., 1967; Hannan and Freeman, 1977; Duncan, 1979). From a contingency theory viewpoint, the environment must be reckoned with and adapted to. The environment-structure imperative indicates that an organization should design its structure in relation to its environment. Here Bon Goût should react and redesign its structure. Even within this framework, there is considerable controversy. What does one mean by *environment*? How does one describe and measure the environment? With the environment defined, described, and measured, what are the implications for the organization's structure, and how should the organization operate? Besides theoretical interest, managers want answers and recommendations about what they should do. In this chapter, the goal is to examine these issues with an eye to developing design recommendations.

The SAS Environment

Until the beginning of the 1970's, the market for international air traffic was heavily regulated and based solely on bilateral international agreements. These agreements controlled which routes each airline company could fly and how much capacity each company could offer. The forum for international price coordination was the International Air Transport Association (IATA), founded in 1945. IATA had members all over the world, and consequently the price agreed on by IATA became the world market price. When a price had to be decided, airline companies produced calculations of how cheaply they could sell their travels, and based on these calculations, a mutually acceptable price was negotiated.

Finally, prices had to be approved by the national governments, which was normally a mere formality. Like so many other airline companies, SAS could consider the prices fixed between two IATA meetings and consequently price competition was excluded. Governmental control of routes and IATA-fixed prices were further reasons that the environment that SAS encountered in the beginning of the 1950's was very stable. The market showed constant growth with room for everyone. Oil prices were increasing only slightly, and most airline companies had a fair profit every year. This was the situation until the beginning of the 1970's when the first oil crisis changed the whole market situation.

Suddenly oil prices rose explosively, and consequently total costs of airline companies increased dramatically, and there was nothing the airlines could do about it. The oil crisis resulted in a general weakening of the entire world market, and where the market for air travel had been steadily growing it suddenly stagnated. Many airline companies found that their assumption that the market growth of the 1960's would continue had led them to purchase planes that were no longer necessary. Furthermore, jet airplanes were now introduced on the market with larger airplanes. All in all, it resulted in too many seats for too few passengers, so the companies lowered prices to fill their planes. At the same time the US government began to liberalize the American airline policy. This meant increased competition on the already keen market for trans-Atlantic air travel. Oil crises and deregulation in the United States resulted in an increasingly noticeable overcapacity. All were eager to get new customers, but where would they find them? The only possibility was to "steal" customers from other airline companies, and thereby the competition in Europe began in earnest.

During the 1980's, competition became increasingly severe. Both European and overseas airline companies began to prepare themselves for the open borders of the European Union in 1992. The airline business had been protected because each country had given airline companies monopolies on their routes. Through their organization, IATA, the airline companies had tried to avoid potential competitive situations. But it was quite clear to everyone that EU's open borders would enable anyone to enter and capture new markets. No one knew for sure what the new situation would look like and which factors would be decisive. Therefore, it was important to get as firmly established as possible in the market before 1993. In this connection SAS had its own special problems. The fact that SAS was situated on the periphery of the EU was not a favorable starting point. The three large European airline companies - British Airways, Air France, and Lufthansa - all had a larger home market, and Air France and Lufthansa, in particular, were situated centrally in Europe.

As mentioned earlier, Jan Carlzon was appointed executive of the part of the new structure called the Airline Company SAS. This began a new era for SAS. Envisioning increased competition from outsider companies due to the deregulation of the industry, Carlzon stressed the importance of SAS to establish a secondary home market. This should be achieved by co-operation with another airline company. Carlzon's successor Stenberg later followed up on Carlzon's idea, and in May 1996, SAS announced plans for a strategic alliance with Lufthansa.

The free market competition resulting from the accomplishment of the deregulation on April 1st, 1997, has further increased the importance of SAS's competitiveness. Even in the Scandinavian market, in which SAS has traditionally held a monopolistic status, the company has experienced increased competition. As a means to take on competition, SAS has expanded its number of flight offerings by making a number of strategic alliances. Through these alliances SAS strives to cover a larger global market. By sharing administrative costs with its allied partners, SAS is furthermore hoping to reduce costs to an extent that will allow the company to reduce prices and hence become more competitive. One of the most important of these is the Star Alliance announced on May 1997 to initially include SAS, Lufthansa, Air Canada, Thai Airways International and United Airlines. The core of this is a code share agreement, enabling all member companies to offer more frequent departures to their customers.

Other environmental factors of importance to SAS are governmental actions. In 1997, SAS's costs were increased when the European Commission's introduced a mineral oil directive, increasing environmental duty on fuel for planes. Other governmental actions aiming at a reduction of the allowed noise level of aircrafts and leakage of CO_2 has stressed the importance of a replacement of SAS's existing aircraft fleet, as SAS may otherwise be excluded from landing at some airports.

The situation in 2003 has changed from severe competition to a hostile environment. The September 11 event, the SARS epidemic, and the general downturn of the economy has created a substantial overcapacity in the industry. Airlines like Swissair and Sabena did not make it and SAS's US partner UNITED went into Chapter 11 in 2003. In the message from the CEO in Scanorama January/February 2003 issue, Jørgen Lindegaard states that the EU court in the autumn of 2002 ruled the bilateral open skies agreements illegal and that will promote further deregulation and consolidation within the next 4-5 years. Basically these bilateral agreements prevent mergers of airlines because the merged airline will loose landing slots. Thus, strategic alliances like the Star alliance are the answers/solutions in an industry where regulations and agreements make mergers very difficult. That may change the future, increasing the environmental equivocality.

Over the history of SAS, the environment has been relatively certain with periods of great uncertainty.

1. How did SAS react to these periods? Was it well prepared?
2. What would the effect of these changes be on the recommended structure (configuration, centralization, complexity, formalization, incentives, and coordination and control)?

LITERATURE REVIEW ON ENVIRONMENT

The concept, description, and measurement of an organization's environment remain difficult and are the basis for continuing controversy (Milliken, 1987). Additionally, the effect on the organization is not fully agreed upon (Koberg and Ungson, 1987). We review the most important contributions to our understanding of the environment-structure relationship. Further, we argue that dimensions of the national culture in which the organization operates is part of the organization's environment.

Measures of the Environment

In the literature the environment has been described using many descriptors and has been categorized in many different ways (Scott, 1992; Buchko, 1994). For the contingency theory of organization, one or two variables related to an uncertainty measure have been the norm in most empirical investigations. However, despite the fact that some measure of uncertainty has been used, it is not clear whether they measure the same thing or how they are related. Lawrence and Lorsch and Duncan are classic examples and are discussed below (Lawrence et al., 1967; Duncan, 1972).

Figure 6.1. Description of the Environment

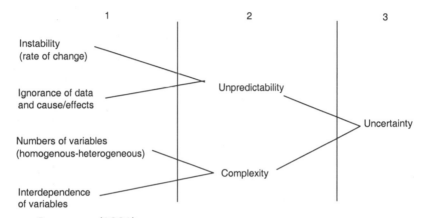

Source: Lawrence (1981).

In their now classic study, Lawrence and Lorsch employed an environmental measure of uncertainty (Lawrence et al., 1967). Uncertainty was a perceptual measurement of the clarity of information, certainty of causal relationships, and time spans of definitive feedback (Lawrence et al., 1967, p. 29). Lawrence maps the instability (rate of change), ignorance of data and cause and effects into unpredictability (Lawrence, 1981, p. 316) (Figure 6.1). He also maps the number of variables, homogeneity or heterogeneity, and interdependency of variables into complexity. Unpredictability and complexity are then mapped into uncertainty. These environmental mappings are thus aggregate measures of more basic descriptors.

Duncan developed a two-dimensional environmental measure to capture the notion of environmental uncertainty: environmental change or dynamism (unstable to stable) and environmental complexity (simple to complex) (Duncan, 1972). The environmental change dimension encompasses the possibility that the environment can change and can change in unpredictable ways. This fits well our intuitive notion of uncertainty; we simply may not know what will happen.

Environmental complexity is another aspect of the environment–namely, the number of elements in the environment that are important to the organization. Duncan's two measures of environmental uncertainty are clearly aggregate measures; each captures a different aspect of the environment: change and complexity. Downey, Hellriegel, and Slocum critiqued Duncan's environmental uncertainty measures and compared them with the measure used by Lawrence and Lorsch (Downey et al., 1975). Downey, Hellriegel, and Slocum found problems with Duncan's measure. In the original study, Duncan directly summed the subscales without standardization, creating a biased weighing (Duncan, 1972). The bias made perceived complexity dominate and "only allow a minute weighing for the ability to assign probabilities" (p. 628). Downey, Hellriegel, and Slocum found that when subscales were standardized, environmental change was the more important aspect of environmental uncertainty (Downey et al., 1975). Comparing the two measures of environmental uncertainty by Duncan and Lawrence and Lorsch, Downey, Hellriegel, and Slocum concluded that the measures actually were two different instruments. Downey, Hellriegel, and Slocum also investigated the Lawrence and Lorsch measure and found, like Tosi, Aldag, and Stoney did, the uncertainty measure deficient in terms of validity and reliability (Tosi et al., 1973).

The importance of the above discussion becomes clear when we discuss the environment and structure relationship because Duncan and Lawrence and Lorsch obtain different results with respect to the environment and structure relationship. We argue that the difference is a result of their different uncertainty instruments. Criticism aside, the two measures, environmental change and environmental complexity, do capture important dimensions of the environment.

Despite methodological criticisms of these classic studies, the organization's environment remains an important contingency. Downey, Hellriegel, and Slocum conclude that "beyond the methodological adequacy of specific research instruments, the inconsistent results obtained in this study between Lawrence and Lorsch's and Duncan's uncertainty measures raise even more serious questions." Uncertainty concepts as presently used in organizational theory involve much ambiguity. This does not mean all contingency theory need be restricted to one meaning for uncertainty. Moreover, it does not mean that contingency theory must wait for the development of the one meaning of uncertainty (Downey et al., 1975, p. 628).

The research developments of the last few years confirm their assessment. We have not settled on *the* measure of environmental uncertainty. Indeed, there are many others. Contingency theory continues to be a well accepted paradigm where the environment, however measured, is important. Other environmental measures have been created for particular purposes.

Bourgeois and Eisenhardt defined a high-velocity environment as one where "changes in demand, competition, and technology are so rapid and discontinuous that information is often inaccurate, unavailable, or obsolete"(Bourgeois and Eisenhardt, 1988, p. 816). This definition incorporates both the reality of the environment (rapid, discontinuous) and what is known (accurate, availability, timely) about the environment. In other frameworks, high velocity could be characterized as dynamic and uncertain. Eisenhardt summarized the implications for management (Eisenhardt, 1989); high-velocity environments call for fast decisions. Fast decision-makers use more information, not less; they develop more alternatives, not fewer. And further, fast decisions based on this pattern lead to superior performance.

Tung (1979) used a three-dimensional typology to measure the environment: rate of change, complexity, and routineness. She showed that these three measures could be mapped on to a measure of perceived uncertainty. Milliken also used a three-dimensional measure (Milliken, 1987). He divided the environmental uncertainty into state, effect, and response uncertainty. State uncertainty means that one does not understand how components of the environment might be changing. Effect uncertainty is the inability to predict whether a change in the environment will have an effect on the organization. Response uncertainty is the lack of understanding of what response is appropriate for a given change in the environment.

In their analysis of the informational requirements in an organization, Daft and Lengel used environmental uncertainty and equivocality (Daft and Lengel, 1986). Uncertainty is lack of knowledge of the value of a particular dimension, while equivocality is the "existence of multiple and conflicting interpretations about an organization." High equivocality means confusion and lack of understanding. Equivocality is a "measure of the organization's ignorance of whether a variable exists in the space" (Daft et al., 1986, p. 567).

Much earlier, Ashby utilized variety as a concept to describe and measure the environment and the capacity of an organism to adjust (Ashby, 1956). Variety is defined as "the number of distinct elements" (p. 126). The environment can be described by a finite list of variables that can take on either discrete or continuous values. The size of an organization's environment depends on the number of variables and their possible values. An environment with four variables, each of two values (yes, no) has 2^4 or sixteen possible states. An organization's environment can be rather large. Ashby's notion of variety of the environment and the organization (or organism) leads to the famous law of requisite variety: "only variety can destroy variety" (p. 207).

Kuhn further interpreted the law of requisite variety: "Environmental disturbances can be kept from causing performance deviations only when the environment's variety is exceeded (or equalled) by the firm's variety, which in turn, is exceeded (or equalled) by the management's decision-making capacity, or variety. To block any disruptions caused by the uncontrolled (i.e., the environment's) variables, then, the controlled (i.e., the firm's) variables must encompass sufficient latitudes"(Kuhn, 1986, p. 3).

Intuitively, the organization must be able to adjust to its environment in its many possible states. Although very appealing, the law of requisite variety requires careful attention to operationalize and to derive implications for organizational theory.

The problem of defining, describing, and measuring an environment is difficult and remains contentious. Somewhere, there is a balance of simplicity and complexity that meets our purpose for a design-oriented contingency theory. Lawrence called for such a balance: "The key problem is whether we can treat this idea as a unitary dimension of environment for the sake of simplicity, or whether we must treat it as a set of separate but related factors for the sake of accuracy. The factors and terms, that have been discussed as related to uncertainty, are complexity, number of variables, predictability, ignorance of relevant data and cause-and-effect relations, interdependence of variables, and rate of change"(Lawrence, 1981, p. 317).

So far we have treated the environment as a single entity. Lawrence and Lorsch recognized that different parts of the organization might be facing different types of environments (Lawrence et al., 1967). They focused on the market, science, and technoeconomic environmental sectors. Fink, Jenks, and Willits discuss multiple environments of economics, technological, political-legal, social-cultural, and external physical environments (Fink et al., 1983). Fink, Jenks, and Willits (pp. 278–284) discuss static attributes of complexity, routineness, interconnectedness, and remoteness, as well as dynamic attributes of rate of change and predictability of change. They present contingency relations between these environmental factors and the organization's structure, goals, and rewards, among others (pp. 284–290). It is a very elaborate set of contingency relations concerning the design of the organization.

The two environmental dimensions of complexity and uncertainty (Robbins, 1990, p. 219) and complexity and change (Daft, 1992, p. 52) are popular textbook concepts. They remain intuitively appealing, if not methodologically rigorous, and can be operationalized as important contingencies for the organization's structure. They are also still the basis for measuring environment in current research (Staber, 2002; Lenox, 2002; Sorenson, 2003). The debate on the proper measure of the environment, however, continues (Buchko, 1994).

One additional measure has to be mentioned here–degree of environmental hostility. Bon Goût was threatened by the actions of Samsonite and had to react appropriately. Hostility is not uncertainty. Extreme hostility implies that someone or something threatens the existence of the organization. *Hostile* environments are characterized by precarious settings, intense competition, harsh, overwhelming business climates, and the relative lack of exploitable opportunities. In 1990, a Danish hairdresser announced publicly that use of the combined shampoo and conditioner Wash-and-Go caused people to lose their hair. The news media jumped on the story, and sales on the Danish market dropped from a 24 percent market share to 3 percent basically overnight. Wash-and-Go is sold worldwide, so the situation was a real threat to Procter & Gamble. Madsen and Jensen (1992) and Jensen (1993) categorized the environmental situation for Procter & Gamble in Denmark as a hostile one. Low hostility implies benign environment (Covin and Slevin, 1989). As we argue later, the appropriate organizational response to high hostility is different from its response to high uncertainty (Robbins, 1990).

We also argue, that the national culture is an important dimension of the environment as it signals which organizational structures employees are most likely to accept. This will be further elaborated in a separate section.

The Environment-Structure Relationship

Environmental-organizational relations are dominated by three perspectives: the environment is dominant; there must be a good fit between the environment and the organization; and there are organizational principles that operate for any environment. Population ecology in organizational theory is an example of an environment-dominant argument. The environmental-organizational fit thesis captures the contingency theory, open systems, and other approaches. We will review briefly the first two themes and then concentrate on implementation of the environmental-organizational fit themes.

Population ecology considers the environment as dominant. It is the "environment which optimizes" (Hannan et al., 1977, p. 939). Through competition, the environment separates the successes from the failures (Pfeffer, 1982, p. 186). The organization must be linked with its environment in order to survive. In the Bon Goût case, EU deregulation may imply that it may not be feasible for a luggage manufacturer to sell through national agents that have exclusive rights. The environment may optimize so there is no room for local agents. The environment, with its changes in information technology and regulation, determines who will survive and prosper and who will die. It is less clear how one can develop approaches for viability– if it is, indeed, possible. The environment selects; whether organizations can adapt consciously is an open question.

Yet, the organization has two choices of design: specialism and generalism. Hannan and Freeman (p.953) offer insight on the choice, "when the environment changes rapidly among quite different states, the cost of generalism is high. Since the demands in the different states are dissimilar, considerable structural management is required of generalists. But since the environment changes rapidly, these organizations will spend most of their time and energies adjusting structures. It is apparently better under such conditions to adopt a specialized structure and 'ride out' the adverse environments." Of course, this choice depends upon a number of factors: how long will the "storm" last; how fast can the structural adjustment be made and what is its cost; and what is the opportunity cost of having a nonoptimal structure. Some organizations will fail in the storm. Population ecology focuses on selection by the environment, whereas the contingency approach focuses on the adaptation of the organization to the environment.

Environmental-organizational relations and the necessity for fit are explicit in much of modern organizational theory. The organization's structure is contingent on the environment and the technology, where the organization's size is also a consideration. Lawrence and Lorsch (1967) argue that increased environmental uncertainty leads to increased organizational differentiation, making organizational integration more difficult. They define *differentiation* to mean that the organization has departments that are different in both tasks and orientation. This relates to horizontal specialization and our measure of organizational complexity. Lawrence and Lorsch studied three well-defined industries that they categorized as ranging from low to high uncertainty. They found that increased uncertainty in the environment required increased differentiation in the organizational structure for the organization to be efficient. Then integration is required to make the different departments work in cooperation. Integration devices typically include rules and procedures, configurational plans, the authority of the hierarchy, and decision-making committees.

Duncan (1979) begins with an environmental classification along a static-dynamic dimension and a simple-complex dimension. Functional, divisional (decentralization), and matrix models are then analyzed for strengths and weaknesses. Finally, the environmental-organizational relations for fit are developed. For example, a simple static environment calls for a functional organization. Galbraith (1973) focuses on the firm's information processing choices as alternatives to cope with environmental uncertainty (Chapter 1). In this stream of research, environmental uncertainty requires the firm to adapt.

As the uncertainty of the environment increases, so does the demand for information processing. A company in a very uncertain competitive market situation has to monitor the market and make quick changes, and doing that correctly requires a large amount of information. An alternative approach is to insulate the organization from its uncertain environment. The classic approach is to create a buffer inventory to absorb environmental shocks. For example, a firm may keep extra raw materials inventories on hand, decouple its sequential processes with intermediate inventories, and use finished goods inventories to coordinate with customers. More generally, slack resources (Cyert and March, 1963) can be created as excessive cash balances, number of employees, and capital equipment. Buffer inventories and slack resources reduce the need to process appropriate information quickly and efficiently.

The open systems theory also emphasizes the importance of environmental uncertainty on the organization's structure (Thompson, 1967, proposition 6.2C, p. 72): "When the range of task-environment is large or unpredictable, the responsible organizational component must achieve the necessary adaptation by monitoring the environment and planning responses, and this calls for localized units." Localized units are similar to the idea of self-contained units in Galbraith's (1973) information processing framework.

Burns and Stalker (1961) developed two types of organizations– the mechanistic and the organic. The mechanistic structure was characterized by high complexity, formalization, and centralization, while the organic structure was characterized by low formalization and decentralization. Additionally, the task definitions were flexible, leading to low differentiation and low complexity. Burns and Stalker believed that the organic structure was needed to be efficient if the environment was turbulent. If the environment was stable, then the mechanistic structure was the best.

Compared to the views of Burns and Stalker, Lawrence and Lorsch give almost opposite advice on how organizations should respond to environmental uncertainty. Burns and Stalker recommend that the proper response to uncertainty is low organizational complexity, while Lawrence and Lorsch recommend a high degree of horizontal differentiation - a high organizational complexity. The empirical studies of the environmental-organizational structure relationships have resulted in mixed results (Tung, 1979; Koberg et al., 1987). Miller (1992) finds that "attempts to achieve fit with the environment uncertainty can prevent or destroy internal complementaries." Basically, he reaches his conclusion by combining various results from the literature.

A careful review of the many studies and arguments reveals that many of the different results may be the result of different use of the term *environmental uncertainty* and that describing the environment using a single measure may be too simple; a proper balance has to be obtained, but there are many pitfalls. Jurkovich (1974) developed five measures of the environment, each measured on a high-low scale that leads to sixty-four different environmental categories.

Such a measure is both too simple and too complex. It is too simple because a high-low score usually is too gross, and it may be too complex with respect to the number of dimensions. In the next section we put the various pieces together and develop a four dimensional measure of the state of the environment for design purposes.

DESCRIBING THE ENVIRONMENT

Environment has been defined, described, and measured in many ways–from unidimensional uncertainty to multidimensional measures. We search for a balance; the definition should be reasonably precise and useful, but not totally descriptive and detailed beyond utility. The goal is to specify the environment in measurement terms that can be of help in the design of an organization - that is, to assess environment in a contingency design framework.

We propose that an organization's environment be described and measured in four dimensions in a four-variable list of characteristics or attributes. Those environmental characteristics are: equivocality, uncertainty, complexity, and hostility. Equivocality is the "existence of multiple and conflicting interpretations" (Daft et al., 1986). Uncertainty is related to earlier discussed measures of uncertainty, but is more restrictive. It is uncertainty of all specific parameter values such as prices, cost and so on. Environmental complexity is similar to earlier measures of the number of factors in the environment that affect the organization. Hostility is the level of competition and how malevolent the environment is.

Equivocality means confusion and lack of understanding. Equivocality is a "measure of the organization's ignorance of whether a variable exists in the space" (Daft et al., 1986, p. 567). We use the notion that equivocality is the organization's ignorance about whether the variable exists in the environment. That is, the variable is ill specified and unknown to the organization or its importance for the organization is not known. Referring back to Figure 6.1 and Lawrence and Lorsch's environmental scheme, equivocality incorporates the "ignorance of data and cause/effects" to the ignorance about what the important variables may be. Of course, the organization must have some information about its environment for equivocality to exist– namely, the organization must be aware of its ignorance. Equivocality is related to the concept of the agenda of the organization (Arrow,

1974). If the agenda of the organization is known and set, then equivocality is low. If the agenda is unknown or not set, then equivocality is high.

Deregulation, as discussed previously, may introduce equivocality. Under regulation, the dimensions on which to compete are known but after deregulation it may not be clear how competition will evolve. EU deregulation and the actions taken by Samsonite increased the equivocality of Bon Goût's environment. In the telecommunication industry, competition has been based on price and service, and competitors have been other telecommunication firms. The entrance of digital switches enabled the computer industry to enter the telecommunications market. In Canada the telecommunication industry has also been challenged by TV cable companies, and it is not clear what the multimedia business will do in the future and how wireless communication will evolve. Denmark is experiencing similar changes. The degree of equivocality has increased. During the last decade the telecommunication industry has faced various degrees of equivocality, and it remains rather high.

The equivocality measure is related to many of the measures discussed above. For example, the routineness measure used in Tung (1979) and Jurkovich (1974) and the effect uncertainty in Milliken (1987) are closely related to our equivocality measure. As Downey, Hellriegel, and Slocum (1975) have pointed out, it is also closely related to Duncan's measure of uncertainty. Additionally, Burns and Stalker's measure of turbulence may be closer to equivocality than to uncertainty.

Uncertainty is our second environmental variable. Uncertainty is not knowing the value of an environmental variable or descriptor. Before the move by Samsonite, Bon Goût knew what suitcases to order and sell. But it might not have known how many it could sell at a given price. Likewise a telecommunication company may not know how many calls it has to handle at certain points in time. In both situations the companies face uncertainty. They do not know the exact value of a particular variable that is important to the company. However, they knew how to handle the uncertainty. This definition is close to the common sense definition and resembles the Lawrence and Lorsch (1967) definition of uncertainty, but more directly to the instability of the environment.

Equivocality and uncertainty are conceptually distinct, yet the distinction is not always easy. Basically, uncertainty requires that at least in principle a probability measure can be specified. Additionally, uncertainty is in most cases related to issues that the organization has experienced previously. Fluctuation in demand of the organization's goods and services is a good example.

When something very new may happen, such as new regulations, new technology, and so on, this is related to equivocality. In this case, a probability measure may not be the right way to describe the lack of certain information. Equivocality is related to something the organization has not experienced before and therefore may not know the exact way of handling the new situation.

The *environmental complexity* is the number of variables in the environment and their interdependency - Bon Goût sales, for example, depended not only on prices but also on fashion. This is a more complex environment than one in which sales are dependent only on price. Sales also may depend on actions taken by competitors, and the variables may be related. The competitors' actions may depend on the price asked by Bon Goût. Managing interdependent variables is more complex than managing independent variables. Therefore, an interdependent environment is more complex than one in which variables vary independently. The environmental complexity measure has been used by a number of authors including Duncan (1972), Tung (1979), and Jurkovich (1974).

Hostility is a measure of how benign or malevolent the environment is. Hostility can vary from a supporting environment to one that is predatory and out to destroy the organization. Natural causes, e.g. hurricane or flood, can be the cause of this extreme level of hostility, or it can also result from the actions of others, e.g. a hostile takeover or sabotage. Bon Goût had been operating in a relatively benign environment, but the hostility quickly increased and it became threatened for its very existence; it lost its product and image of quality, which threatened its very existence. For high hostility, someone or something is manipulating the environment to deteriorate the performance.

The environment can be more precisely described as a list of descriptors or variables $\{X_1, X_2, ..., X_n\}$ and its effect on the organization can be modeled by $f(X_1, X_2,..., X_n)$. Then we state the following:

- Complexity is higher as n increases and the Xi's are interdependent.

- Uncertainty increases as the value of a variable Xi (or variables) is less well known (that is, the variance of Xi increases).
- Equivocality increases as the organization is more ignorant about what are the important variables in its environment, (that is, the organization does not know which set of variables the f-function is defined upon nor the function itself).
- Hostility increases as a change in one of the variables changes to threaten the organization's performance and perhaps its existence; it is a discontinuity.

The four measures were chosen because they can be related to a vast literature of empirical studies as well as fit well with our information-processing view of the organization. The greater the environmental complexity, the greater the information processing demand on the organization. The organization has to monitor more issues and assess the effect on the organization.

Increased uncertainty also increases the demand for information-processing capacity, but in a different way. While increased complexity increases the number of variables that the organization has to monitor and react to, uncertainty relates to the frequency at which the variables have to be monitored and adapted to.

Equivocality relates to the missing information that may hold important contingencies.

Hostility captures whether the environment is supportive or whether the environment is threatening and potentially destructive.

An increase in each of the environmental dimensions increases the demand for information processing capacity but in different ways. Each environmental characteristic presents a different information challenge, as summarized in Figure 6.2. Greater environmental complexity increases the amount of information to process, as there are more issues of importance to the organization. Greater uncertainty requires greater capability to forecast or adjust to the changing environment. This does not necessarily increase the amount of information, but does require a different organizational response. It must either project what will happen or adjust quickly to the environment. The first is forecasting; the second is adapting to feedback. Many organizations use a combination of both: for example, a firm with uncertain sales will forecast and also adjust quickly to actual sales. Greater equivocality requires broader scanning of the environment for heretofore unknown and unimportant environmental parameters. Product and service quality, as defined by the customer, continues to generate new parameters of importance. We frequently do not know what the customer is saying.

Figure 6.2. Information Processing Implications of Environmental Characteristics

Environmental characteristics	Information processing implication
Greater complexity	→ More environmental parameters to monitor
Greater uncertainty	→ Parameters can take on a greater range of values and change quickly
Greater equivocality	→ Unknown issues to scan and attempt to monitor
Greater hostility	→ A change in relevance of what information is important; new information emerges

The four environmental characteristics are general attributes, not the detailed list of X_i's. It is a great simplification. Larger characteristic spaces are possible. Environmental complexity could be split into two measures: one is n, the number of variables, and another is their interdependency. Of course, each bivariable interdependency or correlation could be a more detailed list of characteristics. Although possible and more precise, there is a cost. Smaller characteristic spaces are also possible. Environmental uncertainty, as a perceptual measure, has been applied, as discussed above. Two-dimensional measures include: (1) uncertainty and complexity and (2) uncertainty and equivocality. The rationale for an environmental description or list of characteristics must be justified by its application. Next, we present environmental contingency propositions that rely on the four environmental characteristics defined above.

The Environment as Five Forces[1]

The organization's environment can be described in many ways, using different frameworks. Michael Porter's five forces model is well known and widely applied in strategic analysis. It focuses on the industry structure and the competitive nature of the environment: buyers, suppliers, substitutes, potential entrants and rivalry among existing firms. More specifically, the firm's competitive situation is analyzed as:

- Bargaining power of buyers.
- Bargaining power of suppliers.
- Threat of substitute products or services.
- Threat of new entrants.
- Rivalry among existing firms.

The buyers' power increases as buyers purchase large quantities, purchase standardized products for which they are ready substitutes and can switch easily, etc. and generally, can then demand lower costs, higher quality products and better service.

The suppliers' power is the other side of the firm. Supplier power is high when the suppliers can demand higher prices and deliver lower quality or service, which is basically when the firm has limited alternatives or only alternatives which are very costly. Suppliers have greater power when there are few, if any substitutes, the product or service is unique or the customer is not very important to the supplier.

The threat of substitutes by other firms is increased when other products or services are ready substitutes and are functionally equivalent.

The threat of new entrants is high when the cost of entry is low and the market is attractive in price and/or size. This threat is less when there are economies of scale, product differentiation, high capital entry costs, or unique and specific industry knowledge which is difficult to transfer.

Greater rivalry among existing firms exists when products are similar, high fixed costs, slow growth, overcapacity and only a very few can be winners in a game of survival.

The five forces describe the market structure and embed other dimensions as uncertainty and risk and other environmental characteristics. These additional environmental measures provide a richer understanding of the organization's environment. The five forces–four environmental dimensions crosswalk matrix provides a way to develop more depth about the

---

[1] Our thanks to Thorkild Jørgensen, Aarhus School of Business, who has applied the five forces framework in his organizational design classes to describe the environment, and to Bo Eriksen, University of Southern Denmark, who suggested the outline for this capsule.

environment. In the table below, the five forces are given down the left-hand side and our environmental measures are shown along the top. The matrix entries are then the crosswalk from the five forces of the industry to our environmental measures. For example, the uncertainty in the environment is composed of the uncertainty from each of the five forces; the uncertainty about the threat of new entrants should be reflected in the uncertainty measure.

	Uncertainty	Equivocality	Complexity	Hostility
Bargaining of power of buyers				
Bargaining power of Suppliers				
Threat of substitute products or services				
Threat of new entrants				
Rivalry among existing firms				

For Bon Goût, an industry structure five forces analysis would highlight the threat from the supplier, Samsonite. Samsonite intended to forward integrate its distribution chain and eliminate the local retailer. The other four forces must be considered, but their importance is not so great here. The buyers' have considerable power and can go to other stores and buy other brands of luggage. Substitutes for luggage depend upon the scope; there are numerous substitutes for suitcases to include bags, sacks, etc. New entrants are possible as the cost of entry is relatively low in retailing. However, Bon Goût and Samsonite have considerable reputation. The rivalry depends upon the local situation for Bon Gout. There is ample supply of luggage from around the world and ready availability in the local market; the rivalry is reasonably intense.

To do the crosswalk from the five forces to the environmental dimensions for Bon Gout, we would begin with the supplier threat and focus on the hostility dimension, i.e., the supplier-hostility cell. It is clearly high and perhaps, extreme. The hostility from other forces is minor in comparison. Looking now to the organizational complexity, there are a large number of buyers, which can be differentiated by buyer group according to income, age, etc. The number of suppliers may not be large, but there are more now than in the past. New entrants are possible. And the rivalry suggests that there is price, advertising, promotions, etc. which help make the market competition. Overall, the environment is rather complex and we judge the level to be high. The environmental uncertainty is judged to be medium; the variation in the five forces may be increasing, but is not high. We can put reasonable bounds around the uncertainty. The equivocality is more difficult as it is essentially a measure of what we do not know. There may be new entrants in the shadows, but we know the buyers and the substitutes for luggage. The equivocality seems low. However, before Samsonite's surprise move, the equivocality was high, but Bon Gout did not know it. That is the essence of equivocality.

As illustrated above, the crosswalk from the five forces to the environmental dimensions can be logical and systematic, but it cannot be reduced to an easy method or simple technique. There is no formula as the influences of the five forces will be weighed for the particular situation. We began with the supplier hostility and it is very important in this case, but this situation is not the norm for most organizations. Further, even if we could find a formula, we want to think through each situation and understand it fully.

Porter's five forces industry structure provides an alternative way to describe the environment and a help to developing the environmental measures for our organizational design approach.

ENVIRONMENT AS A CONTINGENCY

The environment as a contingency is well established (Lawrence et al., 1967; among others; Galbraith, 1973; Duncan, 1979). Yet there remains controversy about the exact nature of those influences, as discussed above.

As we take our point of departure from the information processing point of view of the organization, we try to sort out the controversies by using information processing arguments to find the relationships between contingency factors and organizational structures. In the previous section we established how our four measures of the environment affect the demand for information-processing capacity.

The organizational design response is to create an organizational structure with an information processing capacity to match the demand or to change the demand by manipulating the environment. There are numerous ways to create such a fit. If there are many repeated tasks, it is advantageous to write a rule about how to perform the task the best way. If, however, there is a great uncertainty about what to do, then it may not be possible to write such a rule, but rather, management needs to monitor the results and create an incentive system that will lead the organization in the right direction. The fit between the decision-making process and the control and incentive system is, therefore, important.

Environmental Effects on Formalization, Centralization and Complexity

In this section and the next, we develop a number of propositions for equivocality, complexity, and uncertainty. These propositions are summarized in Tables 6.1 and 6.2. Hostility is treated a little later in this chapter.

In Table 6.1, the environmental characteristics of equivocality, complexity, and uncertainty are given a high or low value, yielding 2^3 or eight environmental possibilities. In practice, a finer scale must be used, as is the case in OrgCon. For the sake of simplicity, we limit ourselves here to a high and low discussion. For each environmental situation, we give recommendations on formalization, organizational complexity, and centralization. These propositions are derived and supported by considerable research (Burns and Stalker, 1961; Lawrence et al., 1967; Duncan, 1979; Tung, 1979). Hostility propositions are included in Table 6.2.

Table 6.1 is divided into two parts: part a shows the complexity and uncertainty effects for low equivocality; part b shows the complexity and uncertainty effects for high equivocality.

If the environment is quite simple, equivocality is low so we know what is important. Complexity is low which indicates that only a few variables describe the environment and the uncertainty is low which suggests that we know what the value of each variable is–that is the variance is small then the information processing requirements from the environment are small. Returning to the Bon Goût example the corporation would be in this environment if it sold only Samsonite luggage in Denmark paying Samsonite in Danish currency at a fixed price and if it made an agreement with the stores it serves that they would order and pay for suitcases well in advance of Bon Goût's ordering from Samsonite.

Table 6.1. Environmental Effects on Formalization, Organizational Complexity, and Centralization

Environmental Complexity	a. Low Equivocality	
High	Formalization: high Organizational complexity: medium Centralization: medium	Formalization: medium Organizational complexity: high Centralization: low
Low	Formalization: high Organizational complexity: medium Centralization: high	Formalization: medium Organizational complexity: high Centralization: medium
	Low uncertainty	High uncertainty

Environmental Complexity	b. High Equivocality	
High	Formalization: medium Organizational complexity: medium Centralization: low	Formalization: low Organizational complexity: low Centralization: low
Low	Formalization: medium Organizational complexity: medium Centralization: high	Formalization: low Organizational complexity: low Centralization: high
	Low uncertainty	High uncertainty

The question then is: how should the organization be structured in terms of its formalization, organizational complexity, and centralization? For this environmental situation, formalization should be high. There are few variables in the environment, all variables are known, and their values can be predicted with a high degree of certainty. In such a situation, it may be efficient to use a standard procedure (Burns et al., 1961; Duncan, 1979). High formalization indicates that the organization's procedures are usually written down and codified. There are standard ways of doing things, and these procedures are well known. For this situation, these procedural rules may or may not be complicated, and there can be many or few of them. Also, organizational complexity (that is, vertical and horizontal differentiation) should be medium. There is no need for an elaborate reporting and control hierarchy, or high vertical differentiation but there may be a requirement for work specialization and high horizontal differentiation (Burns et al., 1961). Horizontal specialization may be on the task level, but very likely horizontal differentiation on the department level will be medium to low (Lawrence et al., 1967). The organizational complexity recommendation will be modified by the size of the organization. Small organizations generally tend to have lower organizational complexity than large firms.

Centralization should be high for this situation. Top management has the capacity to gather, interpret, process information and make decisions, most of which are current operations decisions. There are few strategic concerns. However, the recommended high centralization can be questioned on two fronts. First, with high formalization, is high centralization too much for such a simple situation? Clearly, there are tradeoffs between formalization and centralization (Robbins, 1990, p. 112-113). So a balance must be achieved. Second, for a large organization, centralized decision making and control may be too demanding for top management, who simply do not have the time to do everything required. Rather, top management should develop more formalization and be more decentralized in its operations. Overall, this simple environmental situation calls for high formalization of procedures, and centralization also can be high if it does not create an overload on management or if the organization is not too large to be centralized. Similar considerations have to be made for all the propositions presented in this chapter.

We now consider an environmental variation in which the organizational complexity is high, and equivocality and uncertainty remain low. This is the upper left corner of Table 6.1. For Bon Goût this would be its situation if it imported from many suppliers and conditions remained the same as above.

The higher complexity decreases the recommended centralization. The low uncertainty and low equivocality still calls for high formalization. With a high environmental complexity, the environment has many important variables to consider; however, the variables are well known and their values are relatively fixed as equivocality is low and uncertainty is low. Here the argument for high formalization is very strong. It is efficient to develop standard procedures to take care of the many variables and adjust operating decisions as required. Bon Goût could develop a standard ordering and delivery system that would apply to all suppliers, or it could tailor this system to individual suppliers making sure they were handled correctly each time. Through analysis and experience, formalized procedures can be developed and codified. Procedures also can lead to well-conceived coordinated activities within the organization.

Here again, the organizational complexity should be medium; there is no need for an elaborate hierarchy or work specialization (Lawrence et al., 1967).

Centralization becomes problematic. There are arguments for low and high centralization. Unless the organization is small, the high complexity of the environment makes lower centralization the most efficient solution. A bounded rationality argument suggests that a top manager cannot monitor all the external factors. With a stable environment control of operations is maintained through high formalization. We suggest that the weight of the argument is to recommend medium centralization. Top management can be overwhelmed with making enumerable decisions for which formalized procedures are possible; this is particularly relevant for the large organization. The risk is that top management will become overloaded and the timeliness of actions may suffer (Duncan, 1979).

Returning to Table 6.1, in the lower right corner of part a, we change environmental uncertainty to high and complexity to low, and equivocality remains low. Uncertainty is higher, but complexity is lower. For Bon Goût this would be a situation where it would not have preordering from the stores and the amount it could sell was uncertain. Changing the environmental uncertainty from low to high, the formalization, organizational complexity, and centralization all change. Clearly, we suggest uncertainty has a dominant influence on the organization's structure. With a high uncertainty and low complexity and low equivocality, there are only a few important elements in the environment (low complexity), and they are well known (low equivocality), but their values are not. It is a relatively simple environment for which the numerical values are difficult to predict. So there is a need to react quickly to events, as they become known: generally, quick reactions are important.

Medium formalization suggests a relatively small number of written rules. The rules apply to procedures to deal with the uncertainty. In this case social formalization using professionals may be appropriate.

High organizational complexity provides the organization with a capacity to react on many levels. The situation is difficult but well known. The use of specialists is appropriate to deal with these diffi-culties. Since the number of variables in the environment is low, it may be possible to coordinate the activities via centralized decision making. However, with high uncertainty, a highly centralized organi-zation could suffer quickly from information overload, and there also could be a misfit if a social formalization is used. Therefore, a me-dium centralization would be a proper response to the environmental requirements.

We now turn to the upper right corner of Table 6.1. The environ-ment has high complexity, high uncertainty, and low equivocality. For Bon Goût this could mean no preordering and multiple suppli-ers.

Here the high uncertainty is a primary issue; coupled now with high environmental complexity, the arguments above are even stronger. High formalization and written rules are even more likely to get in the way of needed adjustments. Thus, formalization should be medium and related to procedures and policies. High environmental complexity and high uncertainty require that the organization should react and adjust quickly to the many factors in the environment, as their values become known. The force of the argument is increased, as the highly complex environment has even more elements that re-quire quick response. A highly centralized organization is even more likely to suffer from information overload, and low centralization is appropriate. A high degree of differentiation is required (Lawrence et al., 1967), which implies a high degree of organizational complexity.

In these four situations, we have kept equivocality low and varied environmental complexity and uncertainty. Now, we consider the lower half of Table 6.1, where the equivocality is high–that is, the environmental parameters are not well known and there is consider-able ambiguity.

If the environmental equivocality is high, the organization does not know which variables in the environment are or will be important for the organization. The particular set of variables may not be known, but if a given set is realized to be the important set, the number and the associated uncertainty may be high or low. If the equivocality is high and the uncertainty is low, one appropriate re-sponse by the organization could be scenario planning. If the uncer-tainty is high, scenario planning is difficult, and the organization has to be ready to adapt quickly. Looking back at the literature, one may argue that Lawrence and Lorsch (1967) did their study under low equivocality, while the Burns and Stalker (1961) definition of turbulence is close to the high equivocable situation. Therefore, their arguments about turbulence fall under the category of high equivocality. These observations lead us to four propositions for a high equivocality environment.

Referring to the lower right corner of Table 6.1, not much is known about the environment, except that its complexity is low–that is, there are likely to be only a few important parameters in the environment. This is much the situation for Bon Goût, if Bon Goût only had Samsonite as the single supplier. High formalization with an elaboration of rules is not likely to work well. High organizational complexity with a high horizontal differentiation and tasks specialization, a high vertical differentiation with a tall hierarchy, and slow response is not likely to respond appropriately to the environment, as it becomes known. Since the set of variables that will be important is not known, no rules can be developed. Thus, the formalization and organizational complexity should be low. However, actions must be taken and taken quickly. Since the environment has low complexity, a centralized head can deal with the information and react and adapt to the environment in a timely manner. In fact, this type of environment may develop into a hostile environment. In brief, it is a rather informal organization with a centralized head who reads the environment and directs the organization.

If the environmental complexity increases formalization remains low, organizational complexity should be low, but in contrast, centralization should be now low. With the complex environment, there is a lot of information to read, interpret, and translate into actions. Given that this should be done quickly, a centralized head is likely to become overloaded, resulting in costly delays. Consequently, a decentralized organization is most likely to be able to cope with the large amount of data. But if a centralized head can handle the information appropriately and quickly with appropriate decision-support systems, a centralized head is possible. However, given the equivocality of the environment, this decision-support system must be able to handle non-standard situations and not just resolve high uncertainty, which may be handled with large numerical data processing. Given today's information technology for equivocal situations, decentralization is more appropriate. Multiple suppliers that terminated their contract would put Bon Goût in this situation.

In this situation both centralization and formalization are low. This implies that no means for integration is recommended. An appropriate incentive system has to be in place to coordinate the various activities. This issue will be treated in more detail in the next section.

Now let us move to the low complexity and low uncertainty situation. Low uncertainty and low complexity make it likely that a central head can cope with the situation where the organization knows that only a few factors will affect the organization. However, it does not know which ones but only that each possible set of factors has a well-known value. This is a rare situation, but it is possible. Some public organizations may not know their future domain but know only that legislators in the future will decide the domain from a well-specified set of possibilities.

The argument is similar if the environmental complexity is high, but here a bounded rationality argument decreases the centralization from high to low for the higher complexity situation.

Our propositions in Table 6.1a are consistent with Lawrence and Lorsch (1967, pp. 96-97) who found that the required intensity of the integration for the three industries studied was almost the same. However, the type of integration was different. In the low environmental uncertainty case, the integrating devices were more related to the hierarchy of authority, and routine devices, while in the high uncertainty case, were less routine and less centralized. This is all consistent with Table 6.1.

Burns and Stalker's (1961) high-turbulence situation can be described in upper right corner of Table 6.1b. Their stable environment situation would be in cell of the right column of Table 6.1a. Our propositions are consistent with their arguments as well. Comparing the lower left corner of Table 6.1a the upper right corner of Table 6.1b, there is an increase in turbulence. Formalization, centralization, and complexity all move in the right direction.

In many of the empirical studies, environmental measures and organizational structures are lumped into a single measure. It is easily seen that putting high and low equivocality measures into the uncertainty measure could produce almost any result. For example, comparing the two cells in the upper row of Table 6.1a would result in a positive correlation between an aggregate measure of environmental uncertainty and organizational complexity. A negative correlation can be found by comparing the lower left corner of Table 6.1a with the upper right corner of Table 6.1b. Comparing the lower left corner of Table 6.1b with the lower right corner of Table 6.1a would not give a significant difference for formalization, while a negative correlation can be found comparing the upper left corner of Table 6.1a with the upper right corner of table 6.1.a The above observations may explain the non-conclusive results obtained by Pennings (1987) and Koberg and Ungson (1987), and also the internal misfit relations discussed by Miller (1992).

Tung (1979) aggregated her organizational measures into one. She mapped specialization, standardization, formalization, and participation on to mechanistic and organic structures. Despite the fact that standardization and formalization are not identical measures, they both relate to our measure of formalization. She therefore put extra weights on the formalization measure. Tung used environmental measures very similar to our three measures. Comparing her results (Tung, 1979, p. 689) with our proposition, we find that our results are not inconsistent with her results.

The general results in the literature have proposed that the relationships are of a monotone type (Schoonhoven, 1981), i.e., an increase in size will result in an increase in decentralization. This may not be the case and could be the reason for some of the conflicting results. The general results also show that environmental equivocality and uncertainty affect particularly formalization and organizational complexity, while environmental complexity affects centralization.

It is, however, generally agreed that if the environment reaches a level that threatens the life of the organization, then the organization should have a centralized organization regardless of the complexity of the environment (Mintzberg, 1979). An organic structure with low formalization and low organizational complexity may also be appropriate (Covin et al., 1989). Here, the environment is said to be extremely hostile and management must intervene directly and decisively in the details of the situation. Generally speaking, if the environmental hostility is extreme, then hostility overrules, maybe only temporarily, other factors of the appropriate organizational design. Hostility is then a discontinuity of the environment and suggests quick and dramatic organizational changes, if only for a short time. For Bon Goût, the Samsonite change created a high to extreme hostile environment and threatened its very existence–at least, its existence as Bon Goût had known. Management had to intervene directly in many decisions that it otherwise would not have been involved in. In the Wash & Go case described previously, members of the central management of Procter & Gamble came from the United States to deal with the Danish problem. When the problem was solved, regular organizational decision procedures were re-established. In the United States, Johnson and Johnson have experienced similar hostility when the over the counter drug Tylenol was tampered with. A few individuals died. J&J immediately withdrew its entire inventory and re-established the brand. Only top management could take such immediate and far-reaching decisions. As we shall argue later, the organization should establish a simple configuration, if only temporarily.

Extreme hostility calls for extreme management action, but the response is likely to be costly and inefficient. Top priority is focus on the threat and act on it. Many, perhaps thousands of other issues are not considered and many other issues are poorly dealt with. It is unlikely that Bon Goût dealt well with the normal business decisions during the crisis. We know that the Tylenol crisis cost millions of dollars. In order to deal effectively with the crisis, efficiency is not an issue. In hostile situations, we could also say the risk distribution has a high probability of large losses. March (1994) argues that individuals act as risk seekers, where in more normal circumstances the individual would be more risk averse.

For a highly competitive business environment, Richardson (1996) found that successful apparel firms decreased the level of vertical integration for a rapid response required by shortened product life cycles. This indicates that the organizational complexity should be lowered for a high environmental hostility.

Propositions on Environmental Effects on Formalization, Complexity and Centralization

If the environment has low equivocality, low complexity and low uncertainty, then the formalization should be high, organizational complexity should be medium, and centralization should be high.

If the environment has low equivocality, high complexity and low uncertainty, then formalization should be high, organizational complexity medium and centralization should be medium.

If the environment has low equivocality, low complexity and high uncertainty, then formalization should be medium, organizational complexity should be high, and centralization should be medium.

If the environment has low equivocality, high complexity and high uncertainty, then formalization should medium, organizational complexity high and centralization low.

If the environment has high equivocality, low complexity and high uncertainty, then formalization should be low, organizational complexity should be low and centralization should be high.

If the environment has high equivocality, high complexity and high uncertainty, then formalization should be low, organizational complexity should be low, and centralization should be low.

If the environment has high equivocality, low complexity and low uncertainty, then formalization should be medium, organizational complexity should be medium and centralization should be high.

If the environment has high equivocality, high complexity and low uncertainty, then formalization should be medium, organizational complexity should be medium and centralization should be low.

If the hostility is extreme, then formalization should be low, organizational complexity should be low, and centralization should be very high. If the hostility is high, then the centralization and organizational complexity should be lowered from its level determined by other factors.

Environmental Effects on Configuration, Coordination, Media Richness and Incentives

The propositions presented in the previous section were based on the organizational theory literature, a number of empirical results and argued from the information-requirement perspective. When environmental complexity, equivocality, and uncertainty increase, so does the requirement for information processing capacity. Each of the environmental dimensions requires different means to deal with the increased information processing. This was, to a certain extent, dealt with in the previous section; formalization, for example, is one means to increase information processing capacity in the case with low scores on all of environmental dimensions, as was also discussed in Chapter 2.

Here, we consider the effect of the environment on other structural elements: organizational configuration, coordination mechanisms, media richness and incentives. Organizational configuration is the fundamental organizational architecture - that is, functional, divisional, simple, matrix, and ad hoc, as described in Chapter 2. Coordination mechanisms include operating rules and procedures, integrating and liaison activities, group meetings, and planning. There are many choices, and the choice of a coordination mechanism depends on the particular environmental situation. Media richness indicates the form, amount, and kind of information. Daft (1992, p. 286) defines *information richness* as the information-carrying capacity of data. He orders media encounters from the richest medium to the least rich: (1) face-to-face, (2) telephone and other personal electronic media (3) letters, notes, and memos, and (4) bulletins, computer reports, and data reports. Face-to-face conversations go two ways and involve many clues that convey feelings, attitudes, and interpretations. In contrast, long computer printouts of numeric data lack interpretation. A richer medium can deal with more complex information. For media richness recommendations, we categorize richness as high, medium, or low and give a recommendation about the amount or quantity of information. Incentives can take on many variations, but two fundamental dimensions are: (1) individual or group and (2) procedure oriented or results based (that is, whether procedures were applied correctly or whether results turn out well). The challenge is to select an incentive consistent with organizational goals as well as with other organizational design parameters. Here we limit discussion to procedure- or results-oriented situations only. In Table 6.2, the environmental effects of equivocality, complexity, and uncertainty are summarized for organizational configuration, coordination, media richness and incentives. It has the same format as Table 6.1; low equivocality is shown in the upper half, and high equivocality is shown in the lower half. The arguments used to justify the propositions are based on information processing capacity. The conclusion on the configurations follows results by Galbraith (1973), Duncan (1979), and Mintzberg (1979). They also follow general recommendations provided in textbooks like Daft (1992) and Robbins (1990). The results derived on the coordination and control and liaison devices are consistent with the results in Mintzberg (1979). In particular, this treatment of the information processing requirement and the concept of media richness are developed from Daft and Lengel (1986). Some empirical support for the media-richness arguments can be found in Rice (1992). The results on incentives are derived from the discussion on planning in Mintzberg (1979), Burton and Obel (1988), Eisenhardt (1975) and Ouchi (1979). The design of incentives is also closely related to the design of control mechanisms.

We begin with the lower left corner of Table 6.2, part a. This is the simplest environment, with low equivocality, low complexity, and low uncertainty, which suggests a simple or functional configuration, direct supervision and planning for coordination, moderate media richness, small amount of information, and procedural incentives. This fits well with the conclusion from Table 6.1 with its high formalization and medium organizational complexity. Centralization is high, although it could be modified by size. If size is small, a simple configuration is appropriate. If size is large, a functional configuration is appropriate.

This relatively simple environment is well known and changes little: a simple or functional organization is appropriate. The information media need not be rich nor provide a large amount of information (Daft and Weick, 1984). In Table 6.2 two aspects of media richness are given: (1) degree of richness and (2) amount of information. Direct supervision with some planning is appropriate. Incentives can be procedure based and based on implementation of the rules of high formalization. It is appropriate to see that rules are followed and implemented.

This situation provides a baseline. All other environments are less simple and generally call for a more elaborate organization that can deal with the need for more information processing and complex decisions to resolve.

Moving to the upper left corner the environment now has greater complexity. We previously argued that centralization should decrease, moving the configuration toward a functional configuration. The functional configuration permits some task specialization to deal with the more complex environment. Similarly, an alternative argument could be made for a divisional product organization, if the environment partitions itself into a relatively independent number of products and the internal technology is similarly partitionable (Lawrence et al., 1967). Although arguments exist for alternatives, a functional organization is most likely to be able to deal with this environment in an efficient manner.

Table 6.2. Environmental Effects on Configuration, Coordination, Media Richness and Incentives

Environmental Complexity	a. Low Equivocality	
High	Configuration: functional, divisional Coordination: rules and procedures Media richness: low richness, moderate amount of information Incentives: procedural	Configuration: functional with liaison activities, divisional Coordination: integrators Media richness: medium richness, large amount of information Incentives: results
Low	Configuration: simple, functional Coordination: direct supervision, planning Media richness: medium richness, small amount of information Incentives: procedural	Configuration: simple, divisional Coordination: direct supervision, planning and forecasting Media richness: medium richness, moderate amount of information Incentives: results
	Low uncertainty	High uncertainty

Environmental Complexity	b. High Equivocality	
High	Configuration: matrix Coordination: planning, integrators, group meetings Media richness: high richness, moderate amount Incentives: results	Configuration: matrix, ad hoc Coordination: integrator, group meetings Media richness: high richness, large amount Incentives: results
Low	Configuration: simple, functional Coordination: direct supervision, planning. Media richness: high richness, small amount Incentives: results	Configuration: simple, ad hoc Coordination: direct supervision, group meetings Media richness: High richness, small amount Incentives: results
	Low uncertainty	High uncertainty

Due to increased complexity, a larger amount of information must be considered, although it need not be rich for this low-uncertainty and low-equivocality environment. The coordination and control can be based on rules and procedures. The best incentives remain procedure based as the formalization is high and further procedure based incentives fit well with a functional organization where results are difficult to evaluate equitably for the separate departments. Thus, the incentives should focus on performing activities well.

In Table 6.2, we now consider the lower right corner of part a. The uncertainty is now high with low equivocality and low environmental complexity. For this situation we have medium formalization and centralization with organizational complexity high.

This situation provides a basis for controversy and debate. There are a number of alternative arguments for different recommendations. The higher uncertainty definitely requires reading, understanding, and reacting to a larger amount of information. There are a number of alternative ways to accomplish the same end (Galbraith, 1974). Here, the environment has low equivocality and low complexity, but with high uncertainty it is difficult to predict. Thus, it becomes important to react to the environment, as it becomes known. A simple organization with an excellent information-support system can handle the uncertainty. However, if the size of organization is large, a divisional organization is more likely to be appropriate to deal with the internal complications of coordination and control due to its high organizational complexity. These two recommendations are very different.

With high uncertainty, a moderate amount of information will be required. But with low equivocality and low complexity, there is not likely to be a need for rich information. We know what we need to know. Coordination and control should be obtained via planning and forecasting with direct supervision.

Results are clearly important. With a simpler functional organization and low formalization, as argued in Table 6.1, there are fewer rules. Thus, a procedure-based incentive may not meet organizational goals. Incentives could be based on the results.

We now add high environmental complexity to this situation moving to upper right corner of Table 6.2a. It is more difficult to cope with this environment, and we shall see some changes.

The functional organization is not likely to work well without some mechanisms to tie it together quickly and deal with new information and situations, and liaison activities are mandatory. These mechanisms could range from rather informal information sharing to more permanent project, product, or customer managers. In any event, a purely functional organization is likely to suffer from information overload with decisions not being made in a timely fashion. The low degree of centralization and medium formalization require a more elaborate coordination mechanism.

A divisional organization could be appropriate here, but two additional conditions must be met. First, environmental complexity must be mostly in terms of a large number of parameters, but they must be partitionable along the lines of products or customers. If the environment is highly interdependent, a divisional separation is likely to lend conflicts among the divisions over customers, suppliers, and so on. Second, the technology must be partitionable along divisional lines, so that each division can manage relatively independent internal operations. Without these additional conditions, the divisional configuration will lead to numerous conflicting situations. The functional organization with liaison activities requires a high level of managerial skill to cope with the environment, and the required coordination will require integrators.

Media richness is moderate with a large amount of information required to cope with environmental complexity and uncertainty. Incentives remain results based. Procedures are likely to be changing to meet the need for good results. Therefore, the focus should be on the results. To obtain the desired results incentives cannot be based on the procedures.

We now turn to part b of Table 6.2, where equivocality is high. We begin with high equivocality, low complexity, and high uncertainty, which demand a high level of information processing – the lower right corner. Not much is known about the environment, except that only a few parameters are important: that is, the environmental complexity is low. However, much is unknown; what are these important parameters and their values? Thus, there is a need to read and interpret for this environment and to take action in a timely manner. Formalization and organizational complexity should be low with a high degree of centralization.

A simple organization with a focus on the environment is appropriate. The supporting information system must be broad-brush in its scope but also capable of focusing in on important issues. Top management must then determine what to do and give directions. This is consistent with a highly centralized organization, as argued in Table 6.1. Despite the fact it is centralized, many people should be involved in interpreting information. Group meetings and direct supervision are a means for coordination and control.

An alternative organizational configuration would be an ad hoc organization in which everyone is responsible and responsive to environment changes. Of course, a more decentralized organization must be consistent with an ad hoc configuration. This would be the case if other contingency factors drive the organization toward a more decentralized organization than that prescribed in Table 6.1.

Media richness must be high; information will come from many sources and in many formats. However, there will not be a lot of it as the environment is not complex. Generally, high equivocality requires high media richness (Daft et al., 1986). Kurke and Aldrich (1983) found that managers prefer verbal media over written contact when the environment is dynamic.

The incentives should be results based. Formalization is low, and there are few rules and procedures. The goal is to succeed regardless of the uncertainty and equivocality, and there must be considerable improvisation. Incentives on results will encourage individuals to take appropriate actions.

Now let us consider the most complicated situation. Environmental complexity is now high, as is equivocality and uncertainty in the upper right corner of part b of Table 6.2. This is the least understood of all environments.

High environmental complexity generates additional requirements for information processing. A simple configuration of organization is not likely to be able to cope. A matrix organization or ad hoc organization is more likely to be successful. A matrix organization is recommended with caution, because it is costly and difficult to manage. However, simpler configurations such as a functional or divisional configuration are too inflexible for high equivocality, uncertainty, and complexity. Formalization, organizational complexity, and centralization are all recommended to be low. This leads toward the ad hoc configuration. However, if organizational complexity has to increase, then the matrix is the next choice.

Coordination should be based on integrators and group meetings. Media richness should be high, with a large amount of information.

This is the highest information-processing requirement of all the situations. Incentives must be results based.

We now consider modifying the environment for low uncertainty – the left upper corner. This situation is very similar to the previous situation except that the low uncertainty decreases the amount of information that must be processed. The most significant change is that since uncertainty is low, it is possible to do scenario planning, and some formalization related to process will be possible. Integrators and group meetings are still needed to coordinate the various activities. An ad hoc configuration is too costly with the decrease in environmental uncertainty.

Finally, we turn to lower left corner where equivocality remains high and complexity and uncertainty are low. This may not be an unlikely situation, particularly for non-public organizations, as discussed in previous sections. There are only a few environmental parameters, but we do not know what they are. For example, a hot dog vendor may not know whether it is the weather or the taste of the hot dog that generates low sales.

In comparison with upper right corner, reduction of complexity increases centralization and makes a simple structure possible. For large organizations a functional configuration may be appropriate. It also decreases the amount of information even further. Direct supervision and some planning will be appropriate. Media richness must be high to cope with the equivocality, but a small amount of information is sufficient. Finally, incentives must be results based.

Additionally, extra rules are added relating to the condition of environmental hostility. An extremely hostile environment calls for a simple configuration. Decision-making is centralized and the normal structure is suspended. Here we have a crisis which threatens the very existence of the organization. These threats emerge from either natural or competitive actions. The information is focused and not necessarily large. For Bon Gout, Samsonite changed its policy. For Procter and Gamble, Wash and Go was said to cause one's hair to fall out. For Johnson and Johnson, Tylenol had been laced with poison. The implications are huge, but the base information is very focused. Decision-making is a matter of judgment, not more information. Of course, there are millions of minor issues, but they are indeed minor and can be ignored for the large decision. These crisis situations require fast decisions, judgment, but not large amounts of information.

Propositions on Environmental Effects on Organizational Configuration, Coordination, Media Richness and Incentives

If the environment has low equivocality, low complexity, and low uncertainty, then the organizational configuration should be simple or functional, media richness should be medium with a small amount of information, coordination and control should be direct supervision and planning, and incentives should be procedural based.

If the environment has low equivocality, high complexity, and low uncertainty, then the organizational configuration should be functional or divisional; media richness should be low with a moderate amount of information, coordination and control should be rules and procedures, and incentives should be procedural based.

If the environment has low equivocality, low complexity, and high uncertainty, then the organizational configuration should be simple or divisional, media richness should be medium with a medium amount of information, coordination and control should be based on direct supervision planning and forecasting, and incentives should be results based.

If the environment has low equivocality, high complexity, and high uncertainty, then the organizational configuration should be functional with liaison activities or divisional, media richness should be moderate with a large amount of information, coordination and control should be integrators, and incentives should be results based.

If the environment has high equivocality, low complexity, and high uncertainty, then the organizational configuration should be simple or ad hoc, media richness should be high with a small amount of information, coordination and control should be via direct supervision and group meetings, and incentives should be results based.

If the environment has high equivocality, high complexity, and high uncertainty, then the organizational configuration should be matrix or ad hoc, media richness should be high with a large amount of information, coordination and control should be via integrators and group meetings, and incentives should be results based.

If the environment has high equivocality, low complexity, and low uncertainty, then the organizational configuration should be simple or functional, media richness should be high with a small amount of information, coordination and control should be via direct supervision and integrators, and incentives should be results based.

If the environment has high equivocality, high complexity, and low uncertainty, then the organizational configuration should matrix or ad hoc, media richness should be high with a moderate amount of information, coordination and control should be via planning, integrators, and group meetings, and incentives should be results based.

If the hostility is extreme, then organizational configuration should be simple. If the hostility is high, the organizational configuration could be simple, but other factors should also be considered.

OPERATIONALIZATIONS OF THE ENVIRON-MENTAL MEASURES

We have proposed a four-dimensional measure of the environment. Even though such a measure is more detailed than the ones used in many studies, they are still rather aggregate. Two issues are dealt with in this section. The first one is whether our measures are objective measures, and the second issue is what the environment really is.

The most objective dimension of the four is complexity. It is relatively easy to enumerate the important factors in the environment that affect the organization. Disagreements may arise over the type of effect and the intensity of the effect, but usually a rough count can be agreed on. The situation is quite different with respect to the other three dimensions (Downey et al., 1975; Tung, 1979). Both a high degree of uncertainty and a high degree of equivocality may be attributed to lack of information or the randomness of some events. If the uncertainty can be attributed to lack of information, the degree of uncertainty and equivocality depends on the precise information the organization has. One organization may consider a situation very uncertain and equivocal, while another organization in exactly the same environment considers its environmental uncertainty and equivocality very low. Hostility is both an issue of perception and the reality of whether someone or something is out to get you. The environment is enacted (Weick, 1969). In the telecommunication industry, one company may be on top of the newest technology developments, have good contact with the political scene, and based on this, have a very good notion about what the future brings. Another company may not have the capacity to do the same and may find the future uncertain and difficult to understand and predict. A number of means to reduce the uncertainty and equivocality are discussed in the next section.

The environment consists of many different parts. The industry, including its size and competitions, is a major part of the organization's environment. Actual customers are also included in the environment. Suppliers of resources for the organization are important as well. This includes suppliers of raw materials, services, and the labor market. Financial resources have to be taken into consideration including the stock market, banks, and private investors (see the earlier capsule on the Five Forces).

The evolution of the techniques of production is a part of an organization's environment. The political and general economical factors are a part too, including regulation, taxes, services, and the political system in general. Additionally, inflation rates, exchange rates, and other important economic factors have to be considered.

Finally, more general sociological factors should be accounted for. Values, beliefs, education, religion, work ethic, and special current trends such as environmental movements and women's rights are important.

Briefly defined, the environment is everything that is outside the organization. This relates to the unit of analysis, as was discussed in Chapters 1 and 2 and a later capsule on new forms and strategy in Chapter 8. Some of the factors will be more important than others and more or less uncertain for a particular organization.

Hypercompetition and Flexibility

Hypercompetition, or Schumpeterian competition, goes beyond the normal static competition to include a number of elements: dynamics, rapid change and adaptation, short product life cycles, discontinuities, creative destruction and general Darwinian behavior. D'Aveni (1994, p. 154) describes hypercompetition as "an environment characterized by intense and rapid competitive moves," and the "behavior is the process of continuously generating new competitive advantages and destroying, obsoleting, or neutralizing the opponent's competitive advantage, thereby creating disequilibrium..." High technology, telecommunications, software, aircraft, health-care and biotechnology are a few of the hypercompetitive industries. Hyper-competition is widespread and is not limited to a select number of indus-tries. Unlike extreme hostility, hypercompetition does not end and can be thought of as a continuing situation–a continuing state of disequilibrium, perhaps a new type of equilibrium.

What is the appropriate organizational response to hypercompetition? D'Aveni (1994, p. 172-176) offers a new 7S's framework: superior stake-holder satisfaction, strategic soothsaying, speed, surprise, shifting the rules, signaling, and simultaneous and sequential strategic thrusts. Later (p.196), he recasts speed and surprise capability for disruption. The capa-bility for disruption is most closely related to the organizational design; although speed and surprise are organizational characteristics of potential behavior than organizational design itself. That is, for a hypercompetitive environment, we want an organization, which acts with speed and surprise to create disruptions and temporary advantages for the organization. The organizational design rules to realize speed and surprise are largely yet to be developed; however, D'Aveni offers a number of rich and detailed exam-ples including examples from the film industry.

Volberda (1996) argues that a flexible organizational form is required for a hypercompetitive environment–a contingency rule that states, "if the environment is hypercompetitive, then the organizational form should be flexible." A flexible organization involves a paradox. "Flexibility is the degree to which an organization has a variety of managerial capabilities and the speed at which they can be activated, to increase the control capacity of management and improve the controllability of the organization" (p. 361). Ashby's (1956) law of requisite variety is fundamental; the variety of the potential actions by the organization must be as large as the variety of the disturbances in the environment. Volberda then offers three propositions for the rigid form, the planned form and the flexible form–the latter, Propo-sition 3 (p. 366-367) being of most interest here:

In a fundamentally unpredictable environment, which may also be dynamic and complex (hypercompetitive), the optimal form employs a broad flexibility mix dominated by structural and strategic flexibility and has a nonroutine technology, an organic structure, and an innovative cul-ture. The intelligence-gathering and information-processing aspects of metaflexibility are directed toward enhancing the receptiveness to new en-vironments.

He argues that this form is a good balance between change and the need for preservation. In brief, a hypercompetitive environment requires a flexible organizational form.

Social network forms and cluster forms are examples of flexible forms. The social network form works well for the biotech hypercompetition. It is characterized by redeployable technology and organic structure, but with strong profcssional ties between actors. The cluster form gives strategic range of action, but with structural control and preservation.

Hypercompetition and flexible form call for a new vocabulary for the change and its rapidity that we are experiencing for the new millennium. It is interesting to look for similarities of concepts and juxtapose definitions and design rules.

In our terms a hypercompetitive environment could be categorized as: high equivocality, high uncertainty, high environmental complexity and high competitiveness. The technology is nonroutine. For this environment and technology, we would recommend: low formalization, low organizational complexity and low centralization, matrix or ad hoc form, coordination by integrators and meetings, high media richness and a large amount of environmental information and results-based incentives.

We suggest that these recommendations are not unlike the Volberda recommendations: structural and strategic flexibility, organic structure, innovative culture, intelligence-gathering and information-processing aspects of metaflexibility.

Hypercompetition and nonroutine technology are strategic fit statements in our framework and it is a very reasonable requirement.

Hypercompetition and flexible organization are new words and now a part of managerial vocabulary. They embed the essence of time: change is more rapid and the environment is less benign. Yet, it remains a research question whether hypercompetition is a new strategy paradigm and flexibility a new organizational form, or whether these are particular environments and organizational requirements, which have been incorporated, albeit as end points in existing theory. It may be some time for this research agenda to develop and for answcrs to cmerge.

NATIONAL CULTURE

The discussion of the effect of the environment on the organizational structure has so far in this chapter been "culture-free". Based on the work on (Hofstede, 2001) there has been substantial discussion of the effect of the national culture on the way people behave in organizations (Søndergaard, 1994). The society influences the members of an organizations acceptance of structural dimensions. Thus, there has to be a fit between the national culture and the organization (Baligh, 1994; Lachman et al., 1994). Sharda and Miller (2001) specifically asked the question: "Does 'national' culture influence organizational structure?" Comparing Jordan, Iran, and USA, they find that the national culture does not affect the dimensions describing the organization structure, but does affect the magnitude. Thus, it seems appropriate to incorporate the national culture as a part of the description of the environment in which the organization operates.

The work on national culture by Geert Hofstede has been very influential (Hofstede, 2001; Søndergaard, 1994), Hofstede (2001) developed four dimensions of the national culture:

- Individualism – collectivism
- Power distance
- Uncertainty avoidance[2]
- Masculinity – femininity

Individualism measures the ties between individuals; loose ties suggest a high degree of individualism, while a high degree of collectivism indicates that individuals are strongly integrated from birth into cohesive in-groups which gives protection in exchange for unquestioning loyalty. Power distance is the extent to which the less powerful members of an organization within a country expect or accept that power is distributed unequally. Low power distance refers to an expectancy of equality; high power distance refers to an acceptance of inequality. If the members of a culture feel threatened by uncertainty or unknown futures and express this as a need for predictability then the uncertainty avoidance of the culture is high; individuals utilize short term feedback and avoid longer term forecasting. Low uncertainty avoidance is an acceptance of unknown futures. The last dimension is Masculinity - Femininity, which is also known under the name of Achievement vs. Relationship Orientation. It mirrors the extent to which a society is dominated by so-called masculine values, such as assertiveness, achievement, money and luxury, performance and success on the job. On the other hand, feminine aspects are quality of life, well-going relationships, both in private and business life, and caring for others.

Lachman et al (1994) developed a congruence model of culture and organization based on a hypotheses generating scheme. Their findings are shown in Table 6.3.

---

[2] Here, uncertainty is an element of national culture and an aggregation of individuals' preferences concerning uncertainty avoidance. In Chapter 3, uncertainty avoidance is an element of an individual's leadership style.

Table 6.3. The Effect of Culture on Organizational Choices

Cultural Values	Structure	Processes

Power
High/low power
	Hierarchy:
	  differentiation
	  high/low
	Centralization:
	  high/low
		Decision-making:
		  participative/non participative
		Communication:
		  vertical/horizontal
		Control:
		  tight/loose
		Coordination:
		  vertical/horizontal

Social Orientation
Individualistic/
Collectivistic
  orientation
	Horizontal differentiation:
	  specialization
	  high/low
	Rewards:
	  differential'
	  high/low
		Rewards and incentives:
		  individual/group emphasis
		Communication:
		  specific/diffuse
		Decision-making:
		  contentious/consensus

Work Orientation
Work/nonwork
  centrality
	Span of control:
	  Wide/short
		Rewards and incentive:
		  intrinsic/extrinsic

Uncertainty
High/low
  avoidance
	Formalization:
	  High/low
	Centralization:
	  high/low
		Locus of decisions:
		  hierarchical/diffuse

Table 6.3 shows a number of propositions on the relationship between culture values and organization structure. The work/non work centrality variable corresponds closely to Hofstede's masculinity/femininity dimension. The table thus maps the four culture dimensions by Hofstede onto the dimensions of organizational structure. If the power distance is high then both complexity and centralization should be high. Similarly, if the power distance is low, Lachman et al (1994) suggest that both complexity and centralization should be low. If the individualism is high then specialization should be high and incentives should be individual based. If the culture has a collectivistic basis, complexity should be low and rewards group based. Formalization and centralization follow the degree of uncertainty avoidance.

Krokosz-Krynke (1998) argues that if the power distance is high then formalization and centralization can be high as the members of the organization can accept a centralization of power. This is in partial agreement with Lachman et al (1994) above. Krokosz-Krynke also argues that if individualism is high specialization is the preferred mode of operation and thus then horizontal differentiation should be high, while if individualism is low then horizontal differentiation should be low. If individualism is high then the focus is on the individual decision making and thus decentralization should be high. This is in agreement with Lachman et al (1994).

Hofstede (2001, p. 375) argues that it is predominantly uncertainty avoidance and power distance that have an effect on structure. Power distance and uncertainty avoidance together affect the configurations that are acceptable (Hofstede, 2001). If uncertainty avoidance is weak and power distance is low then configuration should be flexible like an ad hoc or matrix. If uncertainty avoidance is low and power distance is high then the flexibility can be centred on the top management like the simple configuration. If uncertainty avoidance is strong then configuration should be a bureaucracy – a professional bureaucracy if the power distance is low while machine bureaucracy if the power distance is high. If uncertainty avoidance is medium and power distance is medium then configuration should be divisional. A high degree of masculinity is related to achievements and thus the incentive system should be based on results while if the culture has a high degree of femininity the incentives should be related to relations and procedures. The above conclusions can be summarized in the following propositions.

Propositions on National Culture and Structure

If power distance is high, then formalization should be high.

If power distance is low, then formalization should be low.

If power distance is high, then centralization should be high.

If power distance is low, then centralization should be low.

If power distance is high, then complexity should be high.

If power distance is low, then complexity should be low.

If uncertainty avoidance is strong, then formalization should be high.

If uncertainty avoidance is weak, then formalization should be low.

If uncertainty avoidance is strong then centralization should be high

If uncertainty avoidance is weak, then centralization should be low.

If uncertainty avoidance is weak and power distance is low, then configuration should be ad hoc or matrix.

If uncertainty avoidance is weak and power distance is high, then configuration should be simple.

If uncertainty avoidance is strong, then configuration should be a bureaucracy.

If uncertainty avoidance is medium and power distance is medium, then configuration should be divisional.

If the culture is masculine, then the incentives should be based on results.

If the culture is feminine, then the incentives should be based on procedure.

If individualism is high, then incentives should be based on individual rewards.

If individualism is low, then incentives should be based on group rewards.

If individualism is high, then horizontal differentiation should be high.

If individualism is low, then horizontal differentiation should be low.

MANAGING THE ENVIRONMENT

When Bon Goût learned that Samsonite had canceled its contract it took action. First, it managed to obtain an appropriate inventory. Second, it sued Samsonite and started to react to the fierce competition from Samsonite–in the stores. This shows that actions can be taken to change the value of all four measurements of the environment.

We argue that there has to be a fit between environment and organizational structure. The fit can be obtained either by adjusting the structure or by changing the environment. If high uncertainty and equivocality come from lack of information, actions that provide better information will reduce the scores. Market analysis, participation in technical conferences, and so on are all means to increase knowledge about the environment. The organization may even change its domain, or it may have to. For Bon Goût there may be no other choice than to find something else to do or die.

Buffering may reduce the effect of uncertainties. Bon Goût managed to obtain an inventory. That is a buffering strategy. If the demand for the output is very irregular, the organization may have to even out the demand. Price differentiation and rationing are strategies that help in that direction. Advertising is perhaps the best-known way to try to influence the environment. Vertical integration and lobbying are further examples, but many more exist.

Some strategies may be available to some organizations and some may not. It may depend on size (large organizations can do more advertising than small organizations), culture (in some cultures bribing is customary, but in other settings it is not), or legal setting (public organizations are usually restricted in such actions).

For the organization it is important to analyze the source of equivocality, uncertainty, environmental complexity, and hostility. Table 6.4 shows a chart to begin. For each of the thirty-two entries in Table 6.4, equivocality, uncertainty, complexity, hostility are to be assessed. A complete analysis then will provide a complete picture on the four dimensions. Interdependence between the various sectors has to be taken into account. For example, decisions with respect to the tax laws may affect economic factors like the inflation.

Table 6.4. Environmental Sectors and Dimensions

Sector	Equivocality	Uncertainty	Complexity	Hostility
Market and industry				
Raw material				
Human resources				
Financial resources				
Technology				
Economic system				
Political system				
Social system				

The organization then has to establish a fit with the environment. The adaptation of the organization to the environment was the subject of previous sections. Fit also can be obtained by changing the environment. However, the manipulation depends on which entry in Table 6.4 can be changed and how it depends on other sectors. The market and industry can be changed by a domain change, vertical integration, advertising, mergers, market analysis, and so on. Contracting and buffering strategies are useful with respect to suppliers. Negotiations with the union, educational activities and recruitment are issues in the human resources sector. Decisions about leverage, choice of bank, and timing of issuing new stocks are choices that affect the equivocality, uncertainty, and complexity of the environment. Investment in research and development can reduce equivocality and uncertainty in the technology sector.

Not all factors can be manipulated easily; general economic factors cannot. Organizations have to respond properly to inflation rates and unemployment rates without having many possibilities to affect these factors. Economic factors may be affected through *lobbying*, the general term for activities that seek to influence political decisions.

Organizations are part of a general social system. It is interesting to contrast the evolution of organizational designs with the evolution of society. The more democratic, less centralized organizations reflect similar trends in our societies. For international organizations, social and cultural factors are particularly important. One organizational design may work in one country, but not in another; national culture has an influence. The global organization has a choice of which activities to do in which country.

The organization has its choices. It can adapt the organization to its environment, or it can try to manipulate the environment to fit the organization.

SUMMARY

Unequivocal, simple, certain environments call for relatively uncomplicated structures with an emphasis on formalization, rules, procedures, specialization, and rewards for doing a job well. As the environment has increased variety, then the organization's structure must have increased variety. The fundamental concern is to be able to process the requisite information–that is, to read, interpret, and take appropriate actions in this more unknown situation. Ashby's law of requisite variety posits that variety in the organization must exceed variety in the environment for viability (Ashby, 1956). In design terms, if the environment has high variety, then the organization should have high variety to cope. The elaboration of the environment into three characteristic descriptions and the organizational structure into seven elements is a way to give precision to the general proposition and also incorporate what we know from the literature and experiences.

In this chapter, we have related the environment of an organization to the structure of the organization. The environment was described in terms of equivocality, complexity, uncertainty, and hostility. The organizational structure was considered in seven ways: formalization, complexity, centralization, organizational configuration, media richness, coordination and control, and incentives. The structure must match the environment, but the structure itself *must* be internally consistent. For example, consider the situation with low equivocality, low uncertainty, and low complexity. From Tables 6.1 and 6.2 it follows that the organizational structure that fits this environment is a structure with high formalization, medium organizational complexity, and a high centralization. The configuration should be simple or functional with some planning. Coordination and control are based on direct supervision. Richness of the media is medium with a small amount of information. Incentives should be based on procedure because the centralized decision-making has all the responsibility about results. All the dimensions have to fit as they do.

The other extreme situation with high equivocality, high uncertainty, and high complexity is best met with low formalization, low organizational complexity, and low centralization. The configuration should be matrix or ad hoc. Coordination should use integrators and group meetings. Media richness should be high and have the capacity to process large amounts of data. Since procedures cannot be established, incentives should be based on results.

The propositions are normative, by which we mean that they are our best recommendation to managers for practice. The support for the proposition is a synthesis: the literature in organizational theory is rich and empirically abundant, our validation of OrgCon (knowledge base) led us to re-examine what we know, ask new questions, and propose tentative answers, and finally, the propositions must be intuitively appealing and arguable to the practicing manager and student.

Chapter 7

Technology

INTRODUCTION

Med Electronic, Inc. is a medium-size company that specializes in electronic apparatus that is used in the treatment of pain and in other electronic devices used by hospitals. The machines have been custom made to the particular needs of the user department or physician. Some basic components are used in all its devices, but no two machines are similar. This has caused problems when machines come in for repair because documentation and specifications for a particular machine may be difficult to find. Med Electronics' performance has been stable for some time, and the owners have been pleased with its performance. The employees are either engineers or highly trained technicians. Med Electronic has had a stable share of the world market for its particular products.

Hospitals everywhere now face difficult economic times. Governmental and private hospitals as well as health insurance companies are trying to control costs. It is more difficult to maintain their usual revenues. On the other hand, consumers spend more money on alternative health care. There are increasing sales in do-it-yourself medical instruments, like blood pressure measurers and pregnancy tests.

The planning committee of Med Electronic has proposed that the company enter the consumer market. One pain-treatment machine that has been produced in a number of different versions is considered to have significant potential in the world market. It could be produced in high quantities in a standard model, and the price could be relatively low. The main problem would be to maintain high quality. The management thinks favorably of this proposal, but has some concerns that the internal structure of the corporation will not fit the new situation.

In the past, there had been no rules. The engineers were both designers and salespersons. They dealt directly with hospitals and physicians. To maintain quality, the production process has to be disciplined. Documentation has to be produced and maintained. Entering the consumer market demands organizational changes. A separation of sales and production has to be considered. Since the price is low, margins are low as well, which requires cost control and efficient operations. Procedures should become more formalized, and activities better coordinated. Design and documentation have to be fixed. The technology of the company would change and could cause a change in the organizational structure.

Technology is the information, equipment, techniques, and processes required to transform inputs into outputs. This very general definition is widely accepted. The skill level and capacity of employees help define the techniques and processes and must be compatible with the equipment and physical plant. Technology as a contingency for the organization is the topic of this chapter. We will also specifically look at the issue of the effect of information technology which is often confused with the technology concept within organization theory.

In this chapter, we review the literature on the technology imperative, i.e., the organization's technology affects the best choice of the organizational design. Information processing provides the framework for developing propositions on the technology and design. We visit again the SAS and examine the effect of its technology on the organizational design.

The Technology of SAS

When SAS was established in 1951, the managers were engineers, and in the following years engineers still occupied a majority of the company's leading positions. Managers were recruited internally because they knew the company from the inside and were in a good position for making decisions. But new ideas were scarce.

From its beginning until the 1970s, SAS's goal had been to maximize the number of passenger kilometers, which meant that the main focus was on the transaction of moving a person from one place to another. Service, as we know it today, was not a major concern, and therefore the passengers' wishes were only secondary. But market demands changed. The passengers wanted more direct departures, and it appeared that passengers were not as concerned about plane types as employees were.

Airplane purchases had been based on the belief that the market would be growing throughout the 1980s. But due to the oil crises in the 1970s, market growth vanished, and overcapacity was an increasingly noticeable phenomenon.

As a result of market overcapacity, utilization of capacity decreased. With Carlzon's appointment in 1982, something had to be done. Carlzon wanted to make SAS more market oriented, which meant that the production apparatus should be adapted to customers and not the other way around. SAS had to use smaller planes on routes to increase the utilization of capacity.

Carlzon's new strategy of making SAS the businessperson's best alternative (Chapter 8) resulted in the introduction of Euroclass–a special service for the business market segment. Euroclass customers had separate check-in counters, separate lounges, and hotels on the ground. In the air, they had roomier seating with more space between the seats in the front of the cabin.

Carlzon's philosophy was to give customers individualized care (SAS, 2002) and problems were to be solved as they arose (SAS, 2002). Therefore, personnel had to be able to make a quick decision to satisfy the customers. A new control system helped the front-line personnel with various solutions to various problems. The control system could be developed because service consisted of certain fixed components, and consequently, customers' problems were mostly well defined within fairly narrow categories. Within the framework of this control system, each employee was responsible for finding a solution that would satisfy the customer in the given situation (SAS, 2002). The service SAS provided was no longer just transportation but also a good experience for passengers. From a technology point of view the major change was the change from a production orientation to a service orientation. The product was now a satisfied customer.

Overall, the technology at SAS is routine; there are few exceptions, and the operating problems are well defined, if not easy to analyze. There are few surprises from the technology. The technology is non-divisible; the flight schedules themselves require integrated operations across continents with narrow time frames. And of course, SAS is a service organization, utilizing skilled personnel and high-technology equipment–both airplanes and information systems. Airlines have very sophisticated information systems.

SAS has a non-divisible technology within its airline company. Its attempt to create a geographical or a customer divisional configuration within the airline company has been difficult. The interdivisional externalities are real and difficult to manage.

As mentioned earlier SAS has from 2001 been threaten by low price airlines like Ryan Air and Virgin. These companies were able to sell air service at a very low price. The basic idea was automation in reservation and ticketing using the internet as much as possible. Further, they try to cut service that did not seem important, such as serving meals on short haul flights. SAS started fighting back with a new service and ticket concept for its European routes, so the customer could buy the flexibility and service needed. In Scandinavia, SAS would only have one class.

From a technology point of view, the EuroBonus system was not well integrated for customer service. Until November 2001 it was not possible to use EuroBonus points to get upgrades to business class at the point of check in, unless one had received a special voucher. Because only few passengers used the voucher system, it had the effect that many EuroBonus gold- and silver-card holders got free upgrades. From 2002 passengers in principle can only be upgraded if they are using EuroBonus points. The check-in counter can now access the passengers' EuroBonus accounts. The EuroBonus card can also be used as an electronic ticket minimizing cost and providing more flexible service.

Further, like other airlines SAS is pushing passengers to use the internet as much as possible.

To compcte head on with the low price airlines, in December 2002 SAS established its own low price airline – Snowflake. It was introduced as non stop, no nonsense, low fare airline (www.flysnowflake.com)

SAS, despite its customer-based strategy, has a highly routine technology using advanced IT systems; in fact, it has utilized its routine technology to make it customer based to meet customer needs but within well-defined parameters. SAS has routine technology and is customer oriented.

1. What is the effect of the service oriented, non-divisible, and routine technology on the choice of an organization structure for SAS?
2. Will the use of advanced IT systems make the effect stronger or weaker?

LITERATURE REVIEW ON TECHNOLOGY

The idea that the technology of the organization affects its structure was a discovery of Joan Woodward (1965). Her study was the first to link technology and structure. In her study of the industry in the south of England, she categorized technology into three types:

- Unit: Custom made and the work is nonroutine (e.g., a craft - shop or a job shop).
- Mass: Large batch or mass produced, usually standard products (e.g., an assembly line).
- Process: Highly controlled, standardized and continuous processing (e.g., a refinery or a brewery).

The scale reflects increasing technological complexity. Woodward linked these technologies with structure. The administrative component, or non-direct worker component, increased with increased complexity.

From an information processing point of view, the categories may be ordered: mass, unit, and process. Mass production may require the lowest information processing capacity, if proper formalization and coordination mechanisms are used. Less formalization can be used in unit production, and more complex information processing is needed to obtain coordination. If the organization is small, coordination can be obtained by centralization. Professionalization also may be used. Process production with its automation requires complex information processing. The amount of information may, however, be lower than for unit production for a given size of the organization. Robbins (1990, p. 180) summarizes Woodward's findings (see Table 7.1).

Table 7.1. Summary of Woodward's Findings on the Relationship between Technological Complexity and Structure

	Low	Technology	High
Structural Characteristic	Unit Production	Mass Production	Process Production
Number of vertical levels	3	4	6
Supervisor's span of control	24	48	14
Manager per total employee ratio	1:23	1:16	1:8
Proportion of skilled workers	HIGH	LOW	HIGH
Overall complexity	LOW	HIGH	LOW
Formalization LOW	HIGH	LOW	
Centralization LOW	HIGH	LOW	

Briefly, as the technological complexity increases,

- Vertical differentiation increases.
- Span of control is curvilinear, with mass production having the widest span of control.
- Ratio of manager per total employees increases.
- Proportion of skilled workers is curvilinear, with mass production requiring the least skill.

- Organizational complexity is curvilinear, with mass production being the highest.
- Formalization is curvilinear, with mass production being the highest.
- Centralization is also curvilinear, with mass being the highest.

Complementary to information processing arguments, an alternative explanation begins with the nature of the work. Unit production is more craft oriented; individual workers are not closely linked and can work relatively independently. Hence, with close supervision, formalization and centralization are not needed. Mass production is driven by the regularity and interdependency of the worker and the machine or assembly line. The machine itself drives the work and provides coordination of the interdependency. Workers must conform to formalized work rules with high centralization; but a supervisor can supervise a large number of workers due to the formalization and standardization.

Process production is yet a different issue; refineries and breweries run themselves. The task is to keep them running smoothly and efficiently. Control and maintenance of the process are major issues. These require high skill, but there is a good deal of discretion about the exact timing of the work. Thus, the complexity, formalization, and centralization can be low, but many managers are required in order to ensure the reliability of the complex technology. Briefly, mass production is clockwork for workers driven by the machine; unit and process production permit greater discretion for workers to accomplish tasks at their own pace. Woodward's conclusions have been criticized for not taking size into account. Usually, mass production organization will be larger than unit and process organization.

From an information-processing complexity point of view, one may argue that the order is mass, unit, and process. Based on the amount of information, it may be ordered unit, mass, and process. The number of vertical levels may be related to the managers per total employee ratio. However, span of control and skill level of employees are linked to complexity of information. The more complex the information, the higher the demand for information processing capacity, and thus, the structure has a higher number of vertical levels. This is a bounded rationality argument. The relationship, however, is contingent on the skill level of employees in two ways. If employees have a high skill level, they can process complex information. If the manager has a high skill level, he or she also can process complex information. Therefore a high skill level will, ceteris paribus, decrease the vertical differentiation. Modern information technology can decrease the vertical information system and the number of vertical levels. Computer information systems, e-mail, and so on increase information capacity and eliminate the need for a large middle management. That is, modern information systems can process information previously handled by middle management. This is one explanation for the flattening occurring in many organizations (See capsule on delayering in Chapter 2). Within this view, Woodward's results and the implications for organizational design make perfect sense.

Perrow (1967, p. 195) viewed technology differently: "By technology is meant the actions that an individual performs upon an object with or without the aid of tools or mechanical devices in order to make some change in that object." He developed a two-dimensional scheme–task variability and problem analyzability. Task variability could have few exceptions or many exceptions. Problem analyzability could be well defined (a logical analytical search process exists) or ill defined (no search program exists). The measure of "ill-definedness" is related to the concept of equivocality (Daft and Lengel, 1986). It may not be known which resources are needed to accomplish the task.

There are then four categories that are labeled routine, engineering, craft, and nonroutine. See Figure 7.1, which places the categories on the two dimensions.

Routine technologies are well defined and have few exceptions. At least one of formalization and centralization is high, and perhaps both are. Tasks are well defined and understood. Control is obtained through the application of rules. Perrow (1967, p. 204) notes, "Given a routine technology, the much maligned Weberian bureaucracy probably constitutes the socially optimum form of organizational structure." Standardized mass production is a well-known example

Engineering is also well defined but involves many exceptions. Formalization is high and centralization is low. Tasks can vary in detail, but they are readily dealt with. Control requires communication through reports and meetings. There is a good deal of task variability, which requires that adjustments be made. Engineering problems tend to be in this category, as well as accounting and perhaps medicine.

Figure 7.1. Perrow's Technology Dimensions

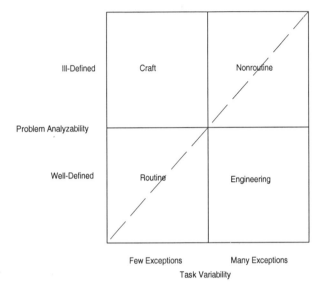

Source: Adapted from Perrow (1967, p. 196).

Craft contains few exceptions but is ill defined. The tasks themselves are difficult and involve few alternatives. Individual skill and experience are quite important to deal with this technology. Formalization is not high, and centralization is low. Initial training is quite important. If coordination is required, it usually is attained through meetings where complexity can be dealt with. Ed May renovates old cars in Durham, North Carolina. He works within a narrow set of tasks, but each car needs special treatment and exactly what should be done is not known in advance. Ed has great skills, many years of experience, and Burton's sports FIAT just looks great after the renovation.

Finally, nonroutine technology has many exceptions and is ill defined. Highly skilled professionals are needed. Formalization and centralization are low. Coordination is obtained through meetings, and group norms provide the control. Basic research is one example.

Technology can be reduced from routine to nonroutine–roughly following the diagonal in Figure 7.1. The information-processing demand increases as we progress from routine to nonroutine. Routine technologies require higher formalization and centralization than nonroutine technologies. Similarly, rules can be used for coordination for a routine technology whereas the give and take of meetings is required for nonroutine technologies. This was the concern of the management of Med Electronic mentioned in the introduction. It planned a move from a nonroutine technology to a routine technology along with an increase in the production quantity. To process the required information and maintain high quality in standard production, rules and standardized behavior is necessary. Centralized decision-making at the overall level will also be appropriate.

Perrow (1967, p. 207), in comparing his categories to Woodward's, speculates that unit production is probably nonroutine production; mass involves routine production, and process is not easy to categorize.

Thompson (1967, p. 14-16) considers technology in terms of long linked or serial interdependence, mediating or intensive. Interdependency of technology (p. 54–55) is described as pooled, sequential or reciprocal. Coordination is achieved, respectively, by standardization, plan, and mutual adjustment.

Scott (1998, p. 228-230) describes and measures technology along three dimensions:

- Complexity, diversity: Number of different items requiring simultaneous consideration.
- Uncertainty or unpredictability: Can the work be predicted in advance?
- Interdependence: Does a change in one item necessitate a change in a different item?

As Scott (p. 257) summarizes: "Complexity, uncertainty, and interdependence are alike in at least one respect: each increases the amount of information that must be processed during the course of a task performance. Thus, as complexity, uncertainty, and interdependence increase, structural modifications need to be made that will either (1) reduce the need for information processing–for example, by lowering the level of interdependence or by lowering performance standards - or (2) increase the capacity of the information-processing system, by increasing the channel and node capacity of the hierarchy or by legitimating lateral connections among participants."

In brief, greater variability in technology requires greater information processing capacity. This increased capacity can come from highly skilled professionals. But modern production technologies also may be used. CAD-CAM, flexible production methods, and robotics are all technologies that can be introduced.

It is interesting to juxtapose Scott's three characteristics of technology (complexity, interdependence and uncertainty) with our earlier environmental characteristics (complexity, uncertainty and equivocality). We suggest the correspondence shown in Table 7.2.

Table 7.2. Environmental-Technology Characteristics

Environmental Characteristics	Technology Characteristics
Complexity	Complexity, interdependence
Uncertainty	Uncertainty
Equivocality	Uncertainty

Environmental complexity definition captures the same aspects as technology complexity and interdependence. Technology uncertainty covers the concepts of both uncertainty and equivocality. Contrary, however, to the measurement of the environment, the empirical research on technology-structure relationships does not make it easy to separate results into the two groups. The underlying concept behind the correspondence is information processing; both the environment and technology require management to cope with the requisite information to meet the organizational goals. Galbraith (1974) argued that "the greater the uncertainty of the task, the greater the amount of information that has to be processed." The task is shorthand for the organizational goals of delivering a product or service in an efficient manner. Uncertainty can arise from the environment, or technology; whatever the origin of the uncertainty, management must deal with it. Stated differently, it is the lack of information that is fundamental–not its origin in the environment or technology. Thus, at a more abstract level, environmental and technological uncertainty poses the same organizational challenge.

Technology also can be defined as the transformation of inputs into outputs. This general definition also is very operational and provides a guide on what to measure. The input-output matrix captures all three elements: inputs or resources, outputs or results, and transformation (or how inputs are utilized to yield outputs). An activity model of the economy, the firm and organization contains an in-put-output matrix of the technological relations. The input-output matrix is illustrated below in Figure 7.2, where the inputs are listed down the left side: employees by skill level, resources by kind, and other required inputs. The outputs are listed along the top of the matrix: automobiles, trucks, accounting services, and consulting services. The outputs can be products or services for any organization. Consider the automobile column in the matrix; it is a list of the input resources required to produce one automobile: 30 hours of autoworker time and .3 hours of the assembly time per automobile and no other inputs. Similarly, a consulting assignment may require: 150 hours of a skilled consultant's time, 10 hours of computer time, and 400 hours of technical analysis by specialists. Each column is a list of the inputs required to yield one unit of the output. It is a list of ingredients. Although not explicit in the matrix, there is a complementary statement that gives the process to convert inputs into output; this is also part of the statement of the technology.

Figure 7.2. Technology: An Input-Output-Matrix

		OUTPUTS			
		Automobile	Truck	Accounting	Consulting
I N P U T S	Employees by skill level	30 hours of auto worker time			150 hours of a consultant
	Resources by kind	.3 hours assembly-line time			10 hours on the computer
	Other required inputs	0			400 hours of technical analysis

The input-output matrix can have various properties. Consider two extreme cases. First, assume each and every output utilizes every input–that is, there is an entry in each and every cell in the matrix, which is a very dense matrix. This dense technology is very complex and interdependent. Referring to Scott's (1998) concepts of complexity, it includes diversity and interdependency. Diversity is the number of different items requiring simultaneous consideration and interdependency refers to a change in one item that necessitates a change in another.

Let us assume that the inputs are in limited supply–for example, there are 600 autoworkers. Then a change in the number of automobiles produced decreases the feasible number of trucks. In addition, the assignment of assembly-line time between automobiles and trucks must be adjusted. If the matrix has no zeroes, then all inputs affect all outputs requiring a simultaneous consideration. The interdependency is very high; any change in one output necessitates a change in another.

Second, let us consider the opposite extreme situation. Assume that only one input is used for each output. The input-output row has all zero entries, except that each row and each column have one and only one nonzero entry. This is a very sparse matrix. The technology is very diverse and very independent. There are no simultaneous considerations, and any item can be changed without affecting another item.

Most technologies fall between these two extreme situations. The first very dense technology cannot be broken into pieces or be decomposed; the second can be. The decomposability of a technology is the degree to which it can be broken into relatively separate parts. The very dense matrix is not decomposable; the sparse matrix is more decomposable. We can relate the decomposability of the input-output matrix to Scott's (1998) complexity-diversity and interdependency measures in Table 7.3.

Table 7.3. Technology Correspondence

Input-output Matrix	Scott's Measures	Decomposability
Sparse - many zeros	Diverse, independent	Perfectly decomposable
Dense - few zeros	Complex, interdependent	Nondecomposable

The technology decomposability is a major contingency for organizational design. Burton and Obel (1984) investigated the effect of technological decomposability on the efficiency of two organizational configurations–the divisional and the functional – holding size constant. Divisional configuration is relatively more efficient with respect to coordination than functional configuration for a more decomposable technology. Thus, as technological diversity and independence increase, the divisional configuration is more likely to be recommended. As diversity increases and independence increases, the divisional configuration becomes relatively more efficient; when complexity increases and interdependence increases, the functional configuration is more likely to be recommended.

In the input-output description of technology, technology uncertainty is reflected by the uncertainty of the input-output elements in the matrix. Consider the time required for an autoworker to make one automobile. If it is known and certain, the uncertainty is low. If this time has high variance, then the uncertainty is high. Technological uncertainty is high when the input-output elements have high variance, and uncertainty is low for low variances.

Technology has been operationalized in many ways. We have reviewed the schemes devised by Woodward and Perrow, and commented briefly on Scott's and Thompson's discussion and its relation to the decomposability scheme. We found that Woodward's unit, mass, and process categories and Perrow's routineness concept are appropriate operationalizations and measures. Scott's notion of diversity is also utilized. Most of the research has directly or indirectly been related to production technology, although Perrow's (1967) measure applies both to production and service technology. From an information processing point of view, it does not matter whether it is production or service. The concepts can be applied equally well as we argue in the next section. However, the information capacity demand and the means to meet that demand may differ in a production technology and in a service technology. Similar arguments can be made for retail and wholesale. To distinguish between the types of technology is therefore appropriate. Next, we consider technology and its effect on the configuration and other design variables.

MEASURING TECHNOLOGY

Technology is the information, equipment, techniques, and processes required to transform inputs into outputs. The above survey of the literature shows, as was the case with both size and environment, that there is not one agreed concept and measure for technology. However, there are common traits in the technology concepts which we will explore below.

We have found that measuring technology along the following four dimensions seems appropriate:

- Manufacturing, service, retail, and wholesale.
- Unit, mass, and process (automation).
- Routine or nonroutine.
- High or low divisibility.

These dimensions capture most of the measures that have been discussed in the literature; and they have proven useful in OrgCon. The combination of these four dimensions gives $4 \times 3 \times 2 \times 2 = 48$ different technology categories. In OrgCon, we use a finer scale for the routine and nonroutine and the divisibility dimensions. Our main dimensions are routineness and divisibility. The two other dimensions are used to further strengthen our hypotheses. From the description below it can also be seen that there is some correlation between the measures. This is particularly true for routineness and unit, mass, and process production as discussed above. The measurement of technology is relatively simple and yet allows for enough variation to be practical.

The first dimension–manufacturing, service, retail, and wholesale - gives a broad type of business indicator. It puts the organization in a technology set. The second dimension follows Woodward's clarification: unit, mass, and process (automation). Automation is like a process technology, as we argue later. Technological routineness or nonroutineness is the third dimension. Scott's uncertainty measure is aligned, but routineness is empirically the dominant dimension in Perrow's framework related both to uncertainty and equivocality. Finally, divisibility is similar to decomposability and the opposite of Scott's interdependence measure. Thus, our technology descriptors and measurements are similar to existing concepts.

The type of business in which the organization operates makes a difference. Manufacturing, service, retail, wholesale are important in determining technology. There is evidence that studies of manufacturing may not be applicable to service firms (Robbins, 1990, p. 198). Every manufactured product ultimately becomes a service, but not vice versa–for example, an automobile is a manufactured product that renders a service. General Motors is a manufacturer; Avis is a service corporation. Some services may, however, involve a few products.

The following definitions have been used:

- *A manufacturing firm.* A firm whose primary activity is the conversion of raw materials, parts and subassemblies into finished products for use by a customer. The manufacturing transformation requires equipment and personnel. It may be either capital or labor intensive. Firms that make automobiles, toys, and computers are examples.
- *A service firm.* A firm whose primary activity is the delivery of a service directly to a customer or client. In the final stages, individuals usually deliver the service. Banking is a service, which usually involves direct face-to-face delivery, but automatic teller machines (ATMs) are an exception. Service requires both capital and labor. Many services are labor intensive, but others are not: telecommunications services, for example, are very capital intensive. Service firms include restaurants, telecommunications, banking, accounting, health care, and dry cleaning.

The distinction between a manufactured product and a service is not always obvious. If you buy an automobile, it is a manufactured product. If you lease the same automobile, it is a service. Consider the distinction between cooking a frozen pizza at home and eating a pizza in a restaurant. The later is a service incorporating a product. Most manufactured products ultimately provide a service, and some services involve a product. Although the distinction between a product and a service is sometimes unclear, the distinction between a manufacturer and a service firm is more apparent. General Motors manufactures automobiles; Avis rents them. The first is clearly a manufacturer; the second is clearly a service firm.

Additionally, we distinguish between retail and wholesale firms.

- *A retail firm.* A firm whose primary activity is the sale of a product or service directly to a customer who is the final user. Usually, retailers do not engage in the manufacture of products. Kroger, Bilka and Wal-Mart are examples.
- *A wholesale firm.* A firm whose primary activity is the sale of a product or service to another firm but that does not sell directly to the final user. Wholesalers do not engage in the manufacture of products. Wholesalers are less well known.

Nonetheless, Coca-Cola sells mostly wholesale as do many famous brand names.

Working only with manufacturing firms, Woodward (1965) based her empirical analysis on this type of categorization and found structural differences between the high and low performers. Her categorization has solely been associated with manufacturing. Mintzberg (1979, p. 258), in his discussion of Woodward's categorization, found that the characteristic process, per se, is not important but that the process industry is highly automated. The unit production is specialized customer-oriented production; the mass production is labor-intensive production of standardized products; and the process production is highly automated with less labor-intensive production. Woodward's categories describe the information demands from various technologies. With this interpretation the categorization can also be used for service, retail and wholesale organizations. Joan Woodward's categories were:

- *Mass production*: Large batch or mass-produced technology.
- *Process production*: Highly controlled, standardized, and continuous-processing technology.
- *Unit production*: Technology in which units are custom made and work is nonroutine.

In the service industry mass production corresponds to standard high-volume services, such as food chains, airlines, and hotel chains. The unit production for the service industry is a service that is tailor made for the customer. Hairdressing, medical services, and specialized portfolio management are examples. The process corresponds to highly automated service industries. These include automated bank tellers, automated telephone service, cable television, and services using advanced electronics and computers without employing many people.

For retail, the standard high-volume retail with high labor intensity would correspond to mass production. An off-the-rack suit is an example of a standard product sold in this manner; a tailor-made suit is an example of unit production in the retail industry. A specialty audio store that combines the set to match the particular needs of the customer is a unit example. The retail process corresponds to stores where the customer does not get advice and where operations are highly automated. Special supermarkets, such as Wal-Mart, are examples of such stores. Similar examples can be given for the wholesale industry. It is clear that Woodward's categories go beyond manufacturing and include these services. This is particularly true with respect to the skill level required by employees, as was shown in Table 7.1.

Routineness is a central concept in technology. Analyzability and uncertainty also have been used. However, a majority of researchers have used the concept of routineness (Robbins, 1990; Miller et al., 1991). Similar to Perrow's (1967) original concept, we define routineness as follows:

- *Routine technology*: Contains easy-to-analyze problems and few exceptions.
- *Nonroutine technology*: Contains difficult-to-resolve problems and many exceptions.

A number of more precise definitions of routineness were discussed by Miller, Glick, Wang, and Huber (1991), and they found that, in general, relationships between routineness and structure do not depend on the particular definition of routineness used.

As we discussed earlier, routineness and uncertainty are similar. There is little uncertainty about a routine technology but a nonroutine technology presents greater uncertainty. Med Electronics plans to move from nonroutine unit production to routine mass production, changing both complexity and uncertainty related to technology.

Technological divisibility is also a central concept. Divisibility is related to interdependency and decomposability. The less interdependent the technology is, the greater the divisibility; the greater the decomposability, the greater the divisibility

Divisibility then is the degree to which tasks can be divided into smaller, relatively independent tasks. Returning to our earlier discussion, a highly divisible technology has a sparse input-output matrix, is decomposable and independent. There are a number of dimensions to consider; technology can be divisible across the functions of research and development, production, marketing, and advertising. Technology can also be divisible across products. For international organizations, the degree of divisibility across various countries is important. The technological divisibility is a measure of these various dimensions. In more technical terms, increasing re-turns to scale in R&D, production, or marketing leads to a low degree of technological divisibility. Satellite television that broadcasts to many countries may decrease the divisibility of the marketing tech-nology when TV commercials are used. A Euro Sport commercial will reach most European countries and require that items advertised are ready for sales in the various countries at the same time. When LEGO introduces a new design in Europe, the design is available in approximately 40,000 stores on the same day. This demands high coordination within the complete logistic operations of the company. Economies of scope can have a similar effect. The R&D of telecom-munications switches, for example, is not divisible due to the high cost of development; there are economies of scale and scope. Some organizations may have high divisibility of technology today, but fu-ture technologies may not.

TECHNOLOGY AS A CONTINGENCY

The various dimensions of technology have an effect on the organiza-tional design. We consider technology's effect on formalization, cen-tralization, complexity, configuration, coordination and control, and incentives.

Technology Effects on Formalization

The more routine the technology, the more the activities are predict-able. Exceptions are few and easy to resolve. Less information needs to be processed. With a high routineness, it is advantageous and efficient to establish rules and a program to regulate and coordinate the work (Perrow, 1967, p. 199-200). This is an efficiency argument. When routineness is low, such rules and programs are likely to be incorrect much of the time. Thus, a good deal of information will need to be processed to schedule and coordinate processes. This is an effectiveness argument. The relationship has been questioned (Robbins, 1990), but it has obtained empirical support by many in-cluding the metaanalysis by Miller, Glick, Wang, and Huber (1991).

Miller, Glick, Wang, and Huber (1991) also found that the strength of the relationship was modified by two factors. First, an organization with many professionals has a mitigating effect on routineness and formalization. The argument is that professionalization and formalization are alternative forms for coordination and control so that when one is in place, the effect of the other vanishes. The important factor is standardized behavior for the organization. There are more means to obtain such behavior.

Second, Miller, Glick, Wang, and Huber (1991) found that one would expect that routineness of the technology was more positively related to centralization, formalization, and specialization in manufacturing organizations than in service organizations. However, their meta-analysis showed exactly the opposite. Jackson and Morgan (1978, p. 196) argue that the reason for such findings may be that in manufacturing the production process is linked to machines and their performance, and these machines indirectly introduce standardization. In service organizations more rules and procedures are needed to obtain the same level of standardized behavior. Additionally, in manufacturing the quality control related to the process may secure high quality of the products while rules are needed to obtain high quality in service organizations. This implies a higher reliance on formalization and centralization. This also fits the view expressed in Mills and Moberg (1982) on the differences between manufacturing and service technologies.

This analysis indicates that highly automated technology should have a greater formalization than otherwise suggested. This is also supported by Child (1973, p. 183), who states that automation leads to higher formalization. However, it is directly opposed to Woodward's original results (see Table 7.1). She found that process production had low formalization. The difference may be attributed to the difference in size of mass production firms and process production firms (Hickson et al., 1979). The mass production firms were generally larger than the process production firms; therefore, from a size argument, formalization was higher in the mass production firms than in the process production firms.

Automation is related to the use of computers and information technology. Zeffane (1989) found that such use would increase formalization. We posit that if the introduction of modern information technology is not followed by standard rules on how to use it, the likelihood of inefficient operations is high. Such standardization also may have negative side effects. We all have received letters from companies urging us to pay their bills when we have done so some days before. If the company's computer has a standard procedure writing letters to all those who have not paid within a given deadline, then it may not be possible to alter procedures. The above discussion can be summarized in the following propositions.

Propositions on Technology Effects on Formalization

If technology routineness is low, then formalization should be low.

If technology routineness is high, then formalization should be high.

If the organization employs many professionals, then the above proposition is not so strong.

If the organization is in the service industry, then the strength of two first propositions is greater than if it is in the manufacturing industry. Retail and wholesale organizations can be expected to fall in between.

If the technology type is process (i.e., high automation) then formalization should be higher than it would be otherwise.

If the organization uses transaction-based information technology, then formalization should be high.

Technology Effects on Centralization

If the organization is small and has a technology that is very routine, then the manager can more easily assess the operations than if the routineness was low. A manager can handle the required information. Therefore, the argument that for small organizations the centralization should be high is further strengthened. It is worth noticing here that our size measure incorporates both the number of people and the skill level of these people.

When the organization is large and has a technology that is routine, then it is very likely that formalization should be high. A control and coordination mechanism, therefore, is in place (Zeffane, 1989).

The size argument presented in Chapter 5 that for large organiza-
tions centralization should be low is therefore further strengthened.
These two propositions are supported by the metaanalysis by Miller,
Glick, Wang, and Huber, (1991).

Propositions on Technology Effects on Centralization

If technology routineness is high and the size of the organization is small, then
centralization should be high.

If the organization is large and technology routineness is high, then centrali-
zation should be medium.

Technology Effects on Organizational Complexity

The relationship between technology and organizational complexity is
not simple either. Size is a moderator again. Generally, the larger
organization with a routine technology is more complex.

The argument is that large organizations can better specialize
and, therefore, use the routine technology to create experts for each
specialty; horizontal differentiation increases. This argument is partly supported by Miller, Glick, Wang, and Huber (1991). However, a
reverse argument also can be made. If the technology is nonroutine,
then the work is very complex, and it is likely that the appropriate
span of control is low. Therefore complexity especially increases vertical differentiation. We therefore state that for a large organization
with a nonroutine technology, complexity should be high–particularly
vertical differentiation. Again remember our special measure of size.

The span-of-control argument says that supervision has limited
information-processing capacity and can deal with a limited number
of issues or exceptions that a nonroutine technology will yield.

The argument is that the more complex the work, the less people
a manager can supervise and control. This is a bounded-rationality
argument and is widely supported (Robbins, 1990). That both nonroutine and routine technology may lead to high complexity, but for
different reasons, may explain why Miller, Glick, Wang, and Huber,

(1991), in their metaanalysis using averages, did not find a significant relationship between routineness and specialization. There may be other technology-based reasons than routineness that may lead to structural conclusions.

Since process organizations use more automation and more skilled personnel, the span-of-control argument suggests that process organizations are more complex than other types. However, Woodward (1965) (see Table 7.1) found the opposite in her research on manufacturing firms. This contrast also may be related to the discussion about the effect of size. Woodward's results have been criticized because it was argued that when controlled for size her results disappeared (Hickson et al., 1979). Because organizations that use a process technology are less labor intensive, they tend to be of smaller size. In most cases the size propositions in Chapter 5 will therefore be activated and recommend a low complexity for a highly automated organization. The balancing of the two propositions will result in the correct recommendation for the particular organization. The balancing issue is treated in more depth in Chapter 10.

Propositions on Technology Effects on Organizational Complexity,

If the size of the organization is large and the organization has a technology that is routine, then complexity should be high - particularly horizontal differentiation.

If the size of the organization is small and the organization has a technology that is routine, then complexity should be medium.

If the size of the organization is large and has a nonroutine technology, then complexity should be high - particularly vertical differentiation.

If the organization has a nonroutine technology, then the span of control should be narrow.

If the organization has a routine technology, then the span of control should be wide.

If the technology type is process (high automation), then complexity is high.

Technology Effects on Configuration

Technology also affects the configuration in many ways. A unit technology is more likely to require a matrix organization. One reason for this is that it may be needed to assign experienced and skilled personnel from one production unit to the next. This sharing of valued and limited resources requires on-line coordination, which can be realized in a matrix structure.

We now turn to a number of mismatches between technology and some configurations. A functional configuration for a nonroutine technology is not likely to be efficient because a functional structure requires high horizontal differentiation, which may be unlikely for a nonroutine technology. And it certainly will require a lot of cross-function coordination, which the functional configuration will not do in a timely fashion.

A divisional configuration and a non-divisible technology is a mismatch. Divisional organizations require that the task be divided and placed in each division. Since these divisions are relatively autonomous, a high degree of interrelationship between them is costly to coordinate. On the other hand, a matrix structure is not needed for a divisible technology, as shown above. The argument is as follows. If the technology is divisible, then the work can be separated into units that are not dependent. A high level of coordination is, therefore, not required due to technological reasons. A matrix structure with its lateral relations for coordination is too costly, and there is little to coordinate. A bureaucracy requires standard behavior either through the use of rules or the use of professionals. Rules are very likely to obstruct needed adjustments for a nonroutine technology.

Finally, adhocracies are costly to coordinate and can operate only where the uncertainty related to the tasks is relatively high. Therefore, an ad hoc configuration cannot operate if the technology is very routine and will not operate efficiently.

Propositions on Effects on Configuration

If the technology type is unit, then it is more likely that the organization has a matrix configuration.

If the organization has a nonroutine technology, then the functional configuration is not likely to be an efficient configuration.

If the technology is not divisible, then the configuration cannot be divisional.

If the technology is divisible, then it is not very likely that the configuration should be a matrix configuration.

If the organization has a nonroutine technology, then it is not likely that a machine or professional bureaucracy is an efficient configuration.

If the technology is not nonroutine, then the configuration cannot be an ad hoc configuration.

Technology Effects on Coordination and Control

We now relate the technology routineness to the recommendations on coordination, media richness and incentives. Generally, with more routine technology, more rule-oriented coordination, and less rich media, the incentives can be procedure based. In contrast, less routine technology calls for coordination by integrators and group meetings using richer media and results-based incentives. The supporting arguments are fundamentally information processing in nature similar to the ones presented in Chapter 6. A routine technology does not change much. Activities are largely known and can be planned. There is little new, detailed, or current need for information. The information is well defined, known for some time, and likely to be numerical: production quantities, product dimensions, and so on. Incentives can be procedure based as procedures are known and well defined and the challenge is to follow them correctly.

The nonroutine technology, in contrast, calls for a large amount of information. There are many issues to decide, implement, and control. The products and the procedures themselves are likely to change often. Galbraith (1974), in his information-processing framework, suggested that the information processing requirements are large for this situation. Integrators and frequent group meetings are appropriate organizational strategies to obtain the required coordination. Daft (1992, p. 290) and Daft and Lengel (1986) argue that relatively rich media will be required to deal with an ambiguous situation when much is unknown and is to be discovered during the decision-making phase. A nonroutine technology could also be described as equivocal and uncertain, terms that we used to describe the environment in Chapter 6. The need for rich media can be realized in a number of ways. Face-to-face is the richest, which is the medium of integrators and group meetings. That is, integrators and group meetings are rich media provided there is truly discussion, joint problem solving, and a give-and-take atmosphere. Integrators who simply tell and meetings that only inform will not work. Finally, incentives must fit the routineness of the technology and the other organizational design recommendations. A routine technology indicates that we know what to do; the incentive is to do it–that is, a procedural-based incentive to follow the rules and implement the plans. The nonroutine technology creates the opposite requirements. The goal is to obtain results in the face of the nonroutine technology. The goal is a working product or a satisfied customer. This result is important, and the procedure is to be developed. Kerr (1975) argues that the best incentive is to reward what the organization wants, and here the organization wants results. Many organizations use many different technologies–some routine and some nonroutine. The various technologies may push the organization in different directions. If that is the situation, technology is not a strong contingency on the overall recommendation relative to other contingencies. Each technology, of course, will be an important factor for the micro design.

Propositions on Technology Effects on Coordination and Control Mechanisms.

If the size of the organization is not small and if the technology is routine, then coordination and control should be obtained via rules and planning, and a media with low richness and a small amount of information can be used. Incentives should be based on procedures.

If the technology is nonroutine, then coordination should be obtained via group meetings, and media with high richness and a large amount of information should be used. Incentives should be based on results.

If the organization does not have a dominant technology, then the technology-structure recommendation should be discounted relative to other contingency factors.

Information Technology[1]

Information technology is often mixed with the technology concept of organization theory. Information technology includes computers, e-mail, voice mail, video-conferencing, databases, expert systems, and other electronic means to store, analyze, move or communicate information in an organization. Information technology is then a means for an organization to process information. The organization itself is an information processing entity and thus, information technology is a means for the organization to accomplish its fundamental work. Of course, there are many other non electronic means for the organization to process information: pencil and paper calculations, face to face conversations, paper memoranda, etc. Therefore we see information technology as a means for coordination and communication. It is thus not a contingency factor, but may be seen as a dimension of the structural design. Huber (1990) found that information technology in itself would not determine the organizational structure, but would drive it towards more extreme positions. If the contingency factors would drive the organizational design towards decentralization, then information technology with decision support systems would allow it to be more decentralized than without the use of information technology. Additionally, if the contingency factors suggest a centralized organization then centralized transaction systems would allow an even more centralized organization. Wang (2001) states that the relationship between IT-related variables and structure might be spurious and affected by certain omitted factors. Further Wang (2003) finds that organizations with high formalization and high centralization may by the use of information technology be rather flexible and adaptive thus influencing the effect of a large size on the choice of organizational form. Basically, this discussion lead to the conclusion that choice of information technology is a central part of the design of coordination and communication and further

---

[1] Starling Hunter contributed greatly to this discussion.

should be seen in the context of obtaining design fit. Thus, we want to explore the implications that the organizational design has for the choice of the information technology.

The connection between the organizational configuration and properties and the organization's information technology has a long tradition and vast literature. As Hunter (1998) suggests, most of the research and studies have focused on the influence of information technology, computers, email, etc., on the structure, properties, behavior and performance of the organization, i.e., the information technology is the independent variable and its effect is the dependent variable (Huber, 1990; Malone and Rockart, 1990).

More recent research has taken a new approach of advanced structuration theory (DeSanctis and Poole, 1994) which focus on the complex interactions between the organizational actors and the information technology. The emergent behavior is uncertain and difficult to predict; the research goal is to describe and understand the interplay between the organization and the information technology without resorting to an independent variable, dependent variable approach. Here, we want to focus on the information technology design or choice question, i.e., what information technology should be adopted by the organization to be compatible with the organizational configuration and the organizational properties. This switches the independent variable and dependent variable so that the organizational design is the independent variable and the information system is the dependent variable.

The organizational design question more directly addresses the managerial question of "What kind of information system do I need to fit with my organization?" Consequently, we want to consider what the information technology should be. This seems reasonable, but most research questions have not been posited in this manner and thus the empirical evidence is wanting. There is a good deal of research on the effect on decentralization when an advanced e-mail system is introduced; however, it is not conclusive.

The main theme of this book is that the organization is an information processing entity, i.e., the organizational task is accomplished by processing information: gathering data, analyzing information, deciding what to do, communicating information, implementing and controlling events, measuring events and results. Individuals talk to each other in the hallway; they write e-mails; they go to meetings, etc. Some organizations have lots of meetings, but discourage hallway conversations. The list of possibilities is long. The organizational design then helps rationalize and organize how the information will be processed. For example, a decentralized organization processes information differently than a centralized organization to accomplish the same organizational task. We suggest that a decentralized organization will use e-mail differently than a centralized organization. "Who makes what decision when" is different and we suggest that the content and frequency of the e-mail would be different. But the e-mail is only a small part of the organization, and indeed, the information, i.e., the electronic information system is only a small part of the organizational information processing. Yet, the electronic information system is an important part of the total organization and it should fit with the rest of the organization.

In Chapter 6 on the Environment, we introduced the concept of media richness: high media richness requires that information comes from many sources, many formats, and probably in large amounts. Low media richness requires much less information from the environment. For high media richness, the information system is most likely to have many elements: e-mail, list servers, the web, news services, trade services, telephones with voice mails and ready access to the outside, multiple formats to receive information and then internally, some information processing capability to make sense of the diverse and large amount of information. In brief, the information system should support the gathering of vast and diverse information and also support its interpretation and meaning for the organization.

Low media richness needs can be met with a focused single information system, e.g., in a low complexity environment, low uncertainty and low equivocality, it may be sufficient to look only at last year's sales. However, this would be very inadequate for a high media richness need.

As we suggested above, the centralization of the organization also affects the choice of the information technology. Most of the studies consider the opposite question of what is the effect of a given information element, e.g., e-mail on the centralization. Huber (1990, p. 57) suggests that the greater use of information technology will be mediating; highly centralized organizations will become more decentralized, and highly decentralized organizations will become more centralized. This is insightful and summarizes what we know; yet, we cannot say definitively what the effect of the information technology will be on centralization of the organization.

But let us consider the design question, if we want to obtain a decentralized organization, what kind of information technology do we need? A decentralized organization means that the decision-making is at a low level in the organization. If the relevant information is local and nearby, then a decentralized organization would not require an advanced information system; the individuals can look at the situation and talk among themselves to make the decisions. Here a centralized organization would require the gathering and transmission of the same information up the hierarchy to make the same decision and hence an advanced information system could be quite helpful. On the other hand, if the relevant information for decision-making is external and widely dispersed, then an advanced information system would support decentralized decision-making. So, the decentralization, advanced information system question requires information about the source of the relevant information as well.

The organizational design question can be stated: for the organizational configuration and organizational properties, what is an appropriate information technology? Here we want to explore some plausible responses to this design question. We will state a few propositions and then argue for their reasonableness. The supporting arguments do have some research support, but their validity rests primarily upon their reasonableness.

A highly formalized organization has a large number of rules and standardized routines. What kind of information technology is appropriate? Without an electronic information technology, the formalization was contained in written documents and the rules were realized through the expertise of the employees who followed the rules. With an electronic information technology, it is possible to incorporate the rules directly into the operations of the organization through quick reference or actual control. Many airline reservation systems incorporate detailed rules and many airline personnel are permitted less discretion than in older paper systems. A highly formalized organization can use information technology to operationalize the rules and control implementation.

A low formalization has few rules. What is an appropriate information technology here? With few rules, information about the environment, customers, competitors, technology, scheduling, etc. becomes information for decision-making and coordination. Would electronic technology such as e-mail, voice mail, bulletin boards, shared databases, etc. be helpful? Yes, it seems reasonable. Here too, an electronic information technology could be helpful.

So we conclude that an electronic information technology can be appropriate for both, high or low formalization. However, there is a difference in detail in what the information technology does: for a high formalization, the rules are incorporated; for a low formalized, information is widely made available throughout the organization. Electronic information technology is a mediating technology; it supports the information processing demands of the organization. In the above discussion, the discriminating variable is the content of the information technology and what we want it to do. So, we must conclude that the information technology is not the primary concern; the main issue is: how does it help (or, hinder) the organizational demand for information processing capability.

A low vertical differentiation means that there are few levels in the hierarchy - top to bottom. There is no large middle management or the organization is "delayered." In information terms, middle management aggregated information as they passed it up, and disaggregated information as they passed it down. And frequently, they simply passed the information on without modification. With low vertical differentiation, the information technology should respond to this information requirement. The specific information technology would also depend upon the organizational degree of centralization as discussed above.

A matrix organization requires a good deal of give-and-take and adaptation to circumstances. The information processing demands are high, both in quantity and type. In addition to face to face meetings, telephone calls, etc., these information technologies can augment the availability and communication of information to support the matrix design. Equally, an information technology which is locked in rules and restricted format may well hinder the matrix design.

We could add a number of other propositions on other organizational design configurations and properties. The supporting arguments must emerge from the information demands of the organization and how the information technology will aid the particular information needs. It is too easy to say that we need more information technology. It begs the question: to do what. Different information technologies are required for different organizational designs.

Whether a given organizational design increases or decreases the need for information technology is not the issue. The question is: what kinds of information technology support the organizational design? Based on the discussion above we can now state design propositions.

Design Propositions

If the organizational formalization is high, then the information technology should incorporate the rules and routines.

If the organizational formalization is low, then the information technology should augment the availability of information through e-mail, voice mail and shared databases.

If the organizational complexity is low and particularly the vertical differentiation is low, then the information technology should facilitate quick hierarchical flow and the aggregation of information.

If the organizational configuration is matrix, then the information technology should be e-mail, voice mail, video-conferencing, and shared databases.

MANAGING TECHNOLOGY

In this chapter, we have presented the rationale for the influence of technology on the organization. However, there is a counterargument that the organization must influence and change the technology. This is managing technology. The organization must choose its technology. From either point of view the organizational structure and the technology must fit as well as technology must fit other contingencies.

Most products and services can be produced using a number of different technologies. Management often faces a choice between a routine and nonroutine technology and between mass production and a high degree of automation. LEGO Systems used to produce its standard plastic bricks in a kind of mass production setting with many workers operating the various machines. The same bricks are now produced in a highly automated factory with numerous robots and almost no workers.

The organization should manage its technology in a comprehensive manner. Total quality management (TQM) is such a comprehensive approach to manage technology. TQM focuses on process and product or service innovation to improve quality continually. Process innovation in TQM includes changing the organizational structure and the organization's procedures to achieve quality improvement and innovation.

Earlier, we defined technology (Robbins, 1990, p. 176) as the information, equipment, techniques, and processes required to transform inputs into outputs. There are many ways to achieve the same output, and consequently managing technology becomes important. The TQM approach assumes that all these technology variables can be changed and should be managed to achieve better quality. Consequently, TQM is a comprehensive approach with an unending quest to do everything better. It requires innovation on processes and products and services to serve the organization's customers and clients better.

Since the early 1950s, quality management has been the mission of Joseph Juran and W. Edwards Deming. One of Deming's fourteen points (Giffi et al., 1990, p. 33) is to "break down barriers between departments; encourage different departments to work together on problem solving." In our model, this recommendation could be a structural change to "increase lateral communications," "use committees," "form teams," and, perhaps, "implement a matrix or an ad hoc organization." Whatever the particular recommendation, the goal is to empower individuals to increase communications, define new issues, and devise ways to implement these changes in ways not currently available to the organization. Quality circles at the worker level and cross-functional teams throughout the organization are frequently used organizational changes. Deming's advice may seem to be too general and lacking specificity. It is a noncontingent maxim. It provides direction for innovation. In a given situation, the particular contingencies of the situation will yield more detailed recommendation and advice.

TQM is a comprehensive approach to quality. In terms of technology and structure, both are to be managed; both are variable and can be changed. New technical processes and products may require a new organization. Even more important, continual changes in technical processes and products require not only changed organizational structures but organizational structures that foster and drive technology innovation. A fundamental tenet of TQM is that quality must be improved continually; consequently, technology must be ever innovated, and organizational structures must drive this innovation. Here, change and innovation are the norm, and the organization should drive innovation.

The need for TQM has grown out of a need to be competitive in international and global markets. In the 1950s, the quality of Japanese products was poor, and they were not competitive. But they changed, and now Japanese products set the world standards for many products. Americans, Europeans, and others have adapted TQM as a way to introduce technological innovations and to answer the competition challenge. The future promises even greater global competition, which demands continual innovation.

TQM is one approach to managing technology but not the only one. Yet, its comprehensiveness suggests that most other approaches could be viewed under the TQM umbrella, adding specificity to a general total view. Med Electronics was considering changing its technology from a nonroutine unit production to a routine mass production. Because the environment had changed, forcing the company to reconsider its strategy, Med Electronics was trying to obtain a strategic fit. A change in technology requires a change in the overall organization. More formalization and maybe specialization may be needed. The quality focus will also change. In the current nonroutine situation, quality is related to creating an apparatus that fits the needs of a particular physician. In the routine technology situation, Med Electronics has to assume that each apparatus is very similar and meets standard specifications with low production cost. More focus has to be put on the procedure. Generally the view of quality affects the organizational structure. If defects are allowed, they have to be handled. A test center has to be created such that only the allowed amounts of defects leave the organization. A department that can deal with complaints may be needed, and possibly a process for replacement or repair has to be established. If defects in principle are not allowed, more structure and processes have to be allocated to prevent defects and fewer resources can be allocated to deal with defects.

SUMMARY

Technology is an important determinant for an organization's design. Technology has been described in numerous ways:

- Production, service, retail, and wholesale.
- Unit, mass, and process.
- Routine or nonroutine.
- Divisible or nondivisible.

Using these technology descriptions, we relate technology to organizational design. Generally, mass, routine, divisible technologies require a low amount of information processing; process, nonroutine and nondivisible technologies require a greater amount of information processing; process, nonroutine, and nondivisible technologies require a greater amount of information for coordination and control. Consistent with an information processing view, the recommendations on formalization, complexity, centralization, and coordination and control schemes are developed. Basically, mass, routine, divisible technologies require high formalization, high complexity, and medium centralization. Process, nonroutine, and non divisible technologies have lesser formalization, complexity, and centralization; but a non divisible technology must be coordinated by some means.

CHAPTER 8

Strategy

INTRODUCTION

When Bon Goût faced its problem with Samsonite, described in Chapter 6, it had a number of ways to deal with the situation. It could fight the decision made by Samsonite, or it could decide to stay in its normal business and find a new supplier to replace Samsonite. One of its major competitive strengths - its ability to deal with manufacturers and retailers - could help it find new products to import and market. Since its problems arose when the single European market was established, Bon Goût could decide to play the European game and import products from East Asia to be sold in the European market. It needed a strategy. Strategy is both means and ends; it includes the definition of the overall end goals, and the means of action needed to obtain these stated goals. Bon Goût must decide what business it wants to be in, and then it has to develop a process to realize what it wants. Environment is important for this choice. The process to obtain certain goals may require a particular orgainzational structure. For example, if Bon Goût decides that it wants to play the European game in the high-end fashion industry, it will need an organization that can operate in several countries. This will relate to both distribution and service to stores as well as to the ability to read fashion trends in various countries. Additionally, it needs an organizational structure that can deal with suppliers.

The fit between the strategy and the organizational structure has crucial implications for the performance of the organization (Miller, 1987a). In the next section we briefly review some of the theoretical developments that relate strategy and structure.

The Strategy of SAS

To review briefly, during the 1950s and 1960s the airline market developed very favorably. It was a stable, rapidly growing, and very profitable market. The challenge was to deliver a given number of ton-miles on every route (Buraas, 1972, p. 113). Problems were realized with the first oil crisis in the beginning of the 1970s. Suddenly, the market was stagnant because of the general economic depression following the oil crisis. It became of utmost importance for organizations to be efficient in order to make a profit. The SAS management recognized the necessity for cost reductions and therefore set up some internal groups to find ways of cutting costs without firing employees.

Management initiated cost cutting where the most savings could be realized. Customer services that were very important for SAS to maintain its market share were cut, whereas less important costs were left untouched. A survey among customers showed that the cuts were noticed and that customers generally felt that customer service at SAS had deteriorated. It was not until 1980 that a major change was initiated.

On March 2, 1981, a strategy outline was presented and approved by the board of directors. The outline consisted of a three-year plan for the organization in which major changes were described.

The main strategy was to be the best alternative for the full-fare passenger who wanted "value for money." The idea was to give special service to business travelers and thereby create an image of being "the businessman's airline."

The first action was to introduce a special category for businesspeople on the planes. In Europe, it was called Euroclass. After 1988, all business-class passengers traveled Euroclass irrespective of their destination (SAS, 2002). Euroclass passengers got roomier seating and better on-board service. In some larger airports special Euroclass check-in counters were introduced for quicker check-in. Special lounges were available for the exclusive use of Euroclass passengers. In this way SAS tried to create an environment specially designed for business travelers.

It was important for employees to be able to give service in their daily contact with customers ("Moment of truth"). As a first sign of a changed attitude toward employees, Carlzon began to inform employees about the changes as soon as the board of directors approved his strategy outline. All 20,000 employees received a red booklet called "Now We Start Fighting." This booklet presented the main concepts of the new strategy. Through this action Carlzon showed that employees should take active part in the changes necessary for the future. This marked the beginning of team spirit among SAS employees, which was an important part of implementing the strategy. Therefore the new strategy focused strongly on the role of front-line personnel in the customer's perception of service.

As part of the increased service for business travelers, SAS had to become a more precise airline company. Many departures had earlier been delayed to await the arrival of other planes so that transit passengers could catch their planes. From now on SAS would compensate transit passengers who did not catch their SAS departure by offering alternative ways of reaching their destination. In that way service for transit passengers remained almost unchanged, and planes could depart on time. Furthermore a serious drive was launched aiming at an improvement of flight schedules and punctuality. In that connection, Carlzon even had a monitor placed at his office so that he could monitor take offs and landings of all planes.

Besides becoming "the businessman's airline" a complementary strategy was devised. The main idea was to utilize available capacity to increase SAS's total revenue within the frame of the main strategy (Obel, 1986). A full airplane generated more revenue. The good years of the 1960s resulted in an overcapacity of planes. Furthermore, the new jet engine airplanes were much larger than earlier planes. The result was too many seats compared to the number of passengers, so, airline companies reduced their prices to fill up their planes. In the 1970s, stagnation really set in, but price agreements and regulation of the airline industry meant that companies were not hit hard. The oil crises, however, hit the entire world economy very hard. At the same time, the U.S. government began liberalizing American airline policy, which meant increased competition on the already tough market for trans-Atlantic air travel. The oil crisis and liberalization in the United States made the impact of overcapacity even more pronounced.

As a consequence a substrategy of capacity utilization was developed to offer the remaining seats on each flight to leisure travelers. These tickets were cheaper, but passengers received less service. On board the plane, this distinction between the customers was particularly important. In order to further distinguish SAS's products from those of other airlines, Carlzon introduced "the total traveling concept." Surveys had shown that businessmen demanded the service of not having to worry about transport and hotel bookings. This was why in 1985 SAS established destination service, whose core product was a ticket, transport and hotel package. Passengers were now able to book airline tickets, ground transportation and confirm hotel reservations with one call. To pursue this vision, SAS now had to develop a hotel network. Because it would take SAS at least 10 years to build up an appropriate amount of hotels, Carlzon decided to buy a 40% share in the Intercontinental Hotels, giving SAS immediate access to 106 hotels worldwide. This investment later turned out to be a very unprofitable, and when SAS finally gave it up in 1992, many judged the investment to be the biggest mistake Carlzon made at SAS.

Because the size of SAS's domestic market limited the competitiveness of the company, SAS would need to co-operate with another airline company in order to gain access to a secondary domestic market. In the eighties several alliances were made, with Thai and the Japanese ANA, and shares were bought in Continental Airlines, LanChile and British Midland. In 1989 SAS formed the European Quality Alliance with Swissair, Austrian Airlines and, in the beginning, also Finnair.

A similar, but larger alliance was the goal of the Alcazar negotiations. Alcazar was projected to include SAS, KLM, Swissair and Austrian Airlines. Because Carlzon was preoccupied with the Alcazar negotiations, Jan Reinås was appointed as new CEO of SAS whereas Carlzon was announced coming leader of the Alcazar group. Reinås' job was to create a "rescue" plan in case the Alcazar negotiations failed. This plan included a harsh rationalization program, aiming at severe cost reductions, extensive firings and a renewed focus on SAS's core business: the airline.

At the end of 1993 the Alcazar negotiations broke down, and Carlzon was left with no position to fill out at SAS. On April 1 1994, Jan Stenberg succeeded Jan Reinås, whose appointment from the beginning had only been temporary.

Jan Stenberg followed up with the focus now being on the airline; several of SAS's non profitable side business such as Diner's club Nordic and SAS Leisure were sold off. Furthermore extensive firings were made, and the life long employment concept was abandoned along with the idea of the total traveling concept.

Stenberg agreed with Carlzon in that the company needed to co-operate with other airlines to avoid becoming a feeder company. By co-operating with other airlines, SAS would not only save administrative costs, but the company would also gain access to more airport hubs. But Stenberg did not want these co-operations to take the form of actual mergers, but rather as strategic alliances. Further the bilateral agreements between countries with respect to landing slots gave a clear benefit to national airlines. This made a traditional merger strategy almost impossible.

As part of its strategy to create a Baltic junction, in 1995 SAS became co-owner of the Lettish company Air Baltic Corporation. In 1996, Lufthansa and SAS signed a strategic alliance, the core of which was a code share of all route networks between Scandinavia and Germany. The SAS-Lufthansa created a strong network between Denmark and Germany, and because Lufthansa has traditionally been strong in Africa and the Middle East, the alliance gave SAS the ability to exploit these areas as well. Further alliances were made with Thai Airways International and United Airlines, giving SAS access to the Far East and the American market. Recently, the so-called Star Alliance was created counting Lufthansa, United Airlines, Air Canada, Thai Airways and Varig (May 1997). The Star alliance and the focus on the airline business were just prior to Jørgen Lindegaard becoming the CEO presented as the SAS growth strategy with new routes and new airplanes.

With the increased competition due to September 11, 2001, the economic down turn after the IT-bubble burst, and the success of the low priced airlines, the SAS strategy had to be with a strong focus on survival - also in the short turn. Airlines such as Sabena, and Swissair went bankrupt. SAS introduced new service concept with lower prices and more flexible tickets on economy class. It bought more airlines like Spanair and Braathens. In particular Spanair which had a much better cost performance than SAS could be used to benchmark against the SAS airline. A cost cutting strategy was announced a reduction of more than 4000 people. With this followed a removal of the old expensive staffing rules, which for example stated than on a fight there should be one Dane, one Norwegian, and two Swedes. This could cause too many flight attendants on a flight and the cost of transporting the crew was very high. Further SAS launched it own low price airline - Snowflake.

The new strategy with majority holdings in other airlines also prepared SAS for the expected change in the bilateral agreements making a restructuring of the airline business a real possibility in 2004 or 2005.

Prior to Carlzon's changes, SAS's strategy was technology based because a growing market and high prices permitted a continuing investment in the latest equipment. During this period, SAS could be described as a defender.

The strategy under Carlzon was to create SAS as an analyzer with innovation. There was a definite emphasis on innovation. SAS clearly moved from a defender to an analyzer with innovation.

However, one of the things that caused SAS's problems was that, over time, the company's strategy became similar to that of a prospector. Once Carlzon had brought SAS "back on its feet" he started focusing on new market opportunities, which were conceptualized in his "total traveling concept." Many of his visions made SAS operate in areas that were very new to the company and most of which were based on emerging environmental trends. Carlzon in his mind saw SAS more of a prospector.

Stenberg's strategy meant that SAS was to focus on its core business: the airline. The company sold off many of its side businesses and subsidiaries, but it still sought selective new opportunities through its many strategic alliances. SAS's strategy under Stenberg can therefore be described as an analyzer with innovation.

Jørgen Lindegaard was forced to have a focus much like a defender fighting both traditional airlines with overcapacity and the new entrants –

the low priced airlines with a different level of service and a different cost structure.

1. Is there a common idea behind the SAS strategies over the years?
2. What are the major effects on structure of these strategies?

LITERATURE REVIEW ON STRATEGY

Structure Follows Strategy

In his historical study of American business Chandler (1962) demonstrated that "structure follows strategy", which continues to be a dominant proposition. Quite simply, it says that the internal structure of the corporation must fit with the adopted corporate strategy. Once a strategy has been chosen, then the appropriate structure follows. Chandler did not argue that the proposition is normative. Nonetheless, the corporations he studied were quite successful, which lends support for using the proposition as a recommendation as well as descriptive. Chandler's proposition has been challenged, supported, and refined; yet it and its derivative forms remain a dominant theme.

Chandler investigated about 100 of the largest American firms. He found that when companies engaged in new strategies, a new organizational structure was required. His basic thesis was that when an organization stayed within a single dominant business, then a functional structure would fit. If the firm diversified into a number of different businesses, then a divisional structure would be required. The argument is basically that when a firm diversifies, co-ordination and control within a functionally organized organization will increase up to the point where it is no longer manageable. A divisional structure is required to internalize some of the coordination and control, an adjustment to obtain a fit between information-processing demand and capacity.

Concurrently, Ansoff's (1965) interest in corporate strategy focused more on the choice of an appropriate strategy. He offered new tools of analysis and rationalized the diversification strategy, which was a widely adapted strategy in the 1960s, 1970s and early 1980s.

He noted the importance of structure and linked it to synergy–a major element in the diversification strategy.

Another classical study is the one by Miles and Snow (1978). They developed a typology of prospectors, analyzers, and defenders and argued that certain organizational structures would best fit each of the categories. Prospectors require a flexible organization, while defenders need a more mechanistic organizational structure.

Using the industrial organization framework, Porter (1980) analyzed the environment in terms of buyers, suppliers, substitutes, potential entrants and rivalry among existing firms (see the capsule in Chapter 6 on the environment). He then developed three generic strategies: differentiation, cost leadership and focus or niche strategy in terms of strategic target and strategic advantage. For a particular market segment target, a focus or niche strategy is called for. A low cost leadership strategy follows for a wide market strategy and a low cost position. And a differentiation strategy follows for a wide market and a uniqueness of product or service as perceived by the customer. Bon Goût had been a niche player with the Samsonite label. Albani was a niche player with its Giraf beer and Jolly Cola in the shadow of Carlsberg and Coke. Emerson and Panasonic are low-cost leaders in electrical and electronic products, capitalizing on economics of scale. Bang & Olufsen and Mercedes-Benz differentiate their products through advanced and reliable technology for which they command a premium in the market place. The appropriate structure for each strategy is the next step.

A recent major study of the relationship between strategy and structure is the one by Miller (1988a; 1987a). He tries to incorporate ideas from Porter (1980) into the views by Miles and Snow (1978). Miller (1987a) creates a typology with the following strategic categories: innovation, market differentiation, breadth-innovation, breadth-stability, and cost control. He investigates the relationship between strategy and structure in a study of Canadian firms. His results are very similar to the results posited by Miles and Snow (1978). For a recent comprehensive review, see Galunic and Eisenhardt (1994).

The Counter Proposition: Strategy Follows Structure

"Structure follows strategy" suggests that the firm chooses a strategy and then chooses a structure that matches the strategy. However, one could hypothesize that the structure constrains what the firm will do and how it will do it. That is, the structure helps determine, or severely constrains, the choice of strategy: strategy follows structure. The proposition "strategy follows structure" was argued by Hall and Saias (1983) and is consistent with Frederickson (1984). Burton and Kuhn (1979) took one of Chandler's companies, General Motors, and presented an analysis using systems concepts showing that the General Motors structure, as developed by Sloan, did indeed constrain what General Motors could do and would do from the late 1920s onward. The theoretical model is fully presented in Kuhn (1986).

Frederickson (1986) began with the structure variables: complexity, formalization and centralization, and developed hypotheses that each will affect strategy. Frederickson finds that increasing complexity makes strategic actions more political and more incremental. Similarly, increasing formalization leads to incremental strategic actions. However, increased centralization leads to strategies that are major departures from existing ones.

Table 8.1. Propositions Regarding the Effects of Structure on the Strategic Decision Process

---

COMPLEXITY

As the level of *complexity* increases, so does the probability that
- Members initially exposed to the decision stimulus will not recognize it as being strategic or will ignore it because of parochial preferences;
- A decision must satisfy a large constraint set, which decreases the likelihood that decisions will be made to achieve organization-level goals;
- Strategic action will be the result of an internal process of political bargaining, and moves will be incremental; and
- Biases induced by members' parochial perceptions will be the primary constraint on the comprehensiveness of the strategic decision process. In general, the integration of decisions will be low.

FORMALIZATION

As the level of *formalization* increases, so does the probability that
- The strategic decision process will be initiated only in response to problems or crises that appear in variables monitored by the formal system;
- Decisions will be made to achieve precise, yet remedial, goals, and means will displace ends;
- Strategic action will be the result of standardized organizational processes, and moves will be incremental; and
- The level of detail achieved in the standardized organizational processes will be the primary constraint on the comprehensiveness or the strategic decision process. The integration of decisions will be intermediate.

CENTRALIZATION

As the level of *centralization* increases, so does the probability that
- The strategic decision process will be initiated only by the dominant few, and it will be the result of proactive, opportunity-seeking behavior;
- The decision process will be oriented toward achieving "positive" goals (i.e., intended future domains) that will persist in spite of significant changes in means;
- Strategic action will be the result of intentional rational choices, and moves will be major departures from the existing strategy; and
- Top management's cognitive limitations will be the primary constraint on the comprehensiveness of the strategic process. The integration of decisions will be relatively high.

Source: Adapted from Frederickson (1986, p. 284).

Strategy and Structure; Fit

There are arguments for both "structure follows strategy" and the counter proposition "strategy follows structure." Amburgey and Dacin (1994) argue that strategy is more important in determining structure than structure is in determining strategy. Either way, there must be fit (Naman and Slevin, 1993). From a normative view, the "structure follows strategy" framework seems the more promising. The organization first sets its strategy, and then it must choose a structure to implement it. The structure may constrain future action; indeed, it should. The organization wants to implement a particular strategy - not all possible strategies. The lack of a strategic focus is not likely to be viable, and thus, a structure that implements the chosen strategy is our design goal. Putting these together, we suggest that strategy and structure should be symbiotic, and fit is very important. This is in alignment with the findings by Ketchen *et al* (1997) in an meta-analysis of 40 empirical studies.

Strategy and New Organizational Forms

New organizational forms illustrate the interrelated nature of strategy and structure. New organizational forms can be viewed as new structures for organizations; but they can also be viewed as new strategies for organizations as well. We want to explore the new forms as strategic choices here.

New organizational forms include: joint ventures, partnerships, strategic alliances, outsourcing, dedicated suppliers, franchising, virtual network organizations, to name a few, and organizational properties, such as flexibility as discussed in the hypercompetition capsule in Chapter 6. Many of these terms and concepts are well known and had been used for some time. So what is new? Recently we had utilized these approaches more frequently and more strategically than in the past. Now, large organizations have significant and major aspects of the organization involved in these new forms and further, the success of the organization depends upon the success of the new forms. Earlier, these forms were peripheral and minor appendages to the organization whose success or failure was often not critical to viability. For many small organizations, they are solely a new form in that the organization is a franchisee, a dedicated supplier, a strategic partner or an element in a virtual network organization. The "newness" results more from its strategic importance than novelty of concept.

To examine the new forms, it is instructive to examine how they differ from the "old" forms of organization. Our traditional notion of organization evolved out of traditional notions of property rights and the privileges that property rights give the owner. The owner of a piece of land, a building, a factory, an office or a business then had the right to command and control what took place in that property within very broad limits of the law and society's norms. An owner could plant or not plant crops, occupy or not occupy the building, choose the products to produce and in what quantities, hire or not hire employees for the office, and generally determine the activities within the organization. Complementarily, the owner could not make these decisions for properties that were owned by others. The owner could set the course for his/her own organization, but not for other organizations. The managerial prerogative followed the property rights for the organization. The organizational boundary was well defined and what is inside/outside of the organization was well demarcated. The property "line" marked what was outside and what was inside and therefore could be commanded. The Porter (1980) model of the firm within an industry is an example where the boundary between the firm and the market is well defined. Of course, there could be disputes about who owned a given property and what the boundaries were; the courts could resolve these issues.

Normal relations between organizations with well-defined boundaries took place in the marketplace where owners would engage as buyers and sellers. An owner could participate in many markets: for supplies, including labor and to sell its products and services. These markets have many forms: today's spot market to buy and sell and through contracts which include longer term agreements, future agreements and contingencies for which the future outcomes depend upon events at a later point in time. Over the last few years, contracts have evolved from relatively simple to very complex instruments

The traditional boundary of the organization also delineates "who" is inside and who is outside, and thus sets the boundary. Employees are inside and individuals who are not employees are outside. In Chapter 5 on size, we defined the size of the organization as the number of employees, which also sets the boundary of the organization as employees as part of the organization and everyone else as outside the boundary. More generally, assets and employees are inside the boundary of the organization. This is the accounting convention as well as the legal notion. March and Simon (1958, p. 89-90) note the legal boundary as fundamental, but also acknowledge its limitation for managerial purposes. Further, they (p. 90) suggest that consumers be considered "in," but argue there must be a limit; they draw the limit as the employees of the organization. The point is clear: the boundary of the organization is not given for all time and purpose, but can be set to meet a particular need or purpose.

The distinction between units in a production-distribution process that are "in" the organization and those that are "out" of the organization typically follows the legal definition of the boundaries of a particular firm. We find it fruitful to use a more functional criterion that includes both the suppliers and the distributors of the manufacturing core of the organization (or its analogue where the core of the organization is not manufacturing). Thus, in the automobile industry it is useful to consider the automobile dealers as component parts of an automobile manufacturing organization.

New forms of organization all share one common aspect; they break down the traditional boundaries and the simple notions of property rights and the associated management prerogatives and responsibilities. In short, the new forms have complex boundaries of ownership, but more importantly of managerial prerogative and complexity. Older notions of authority, responsibility, command and control break down and call for new attitudes and concepts.

Where is the boundary of the organization? Baligh and Burton (1982) argue that the managerial boundary of the organization and the legal boundary are normally not the same. The managerial boundary is a choice for management and should be thought of in information processing terms. The managerial boundary is only limited by the extent to which the management chooses to manage the information and take action. Many managerial issues fall within both boundaries - for example, the management of an owned building or factory. But other information issues fall outside the property rights of the firm; advertising, for instance, is a direct intervention by management beyond the legal boundaries of the firm. (Automobile companies gather information and take action for automobile dealers, whether they own the dealership or not.) The managerial information boundary for the modern organization goes "outside" in many ways: gather environmental information, advertising, influence buyers, influence government and policy, monitor new product and process technologies, monitor and direct supplier activities, etc. It is now a small step to go further with these activities: outsourcing, creating strategic alliances, establishing joint ventures, building franchising, and entering a virtual network organization. From the organization's point of view, the first step is an extension of the activities into the environment, and then the second step is the intervention into the activities of another organization. This relation may be unidirectional as in outsourcing, but it is likely to be reciprocal as in a joint venture or a strategic partnership. In short, both organizations are managing in part the activities of other organizations. This is a very major departure from the traditional concept of the organization with fixed boundaries operating in markets to complex boundary relations where each organization is managing in part the activities of the other organization. The "in part" management does require new notions of authority, responsibility, command, control and how managerial actions will be realized; it is a new form.

Williamson (1975) posed the boundary issue in different terms. Using transaction cost logic, he posited that the organizational size will be determined by the optimal mix of activities inside the organization (hierarchy) and activities outside the firm (market). In this approach, the size of the firm also sets the boundary of the firm, i.e., what is inside and what is outside. Powell (1990) extended the markets and hierarchies analysis to include hybrids or networks - organizations that are neither purely market nor hierarchy, but a complex combination. Here, the boundary becomes more complex, involving relations that go beyond either market or hierarchy. A new form is then an extension of the markets and hierarchies; and hybrids can be thought of as a generic category for new forms.

The concept of new forms has been explored further by Daft and Lewin

(1993). Among a large number of characteristics, they include "permeable and internal and external boundaries" (p. ii) and continue to discuss "organizational collaborations ... organizations team up with others to create strategic alliances and other forms of interdependencies that permit them to find a niche in a turbulent world. Boundaries among organizations are blurring as they explore hundreds of interorganizational connections and joint ventures. Network forms that emphasize interdependence rather than independence may emerge as a distinguishing characteristic of the new organizational paradigm" (p. iv). Besides the blurring of boundaries, they emphasize the "interdependencies" which transcend these boundaries to create connections, or networks.

Yet, the question remains; are new forms new combinations of traditional forms, or are they entirely new. If new forms are extensions and new combinations of traditional forms, then the theory about traditional forms should be relevant for new forms, albeit, modified and extended. If new forms are truly a departure, then traditional notions should be discarded and we should begin de novo to think about managing new forms. Daft and Lewin (1993) entertain this possibility in their call for new research and new research approaches. No doubt, there are arguments for both views, but we suggest that new forms can be fruitfully considered as extensions of traditional forms as we have developed above. In short, the central theoretical issue is the boundary of the organization. New forms have new and different boundaries. These boundaries are managerial and not limited to legal boundaries.

As such, new forms can be thought of as old forms with new boundaries. For a joint venture, we can think of it as a new and different organization from either of its parents and it has its own distinct boundary for which we can now analyze its environment, strategy, technology, etc. to determine an appropriate organizational design. Outsourcing extends the boundary of the organization to include elements of its vendor. Strategic alliances are organizations for which each partner will consider elements of other organization as elements of its own. Virtual network organizations are perhaps the purest of information processing organizations where the property rights are essentially nil and the organization itself only processes information.

So, new organizational forms are more questions of strategy about where managers choose to place the boundary of the organization than a question of organizational form itself. If the organization chooses to engage in joint ventures, partners, outsourcing, etc for strategic reasons, then management has shifted the boundary of the organization or created a new organization and a new organizational design is required; and, the approach developed in this book is relevant.

MEASURING AND CATEGORIZING STRATEGY

Strategy and strategic planning are important issues for managers and researchers. There have been numerous attempts to classify the companies according to their strategic behavior. Our approach is to categorize the strategy so that a given type of strategy fits with organizational structure. In the research, this relationship has received less attention than has the choice of strategy.

To describe strategic choice, Miles and Snow (1978, p. 29) developed a four-category typology: defenders, prospectors, analyzers, and reactors. In their typology, the organization is analyzed as an integral and dynamic whole, taking into account the interrelationship among the strategy, the process, and the structure. Miles and Snow's categories are characterized as follows:

- *Defenders*: organizations that have narrow product-market domains. Top managers in this type of organization are highly expert in their organization's limited area of operation but do not tend to search outside of their domains for new opportunities. As a result of this narrow focus, these organizations seldom need to make major adjustments in their technology, structure, or methods of operation. Instead, they devote primary attention to improving the efficiency of their existing operations.

- *Prospectors*: organizations that almost continually search for market opportunities and regularly experiment with potential responses to emerging environmental trends. Thus, these organizations often are the creators of change and uncertainty to which their competitors must respond. However, because of their strong concern for product and market innovation, these organizations usually are not completely efficient but they are effective.

- *Analyzers*: Organizations that operate in two types of product-market domains - one relatively stable, the other changing. In their stable areas, these organizations operate routinely and efficiently through use of formalized structures and processes. In their more turbulent areas, top managers watch their competitors closely for new ideas, and then they rapidly adopt those that appear to be the most promising.

- *Reactors*: Organizations in which top managers frequently perceive change and uncertainty occurring in their organizational environments but are unable to respond effectively. Because this type of organization lacks a consistent strategy-structure relationship, it seldom makes adjustment of any sort until forced to do so by environmental pressures.

Miles and Snow viewed their categories as being points on a scale going from defenders to prospectors with the analyzers in between. They asserted (p. 30) "we believe that our formulation specifies relationships among strategy, structure, and process to the point where entire organizations can be portrayed as integrated wholes in dynamic interaction with their environments." Indeed, the empirical evidence supports their claim.

To validate the strategy typology, researchers have performed empirical research using cluster analysis. The clusters are generally few in number and match the Miles and Snow typology remarkably well (Smith et al., 1989; Roth and Miller, 1990).

It may be useful, however, to develop the typology further. Nicholson Rees, and Brooks-Rooney (1990) present a typology with the following five categories:

- *Defenders*: Organizations whose strategy is to produce efficiently a limited set of products directed at a narrow segment of the total potential market.
- *Prospectors*: Organizations whose strategy is to find and exploit new products and market opportunities.
- *Analyzers*: Organizations whose strategy is to move into new products or new markets only after their viability has been shown, yet they maintain an emphasis on their ongoing products. Analyzers have limited innovation and the innovation is related to the production process and generally not to the product.
- *Hybrids*: Organizations that combine the strategy of the defender and the prospector. They move into production of a new product or enter a new market only after viability has been shown. But unlike analyzers, they do have innovations that run concurrently with their regular production. They have a dual technology core.

- *Reactors*: A residual strategy that describes organizations that follow inconsistent and unstable patterns.

Table 8.2. Miller's Integrative Framework

Strategic Dimension	Challenge	Predicted Structural Characteristics
Innovation	To understand and manage more products, customer types, technologies, and markets	Scanning of markets to discern customer requirements, low formalization, decentralization, extensive use of coordinative committees and task forces
Market differentiation	To understand and cater to consumer preferences	Moderate to high complexity, extensive scanning and analysis of customer's reactions and competitor strategies, moderate to high formalization, moderate decentralization
Breadth	To select the right range of products, services, customers, and territory	
- innovation,		High complexity, low formalization, decentralization
- stability		High complexity, high formalization, high centralization
Cost control	To produce standardized products efficiently	High formalization, high centralization

Source: Adapted from Miller (1987b, p. 55-76).

Nicholson, Rees, and Brooks-Rooney (1990) tested their categories on only a very limited set of cases, but their categorization fits well with the categories developed by Miller (1988b; 1987a; 1987b). Miller combined the categories developed by Porter (1980) and Miles and Snow (1978). His results are summarized in Table 8.2.

Miller's and Nicholson, Rees, and Brooks-Rooney's categories can be compared as follows:

Miller	Nicholson, Rees, and Brooks-Rooney
Complex product innovation	Prospectors
Breadth–innovation	Hybrids

Breadth–stability   Analyzers
Cost control   Defenders

Smith, Gutherie, and Chen (1989) did a cluster analysis investigating the strategic behavior in forty-seven electronic firms. Their purpose was to test the validity of Miles and Snow's (1978) typology. They identified four clusters and three of the clusters could easily be identified with prospectors, analyzers, and reactors. However, for the fourth cluster, it looked like the defender but with more product diversity. Small "defenders" outperformed large "defenders" in their study. Comparing the empirical results by Smith, Gutherie, and Chen (1989) with the typologies developed by Nicholson, Rees, and Brooks-Rooney (1990), one might argue that there are two types of defenders–those with a narrow market diversity, low cost, and little environmental monitoring and those that have a more broad product diversity, which comes from copying successful markets and products from other companies within the same industry. These latter companies have no innovation.

Table 8.3. Strategy Content Scores

Metavariable: Content	D	A	P	R
Technological progress	64	64	77	20
Product/market breadth	10	70	78	32
Product innovation	11	61	99	17
Quality	81	72	75	20
Price level	5	57	94	55
Active marketing	36	74	92	18
Control system level	100	58	24	18
Resources level	55	70	64	28
Investment in production	86	60	32	43
Number of technologies	1	46	96	37
Professionalization	22	62	95	20

Note: D = defender, A = analyzer, P = prospector, and R = reactor.
Source: Adapted from Segev and Gray (1990, p. 255-256).

Segev and Gray (1990) developed eleven metavariables to describe what they called *strategic content*. They used nine different typologies to describe the organization's strategy and strategic process. For each typology they assigned values for each metavariable that would best fit a category in a given typology. For the Miles and Snow (1978) typology, they had the scores shown in Table 8.3. In Table 8.3 eleven metavariables describe the strategic category of the organization. Scores range from 0 to 100 with 50 the mean.

Table 8.3 shows that a defender is high in technological progress, quality and investments in production. Further the control system has a high priority. A prospector scores high on technological progress, product market breath, product innovation and quality. Further a prospector can achieve a high price level and have high profile on marketing. The professionalization is at the highest level as is the number of technologies. The analyzer falls in between these two strategies. Some of these scores fit the Miles and Snow very well, while others are a little surprising.

Hambrick (1983) analyzed functional attributes similar to those shown in Table 8.3. In general, his results support the scores in Table 8.3, except for the price-level metavariable. Hambrick did not find a significant difference in the price level between defenders and prospectors.

- *Prospector*: an organization that almost continually searches for market opportunities and regularly experiments with potential responses to emerging environmental trends. Thus, the organization often is the creator of change and uncertainty to which its competitors must respond. However, because of its strong concern for product and market innovation, it usually is not completely efficient.
- *Analyzer with innovation*: an organization that combines the strategy of the defender and the prospector. It moves into the production of a new product or enters a new market after viability has been shown. But contrary to an analyzer without innovation, it does have innovations that run concurrently with the regular production. It has a dual technology core.
- *Analyzer without innovation*: an organization whose goal is to move into new products or new markets only after their viability has been shown yet maintains an emphasis on its ongoing products. It has limited innovation related to the pro-

duction process and generally not the product.

- *Defender:* an organization that has a narrow product market domain. Top managers in this type of organization are highly expert in their organization's limited area of operation but do not tend to search outside their domains for new opportunities. As a result of this narrow focus, these organizations seldom need to make major adjustments in their technology, structure, or methods of operation. Instead, they devote primary attention to improving the efficiency of their existing operations.

- *Reactor:* an organization in which top management frequently perceives change and uncertainty occurring in their organizational environments but is unable to respond effectively. Because this type of organization lacks a consistent strategy or structure relationship, it seldom makes adjustment of any sort until forced to do so by environmental pressures.

We have used a five-category typology to describe the strategic behavior of the organization. Prospector, defender, and reactor which follow the usual definitions by Miles and Snow (1978). The analyzer strategy is divided into two categories: (1) analyzers with innovation, which is defined as the hybrid category by Nicholson, Rees, and Brooks-Rooney (1990) and (2) analyzers without innovation, which is defined as the analyzer category by Nicholson, Rees, and Brooks-Rooney (1990). The analyzer without innovation category can be seen as an extension of the defender strategy. The category also fits the breadth-stability category, which is one of Miller's (1988c) typology categories. Before we move on, we want to compare this framework with other well-known schemes.

Chandler (1962) implicitly used a product based strategy typology. He argued that a functional organization performed well for a firm with few products; a divisional organization performed better for a firm with a large number of products. The strategy scheme is then based on the number of products: a small number and a large number. Through historical studies he demonstrated that the firm got into coordination and control problems when the number of products grew large and the firm continued with a functional organization. The Miles and Snow typology does not consider the number of products as a primary dimension of strategy, but focuses more on the innovation and the attitude toward the market.

Ansoff (1965) focused on synergy and the implications for diversification in strategy. Diversification is related to the number of products - the greater the diversification, the greater the number of products. The Miles and Snow typology is not explicit on the level of diversification. We might argue that the hybrid strategy of blending is a form of diversification. But this is a different notion than Ansoff's diversification of products and markets.

Porter's (1980) strategies of differentiation, cost leadership and niche are more closely related to the Miles and Snow typology. The differentiation strategy focuses on the product or service characteristics. Innovation is one approach to obtain and maintain products which are different. The Miles and Snow categories of prospector and analyzer with innovation also focus on uniqueness of product, which permits the firm to obtain higher profit margins, which is a return for higher risk. Cost leadership is a defender strategy, where low cost permits a profitable low price for the product and fends off imitators and competition. The niche strategy focuses on a small segment of the market with unique customer needs. The Miles and Snow prospector may be a niche player, but from a different point of view. The prospector is more driven by technology and what is possible rather than a direct orientation to the customer needs. Yet, in both, we can see a need for innovation and uniqueness.

Beginning with the prospector and moving to a defender, the prospector is an environment-oriented, risk-taking organization, frequently an inefficient organization with a zealous spirit for the "new"; the defender is an internal-oriented, risk-avoiding, cost-conscious organization that evades the "new." The two analyzers fall in between. The analyzer with innovation is more like the prospector in its external orientation, but it accepts the "new" only after careful consideration. The analyzer without innovation only copies what is known to work; it is not as dogged in its protection of its turf as the defender. Examples abound.

3M Corporation is a prospector. It exists to develop and market new products. It thrives on the new, using technology and developing commercial application.

Traditionally, IBM has been an analyzer with innovation. IBM simultaneously defended its position in the marketplace, carefully developed new products as demanded, but was infrequently first to the market. IBM has a large R&D effort as a base for its innovation.

Matsushita can be categorized as an analyzer without innovation. Matsushita is a Japanese-owned and -managed global manufacturer and distributor of consumer electronics. Panasonic and National are well-known brands. Matsushita enters markets that are established but not exhausted, with high-quality low-cost products. It will undertake process development to reduce costs.

Coca-Cola, in recent years, can be called a defender. It defends its position as the top worldwide purveyor of cola drinks. Frequently, it is pitted against Pepsi. The strategy is nonetheless the same; it uses advertising and promotion to attract customers and maintain market share.

These five categories are used in the next section when we develop the propositions that set the relationships between strategy and structure.

STRATEGY AS A CONTINGENCY

We recommend that an organization adopt "structure follows strategy." Describing and categorizing the strategy is the first of a two-stage process. Specifically, the two stages comprise: (1) describing the strategy in detail and the categorizing of the strategy into one of the modified Miles and Snow(1978) five categories, and (2) recommending a structure for the strategy. The first stage is descriptive; we want to state the strategy. The second stage is normative; we recommend an organizational structure for the given strategy - structure follows strategy.

Figure 8.1. Two-Stage Model on Strategy

Strategy Measures		Structural Properties
Product innovation		
Process innovation	**Prospector**	Configuration
Product and market breadth	**Analyzer with Innovation**	Complexity
Concern for quality	**Analyzer without Innovation**	Centralization
Price level	**Defender**	Formalization
Control level	**Reactor**	Coordination and Control
Technology		Incentives
Barrier to entry		

The two-stage model is given in Figure 8.1. Stage 1 describes the strategy in more basic terms and categorizes that strategy. The second stage takes the strategy and then recommends an organizational structure.

Miles and Snow (1978) argued for a fit among environment, strategy, and organization. They found that a defender would fit best with a stable and simple environment and an organization that is rather mechanistic, while a prospector at the other extreme would fit a dynamic and complex environment and would require a more organic organization. Miller (1987b;   1989) in his empirical investigations basically found the same results. We refer to his results in more detail in the next sections. Doty *et al* (1993) support the Miles and Snow typology relating it to organizational effectiveness. The typology seems to be rather robust and a good way to categorize strategic behavior in many industries such as: banking (James and Hatten, 1994), transportation (Murphy and Daley, 1996), health (Byles and Labig, 1996), service (Rajaratnam and Chonko, 1995), and biotechnology (Weisenfeld-Schenk, 1994) among others.

The modified Miles and Snow typology provides the basic strategy concepts and framework for structural choices. First, strategy is categorized from a description of the organization's innovation, barrier to entry, price level, product and market breadth, technology, concern for quality, and control level. Second, strategy helps prescribe the appropriate structure for the organization.

Each of the strategy descriptors must be assigned a value or category. For example, product innovation may be high or low. An organization, which develops and introduces new products on a regular basis, would be high, whereas an organization that seldom has a new product would be low on innovation.

Process innovation is perhaps more difficult to observe than product innovation, but the same concept holds. An organization, which changes technical and administrative procedures, has high process innovation. An organization, which seldom changes, has low process innovation. In the car industry with the introduction of flexible manufacturing systems, process innovation is high. Product/market breadth is defined according to number of products and number of markets.

Coca-Cola has few products - it sells carbonated soft drinks of various kinds in bottles and cans. 3M has a broad product/market;

a local company with few products does not have a broad product/market.

The concern for quality is an attitude about quality. Does the organizational value focus on delivering high quality products and services? This concern is not only a management issue, but also one that must permeate the entire organization if it is to be realized. 3M, IBM, Matsushita and Coca-Cola - all have a high concern for quality. Today, General Motors and SAS have a much higher concern for quality than they did in the 1970s. In general, concern for quality has gone up in the last decade.

Price level is relative; price level is high if it is higher than the average price in the industry, a low price level is lower than the average. 3M, IBM, and Coca-Cola have consistently commanded high prices for their products in competitive markets. But Matsushita is a low price competitor - good quality at a low price.

Control level is a measure of preferred level of control. A focus on short-term detailed feedback is a high level of control. A focus on long-term general trends is a low level of control. The technology dimension that is used to define strategy is the routineness dimension defined in Chapter 7. And finally, barrier to entry includes monopoly situations, regulations and high capital requirement. A high capital requirement is a high capital-to-labor ratio which is one way the barrier to entry can be high. A steel mill has high capital requirement while a consulting firm does not. 3M also has low capital requirement, whereas Matsushita scores high on the same dimension

There are alternative ways to determine an organization's strategy and categorize it into prospector, analyzer, defender or reactor. Perhaps the most straightforward approach is self-typing; simply ask managers, "What is your strategy type?" Grønhaug and Falkenberg (1989) indicate that managers frequently have a bias in judging their own strategy. The manager may report his own strategy as analyzer, where a competitor sees it as a defender. Second, an outsider–competitor or expert–can also be used to indicate an organization's strategy. Finally, strategy descriptors as discussed above can be used.

In most empirical studies self-categorization or expert categorization has been used. But as managers have difficulties consistently categorizing their own company strategy, we have introduced a choice. The strategy can be categorized using the strategy descriptors or the strategy can be categorized by an ad hoc procedure.

Describing a Prospector

We begin with a description of a prospector. A prospector is likely to have a number of characteristics. At least one of these characteristics must be present, and the greater the number of operative characteristics, the more likely the strategy will be a prospector. A prospector is likely to have high product innovation. Miller (1987b) found a significant correlation, and the Segev and Gray (1990) scores also indicate that a prospector has a high product innovation. The hypothesis is generally supported by the literature. 3M is a product innovator and a prospector. It cannot sustain its new product strategy without product innovation. However, an organization can have high product innovation and not be a prospector. Xerox is an example of a product innovator that was not a prospector. Xerox developed much of the PC basic technology. Yet it failed to exploit the unknown market. Xerox focused more narrowly on the known but changing copier market and missed the personal computer revolution. Product innovation is an activity for prospectors and is a primary characteristic. The high requirement for innovation requires either a prospector or an analyzer with innovation strategy.

An organization with many products is a prospector. A prospector is constantly seeking new product opportunities to serve existing and potentially new customers and usually has an array of products at any one time.

A prospector strategy that is aggressive in product development or market opportunity exploitation requires high capital and also skilled individuals. A nonroutine technology is also consistent with a prospector. A routine technology will not generate the required new products. A nonroutine technology is likely to be costly. If the organization engages in product development where margins are likely to be high, nonroutine technology is very reasonable.

The basis for a prospector is often to exploit an area where competitors have a difficult time to follow. This includes a high barrier to entry situation.

Segev and Gray (1990) put a high score on the price level; there is a high association. A prospector is not likely to survive with a low price as it is unlikely to generate sufficient revenue; other strategies may also require or allow high prices. Hambrick (1983) did not find that price level is a significant descriptor of strategy.

A concern for high quality is consistent with a prospector strategy. This fits the results by Hambrick (1983). However, Segev and Gray (1990) did not have significant scores on the quality dimension. This relation is not as strong as the price relation above. It is possible to innovate with new products that are not necessarily of the higher quality nor produced in a high-quality process.

There are arguments to support the low-control prospector relation. A prospector strategy requires that risks be taken and new ideas be generated freely. This can be obtained only if the management prefers a relatively low level of control. This relation is supported by Segev and Gray (1990).

Thomas, Litschert, and Ramaswamy (1991) found that firms with a greater alignment between strategy and the profiles of top managers generally realized superior performance outcomes. Their definition of manager profiles is not the one we have used, but they are related. As we do, they emphasize that the level of risk aversion is one critical dimension of the manger profile. The proposition that relates product innovation with a prospector strategy is the most important one. It is nearly definitional where the other propositions are more indicators of a prospector.

Describing a Prospector

If product innovation is high, then the strategy is likely to be prospector.

If the organization has many products, then the strategy is likely to be prospector.

If barrier-to-entry is high, then the strategy is likely to be prospector.

If technology is nonroutine, then the strategy is likely to be prospector.

If price level is high, then the strategy is likely to be prospector.

If concern for quality is high, then the strategy is likely to be prospector.

If the preferred level of control is low, then the strategy is likely to be prospector.

Prospector Effects on Structure

We now recommend some organizational design properties that are required for the prospector to be effective. The prospector should have low formalization. Prospector strategy requires a low formalization so that the organization can react quickly to the new situation and not be constrained by rigid rules. Low formalization is required because of the need for innovations (Miles and Snow, 1978; Fredrickson, 1986; Miller, 1987a). Complex information has to be processed, and task uncertainty is high, which from the information-processing argument leads to low formalization.

Complexity should be either low or high when the organization has a prospector strategy. It should either have generalists or specialists to deal with the high requirements for innovation (Miles et al., 1978; Fredrickson, 1986; Miller, 1987a) Again, demand and capacity of information processing have to be balanced.

For an organization that has a prospector strategy, a low centralization is required so that the organization can react quickly. Low centralization is also required because of the need for innovation (Miles et al., 1978; Fredrickson, 1986; Miller, 1987b).

Finally, a prospector strategy should be configured as a simple, ad hoc, or matrix organization but not a bureaucracy–machine or professional. Miles and Snow (1978), Miller (1987b, Table 24.2) and Daft (1992, Exhibit 14.4) all support these propositions. The basic idea is that if the organization has a prospector strategy, then a configuration that will enable it to react fast and engage is innovation as necessary. It may also be possible to support a prospector strategy with a virtual configuration.

Coordination and control will be rather complex with high media richness. Meetings, direct supervision and the use of IT can be used as means for communication. Rewards should be based on results as the product innovation has a high priority.

Prospector Effect on Structure

If the strategy is prospector, then formalization should be low.

If the strategy is prospector, then complexity should be low or high.

If the strategy is prospector, then centralization should be low.

If the strategy is prospector, then configuration should be either a simple, an ad hoc, or a matrix configuration. It may be virtual network.

If the strategy is prospector, then configuration should be neither a professional bureaucracy nor a machine bureaucracy.

If the strategy is prospector, then meetings, direct supervision, and the use IT can be used. Communication should have high media richness.

If the strategy is prospector, then the incentives should be results based.

Figur 8.2. Prospector Strategy: Description and Effect

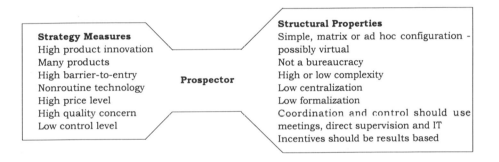

Strategy Measures		Structural Properties
High product innovation		Simple, matrix or ad hoc configuration - possibly virtual
Many products		Not a bureaucracy
High barrier-to-entry	**Prospector**	High or low complexity
Nonroutine technology		Low centralization
High price level		Low formalization
High quality concern		Coordination and control should use
Low control level		meetings, direct supervision and IT
		Incentives should be results based

Describing a Defender

The defender is identified by a number of characteristics. At least one of these characteristics must be present, and the greater the number of operative characteristics, the more likely the strategy is to be a defender.

We begin with low product innovation. The low-innovation requirement fits the defender strategy where the organization can utilize its particular expertise.

A high process innovation is a focus that tends to drive out other potential competitors. This fits well with an organization that is protecting its markets and products: the defender. The realized efficiency and cost reduction can yield a competitive advantage.

A defender normally will have few products. It needs to defend these products well in the marketplace and with an appropriate technology. Viability depends on being successful with these limited activities. For a company with a high capital investment, the ability to adjust its capital base quickly is not likely. Thus, it needs to protect and defend its position; a defender strategy and technology protection is appropriate.

A defender usually has a high capital investment that creates a high barrier to entry. A potential new competitor faces a large capital investment against an established firm that is protected by capital intensive highly efficient production means and an established market position as well.

A defender is characterized by a routine technology. Consequently, new products for new customers are less likely to be possible. It needs to defend its position for the technology it has or copy well-known products or markets. A mass production can keep the costs low.

A defender uses high quality to keep its markets and keep competitors out. The defender will have a high concern for quality.

A low price is a sign of a defender. To defend its position the organization needs generally to be competitive both on quality and prices. The efficient operation makes it possible to defend on prices, and this is normally the case when the organization produces a standard product. If the organization defends a high-quality brand, like LEGO, then a high price can be charged.

Finally a high level of control and a focus on doing things well signals an element in a defender strategy.

Describing a Defender

If product innovation is low, then the strategy is likely to be defender.

If process innovation is high, then the strategy is likely to be defender.

If the organization has few products, then the strategy is likely to be defender.

If the barrier-to-entry is high, then the strategy is likely to be defender.

If technology is routine, then the strategy is likely to be defender.

If technology is mass production, then the strategy is likely to be defender.

If concern for quality is high, then the strategy is likely to be defender.

If price level is low or high, then the strategy is likely to be a defender.

If preferred level of control is high, then the strategy is likely to be a defender.

Defender Effects on Structure

The effects of a defender strategy on the organizational structure are recommendations that follow the general proposition that "structure follows strategy". Given a focus on maintaining the status quo, the demand for information-processing capacity is low and the task uncertainty is low. A defender strategy generally requires high formalization to fit the routine technology. The reason is that a defender needs cost efficiency, which can be obtained through formalization (Miles et al., 1978; Miller, 1987b).

A defender strategy generally requires medium to high complexity. The reason is again that a defender needs cost efficiency, which can be obtained through specialization (Miles et al., 1978). Empirical results have shown that defenders may also have low complexity for small companies (Miller, 1987b).

A defender strategy usually requires a high centralization. Again, the reason is that a defender needs cost efficiency, and that can be obtained through centralized coordination (Miles et al., 1978; Miller, 1987b).

Finally, a defender strategy usually requires a functional configuration. The emphasis on specialization and efficiency is consistent with a defender's goals and its competitive position. Coordination and control should be based on planning and budgeting. Vertical information systems may be appropriate. Incentives are based on efficiency and should thus be procedural based.

The defender two-stage model is summarized in Figure 8.3. The first stage gives the description of a defender, and the second stage states the structural recommendations.

Defender Effect on Structure

If the strategy is defender, then formalization should be high.

If the strategy is defender, then complexity should be medium or high.

If the strategy is defender, then centralization should be high.

If the strategy is defender, then the configuration should be functional.

If the strategy is defender, then the coordination and control should be based on planning and budgeting e.g. by use of a vertical information system.

If the strategy is defender, then incentives should be based on procedures.

Figure 8.3. Defender: Description and Effects

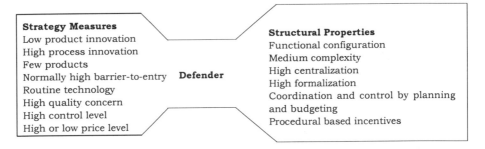

Strategy Measures
Low product innovation
High process innovation
Few products
Normally high barrier-to-entry **Defender**
Routine technology
High quality concern
High control level
High or low price level

Structural Properties
Functional configuration
Medium complexity
High centralization
High formalization
Coordination and control by planning and budgeting
Procedural based incentives

Describing an Analyzer without Innovation

The two analyzer strategy types fall between the prospector and defender. Product innovation is an important element in defining an analyzer without innovation. The firm with this strategy copies proven ideas effectively and does not incur the cost of product innovation. It does, however, require selectivity in the choice of what to copy.

An analyzer without innovation will likely have many products. An analyzer without innovation searches the environment for new opportunities to maintain a reasonable level of product diversity and appropriate match with the market. The analyzer without innovation seeks new opportunities but also maintains its existing profitable position. It is important to handle the old products or services in a routine way to maintain quality. New products will be included in the routine operations quickly. Thus, a routine technology is characteristic of an analyzer without innovation.

Process innovation, however, is quite different. An organization with medium or high process innovation devises new ways to make a product or deliver a service. The focus is the process and making it more efficient. Usually, the product or service is known. It is not likely that an organization can accomplish simultaneously the process development and matching new products. Rarely, a process innovation will change the product. A more likely approach is that the organization sees a product with market potential; it copies the product and attempts to become efficient and competitive through process innovation.

Low or medium capital requirement is consistent with an analyzer without innovation. An organization with a medium capital investment is likely to have some fixed capabilities but can adjust also. Thus an analyzer without innovation usually has a situation with a low to medium barrier-to-entry.

For an analyzer without innovation, cost efficiency is important. A routine technology that fits the mass production is appropriate and may be necessary. When prices are low, the organization must be cost effective to survive. For an imitation strategy, the firm can maintain low costs and be profitable with low prices. This strategy focuses more on cost and quality, and it is, thus, appropriate to require a high level of control.

Describing an Analyzer without Innovation

If product innovation is low, then the strategy is likely to be analyzer without innovation.

If the organization has many products, then the strategy is likely to be analyzer without innovation.

If barrier-to-entry is medium or low, then the strategy is likely to be analyzer without innovation.

If technology is routine, then the strategy is likely to be analyzer without innovation.

If process innovation is medium or high, then the strategy is likely to be analyzer without innovation.

If technology is mass production, then the strategy is likely to be analyzer without innovation.

If price level is low, then the strategy is likely to be analyzer without innovation.

If concern for quality is high, then the strategy is likely to be analyzer without innovation.

If preferred level of control is high, then the strategy is likely to be analyzer without innovation.

Analyzer without Innovation Effects on Structure

An analyzer without innovation looks very much like a defender. An analyzer does not invent new things but copies when it is safe. The Danish charter-tour-operator Spies is a very good example of an analyzer without innovation. The company never developed a new destination. It waited, and when some destinations seemed to be ready for mass-marketing, it moved in a highly formalized way. It marketed low prices and high quality. Compared to a prospector, the demand for information processing was lower and less complex.

An analyzer without innovation should have a medium or high formalization. This organization has a need for limited change and in a reasonable time. Internal efficiency is a goal, and formalization is one means to achieve the economics of regularity in process.

The top management can process the required information and can make decisions that are coordinated and consistent in the organization's goals. Thus a medium or high centralization is recommended. High specialization is required to be efficient as an analyzer without innovation. Additionally, the organization should operate both current and new products and services. This generally leads to a high organizational complexity.

A functional configuration is recommended for an analyzer without innovation. The functional configuration captures the efficiencies of specialization when the need for adjustment is slow and reasoned. For an analyzer without innovation, a functional organization is normally best. If the technological diversity is low due to low economies of scale, then a divisional configuration may be appropriate. However, if the organization is small, a simple structure may do.

Coordination and control should be based on planning and budgeting. It should also involve forecasting and liaison activities. Vertical information systems may be appropriate. Incentives are based primarily on efficiency, but also on effectiveness. Incentives should thus be procedural based with elements related to results.

Analyzer without Innovation Effect on Structure

If the strategy is analyzer without innovation, then formalization should be medium or high.

If the strategy is analyzer without innovation, then centralization should be medium or high.

If the strategy is analyzer without innovation, then complexity should be high.

If the strategy is analyzer without innovation then the configuration should be functional or divisional.

If the strategy is analyzer without innovation, and the organization is small, then the configuration should be simple.

If the strategy is analyzer without innovation, then the coordination and control should be based on planning, forecasting and budgeting e.g. by use of a vertical information system.

If the strategy is analyzer without innovation, then incentives should be based on procedures and to some extent on results.

Figure 8.4. Analyzer without Innovation: Description and Effect

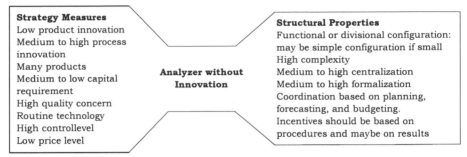

Strategy Measures	Analyzer without Innovation	Structural Properties
Low product innovation Medium to high process innovation Many products Medium to low capital requirement High quality concern Routine technology High controllevel Low price level		Functional or divisional configuration: may be simple configuration if small High complexity Medium to high centralization Medium to high formalization Coordination based on planning, forecasting, and budgeting. Incentives should be based on procedures and maybe on results

Describing an Analyzer with Innovation

The characteristics of an analyzer with innovation are more towards the prospector than the defender. A high product innovation firm is likely to be an analyzer with innovation, but to be efficient it also has to have some process innovation. With many products, the firm can maintain its diversity by developing new products, which is also consistent with an analyzer with innovation.

An organization with a medium capital investment is likely to have some capabilities rather fixed but can adjust also. Thus a medium barrier-to-entry situation is consistent with an analyzer with innovation.

For a medium routine technology, the organization has some flexibility. It is also consistent with an analyzer with innovation strategy.

The price level is medium for an analyzer with innovation. Moderate prices create some flexibility but also require attention to efficiency. An analyzer with innovation strategy both attends to existing profitable markets and seeks selective new opportunities.

A concern for high quality is also consistent with an analyzer with innovation. With a medium preference for control the management wants some influence. This can be obtained via control over current operations. Product innovation should then be less controlled.

Describing an Analyzer with Innovation

If product innovation is high and process innovation medium, then the strategy is likely to be analyzer with innovation.

If the organization has many products, then the strategy is likely to be analyzer with innovation.

If the barrier-to-entry is medium, then the strategy is likely to be analyzer with innovation.

If technology is medium routine, then the strategy is likely to be analyzer with innovation.

If price level is medium, then the strategy is likely to be analyzer with innovation.

If concern for quality is high, then the strategy is likely to be analyzer with innovation.

If the preferred level of control is medium, then the strategy is likely to be analyzer with innovation.

Analyzer with Innovation Effects on Structure

An analyzer with innovation strategy requires that formalization should be medium. There should be high levels of standardization in current activities and high levels of flexibility in new undertakings.

In an analyzer with innovation strategy, the organization is set up to deal with both new and current activities, which leads toward high complexity. An analyzer with an innovation strategy is actually a dual one. The current operations are different from new undertakings.

When the organization has an analyzer with innovation strategy, the centralization should be medium. There should be tight control over current activities and looser control over new ventures.

The analyzer with innovation requires a focus on change and efficiency. A matrix or divisional configuration is recommended except for a small organization where a simple configuration is appropriate. These propositions are developed from Miles and Snow (1978), Miller (1989, Table 24.2) and Daft (1992, Exhibit 14.4). An analyzer with innovation requires a rather complex set of means for coordination and control. Some planning is necessary. Meetings, direct supervision and the use of IT should also be used. Rewards should primarily be based on results as the product innovation has a high priority. Some rewards may be procedural based to maintain efficiency.

Figure 8.5. Analyzer with Innovation: Description and Effects

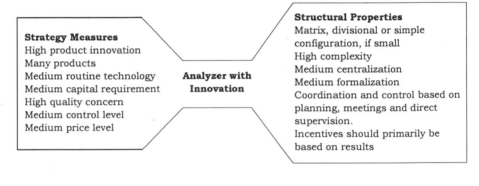

Analyzer with Innovation Effect on Structure

If the strategy is analyzer with innovation, then formalization should be medium.

If the strategy is analyzer with innovation, then complexity should be high.

If the strategy is analyzer with innovation, then centralization should be medium.

If the strategy is analyzer with innovation, then the configuration should be a matrix or divisional configuration.

If the strategy is analyzer with innovation and the organization is small, then the configuration should be simple.

If the strategy is analyzer with innovation, then planning, meetings, direct supervision, and the use IT can be used.

If the strategy is analyzer with innovation, then the incentives should primarily be result based.

Reactor Strategy

In the previous section we described four strategic types. For each of these types we have assumed there is consistency among the descriptive factors. But some descriptors may lead to more than one strategy categorization. For example, a high price level is consistent with both the prospector and the defender strategies. However, a consistent pattern must exist for each strategy to be viable. If that is not the case, the strategy is termed a reactor strategy. This is the case when half the descriptor variables would lead to a defender strategy while the rest would lead to a prospector strategy; consistency is absent.

A reactor is also normally characterized by a low concern for quality. In fact, it may be the only significant factor separating a reactor from the other categories. However, concern for quality does not mean the same for a prospector and a defender. Quality for a prospector means that the organization can react quickly to trends and changing needs and is flexible with an effective product innovation. Cost is normally not a major concern.

For a defender, quality concerns normally mean that the product has to meet standard specification; cost has to be low; and machinery has to be effective and efficient.

Reactor Proposition

If the descriptor factors in Figure 8.1 are inconsistent and quality concern is low, then the strategy is reactor.

INTERNATIONAL DIMENSIONS

The international dimensions of a strategy do not affect the choice of basic structure per se. Of course, if an international strategy means that the organization increases its market or product diversity that affects structural choice.

To further strengthen the hypotheses on the choice of configuration, we add four propositions related to international issues. They are based on whether the organization has a high or low foreign product or service diversity and how intensive foreign involvement is. The propositions are based on Phatak (1992) and Habib and Victor (1991).

A product and market diversity may lead to divisionalization. Again, an information-processing perspective can be used as an argument. The more diversity, the more demand for information processing capacity. The higher the involvement in international markets or production the more important it is to reduce the information-processing demand by creating units that can internalize the coordination requirements. When that is not possible, a matrix configuration is needed. In fact, the choice of an information system with its channels and capacity is a strategic choice.

International Configurations Propositions

If foreign product or service diversity is low and foreign involvement is low, then configuration should be either an international functional configuration or an international divisional configuration.

If foreign product or service diversity is medium and international involvement is medium, then configuration should be a global configuration.

If foreign product or service diversity is low and international involvement is high, then configuration should be a global functional configuration.

If foreign product or service diversity is high and international involvement is high, then the configuration should be a multidimensional, global configuration.

The contingency theory of international configuration has not been researched to the same degree as our basic configurations. The strengths of the proposition on internationalization are therefore offered with caution (Habib and Victor, 1991).

CHOOSING THE RIGHT STRATEGY

Bon Goût had to figure out what to do to solve the crisis; it had to select a strategy. The strategy of Bon Goût could be described by accessing the values in the descriptive part of Figure 8.1 - finding the value for innovation, product and market breadth, quality, control, technology, and capital requirement.

From an organizational point of view the "structure follows strategy" hypothesis, as discussed previously, would be to choose values for the descriptor variables that could result in a strategy type that fits with the organizational structure.

The organization has its choice of the amount of product and process innovation to employ. Additionally, it can choose the number of products and markets. The choice could be the organizational evolution that was previously discussed going from exploiting the domestic market to diversifying into other markets, and finally, to diversifying into more products.

The choice of technology - including quality, control, and standardization - is a part of the strategic choice. Investments in long-term commitments like R&D equipment and market development are also a part of choosing the right strategy.

Generally, managing size, environment, and technology to fit the environment is a part of selecting a strategy. Choosing objectives, goals, and ways to obtain those goals requires choosing and implementing a strategy. The strategy may be chosen to fit the environment and technology, and then a structure is selected, or as discussed above, the strategy is chosen with the limits given by the organizational structure.

The literature on strategic choice is vast (Porter, 1980; Quinn et al., 1988), and a discussion of best choice lies outside the scope of this book.

SUMMARY

"Structure follows strategy" is a basic proposition. An organization's strategy is an important determinant of its organizational design.

A modified Miles and Snow strategy categorization is both comprehensive and discriminating. They are prospector, analyzer without innovation, analyzer with innovation, defender, and reactor. Each strategy calls for a different organizational design.

The prospector requires an open, flexible organization that adjusts quickly. The analyzer adjusts more slowly and deliberately. The defender needs a stable organization. In general, information-processing requirements decrease moving from prospector to defender; however, the defender needs particular information to fight off competitors. Prospectors need a lot of information to deal with an unknown environment and technology; everything can be in flux. Analyzers fall in between. We do not offer organizational recommendations for a reactor strategy, but rather suggest a reactor strategy is not a viable strategy in the long run. No organization can salvage it.

Strategy and structure should fit, and that fit is realized with a structure that processes the right amount of the right kind of information.

Chapter 9

Diagnosis and Misfits

INTRODUCTION

Strategic diagnosis is an assessment of the organization's strategic factors, structure and performance. Diagnosis identifies misfits - situations which yield diminished performance; design is devising structures that fit the strategic situation and result in good performance. Strategic organizational diagnosis is the assessment within the multi-contingency model presented in Figure 1.3 of the strategic factors of the organization, the contingency relations, and its configuration and properties.

SAS Misfits

In the previous chapters, the development of SAS from its creation to 2003 has been described. The first major strategic misfit occurred in 1975 when the environment changed due to the oil crisis. The reaction from SAS was to hire a new CEO Jan Carlzon in 1981. He made changes to bring SAS back to a new fit situation. He then went on to develop the strategy of SAS further creating a new misfit situation. Stenberg in his period worked to bring SAS back into a fit situation. It seemed to work although it was later revealed that unethical and illegal agreements with other airlines had been used to decrease the uncertainty of the environment.

Jørgen Lindegaard thought that he inherited a fit situation but the EU ruled that the secret agreements with some other airlines were illegal and a heavy penalty had to paid, the economic downturn, and the September 11 incident created a very severe misfit situation.

1. Which misfits was Carlzon facing in 1981?
2. Analyze the misfit situation as of 2001. Do you agree with the decisions that Jørgen Lindegaard made? Look at the SAS homepage - www.scandinavia.net to see the major news releases.

Total fit occurs when strategic fit, contingency fit, and design fit are all met. A good fit means better performance (Donaldson, 1987) than with one or more misfits. Fit is a matching process where the organization can be matched with its strategic factors. If any of the fit conditions are not met, then there is a misfit. This is supported by Burton, Lauridsen and Obel (2002) who demonstrated that organizations that have total fit perform better than organizations that have one or more misfits. As we argued in Chapter 6, an organization may choose a flexible structure that fits a turbulent environment, or it may manipulate its environment to reduce the turbulence. The organization wants to avoid or correct for the misfit of a nonflexible structure with a turbulent environment.

In this chapter[1], we present the background and logic of misfit diagnosis, the development of misfit rules, which are diagnostic and can be used by managers as part of the design process.

LITERATURE REVIEW ON MISFITS

In the development of a multi-contingency normative theory of organizational design, we utilized both fit and misfit notions. Fit can be categorized in terms of strategic, contingency and design fit - see Figure 1.3. The strategic fit involves the balance of the leadership style, the organizational climate, the size and skills, the environmental conditions, the strategy, and the technology. The research in strategy, climate, size, leadership, and technology provides the basis for strategic fit, while organizational contingency theory provides the support for the contingency fit rules. The multi-contingency of environment, technology, size, strategy and climate and leadership requires a fit of these factors with the characteristics of the organization, such as its level of decentralization, formalization, etc. Contingency fit is a match "between the organization structure and contingency factors that has a positive effect on performance" (Donaldson, 2001, p. 7, 10). Donaldson indicates that a "misfit produces a negative effect on organizational performance; structural change to a fitting structure is not usually immediate" (p 15). He discusses misfit

---

[1] This chapter relies heavily upon: Richard M. Burton, Jorgen Lauridsen & Borge Obel, *Returns on Assets Loss from Situational and Contingency Misfit*, Management Science, 2002, v. 48, n. 11, p. 1461 - 1485.

as a condition that calls for the organization to move back to fit (p. 165). Fit recommendations are triggered by misfit observations. For example, a fit recommendation that the organization should be decentralized is operative only if it is not currently decentralized. Fit recommendations are "should be" statements; misfit statements are "should not be." Other terms have been used for misfit: deviation, misalignment, incongruence, out of kilter, incompatibility, gap and so on.

Fit is an organizing concept for the creation and development of the knowledge base, i.e., the diagnosis and design rules. The development of the multi-contingency model for organizational design rests upon the proposition that a fit among the "patterns of relevant contextual, structural, and strategic factors" will yield better performance than misfit will (Doty et al., 1993). Our model builds on Miles and Snow strategic fit model (Miles and Snow, 1978), Mintzberg's structuring model (Mintzberg, 1979), Galbraith's star model (Galbraith, 1973; Galbraith, 1995), Nadler and Tushman's congruence model (Nadler and Tushman, 1984), Miller's environment, strategy and structure model (Miller, 1987), and Meyer, et al's configurational model, (Meyer et al., 1993) among others.

There is a large literature that tests alignment and fit relations for a few variables, whereas our model is a comprehensive model involving many variables. Fry (1982) showed that deviations from the fit/norms decrease performance. Gresov (1989), in his empirical study of the work units in employment-security offices, measured explicitly both fit and misfit conditions to test propositions on context-design and performance. He found strong support for the context-design proposition that an organic design is a fit with an uncertain environment, but mixed support that performance decreases from misfits. His results are consistent with equifinality in that functionality is the performance requirement, not the structure per se (Gresov and Drazin, 1997). That is, different structures in a given circumstance can yield performance sufficient for viability, although not necessarily optimal. Naman and Slevin (1993) devised and tested a normative fit model of entrepreneurship. They modeled fit among the entrepreneurship, organicity, and mission strategy and operationalized fit by a misfit measure of the sum of the absolute differences between desired levels and observed levels of entrepreneurship, organicity, and strategy. Using senior executives as respondents from 82 manufacturing firms, they tested the model and found that fit was positively related to performance. They called for additional studies on the synergy of fit. In their empirical study of savings and loan organizations under crisis, Zajac et al. (2000) tested a normative dynamic multi-contingency strategy fit/misfit model, which includes structural fit. That is, the environment, structure, and strategy should fit for better performance. Here, a strategic misfit is realized when needed strategic change is excessive, or insufficient. Studying Savings and Loan institutions (S&L) they found support for a number of contingency hypotheses, specifically, "the greater an S&L's deviation from the contingency model of change, the worse the performance" (p. 444). They identified normative implications on how a firm should change when the environment and structure create a misfit with the strategy. Thus, the understanding of interactions or fit among firm activities is important.discussing external vs. internal fit, Miller (1992) argued that it may be difficult to obtain concurrently the different kinds of fit, and that a sequential approach may be needed to obtain total design fit. This may include "periodical disrupting the harmony" to adjust to changes in the situation while more generally "striving for harmonious alignment." Misfits provide a managerial view of organizational design for a number of reasons. First, deviations from fit are easier to identify and measure than are conditions of fit. Second, misfits are managerial in orientation; managers react to misfits. Third, misfits come in smaller sets than do fit conditions. With equifinality, there are a very large number of fit conditions for a given organization, which yield reasonable performance outcomes. Therefore, it is both empirically and managerially convenient to focus on misfits. Based on the model depicted in Figure 1.3 we developed a set of extreme misfits that were the basis for an empirical analysis (Burton et al, 2002). This article is the basis for the strategic and contingency misfits that follow.

STRATEGIC MISFITS

Strategic misfits are performance-diminishing situations among (between) strategic factors in the multi-contingency model shown in Figure 1.3. In that model, there should be a congruency and balance among the strategic factors. For example, a highly uncertain environment fits with a prospector strategy, but is a misfit with a defender strategy; the defender cannot make the needed adjustments to changes in the environment as they occur.

Environment-Strategy

The first group of misfits deals with environment and strategy relationships. The general proposition is that the environment influences what are viable strategies for the firm. For example, a highly uncertain environment calls for a strategy that can adapt to the environment in a timely manner. These relations have been well studied (Venkatraman and Prescott, 1990; Miles et al., 1978; Porter, 1985). In Chapter 6, we defined four dimensions: complexity, uncertainty, equivocality, and hostility that could be used to describe the environment. Complexity is the number of variables in the environment; uncertainty is the lack of information about probability distribution of the value of the variables; equivocality is confusion and lack of understanding about the environment itself, and hostility is the malevolence of the environment. Strategies are categorized using the modified Miles and Snow categories (Miles et al., 1978; Burton et al., 2002) of prospector, analyzer with innovation, analyzer without innovation, defender and reactor as developed in Chapter 8.

A low uncertainty environment is a misfit with a prospector strategy. A low uncertainty environment can be predicted; the firm can plan, meet the competition, and realize efficiencies. There may be little need for new ideas and products. The prospector is an innovator who projects into the environment with new ideas and products, but these ideas do not necessarily meet current and known needs. The prospector strategy tends to have higher costs from innovation and it is not focused on efficiency. The prospector strategy is high risk, and can fail. Its market niche may emerge and evolve from its projection into the market (Miles et al., 1978, p. 56 - 58); some uncertainty yields this possibility.

The poorly understood equivocal environment is a misfit with a defender strategy. A high equivocality in the environment calls for a capability to vary products and services as the environment becomes clear. Without an innovative capability, it may be very difficult to adjust. A highly equivocal environment calls for a prospector or analyzer with innovation to adjust (Miles et al., 1978; Burton et al., 2002). The defender and analyzer without innovation cannot adapt quickly and perform better in an environment where good predictions can be made and realized - a low equivocality environment. Copying what others have done may be possible, but it is not likely to be viable for the long run. Again, without innovation, the organization is limited to copy what others have done. The organization needs to develop some innovative capabilities to adjust and adapt to the uncertainties in the environment.

A low equivocality environment is a misfit with an innovative strategy. There may be limited need for innovation and adaptation and probably limited opportunities to which to adapt. An analyzer without innovation, or a defender strategy, which focuses directly on the few environmental factors and meets market needs efficiently will usually yield better results (Miles and Snow, 1978, 36-38, 72-73). The prospector and the analyzer with innovation need some opportunity for new ideas and new products to realize success. They cannot meet the efficiency and low cost demands that are needed with known competition. The prospector and analyzer with innovation are not efficient and low cost providers, which is needed where competition is fierce in a low equivocal environment, i.e., low cost strategy (Porter, 1985, p. 12-14).

Propositions on Environment- Strategy Misfits

A low uncertainty environment is a misfit with a prospector strategy.

A highly equivocal environment is a misfit for a defender strategy.

A highly equivocal environment is a misfit for an analyzer without innovation.

A low equivocality environment is a misfit with a prospector strategy.

A low equivocality environment is a misfit with an analyzer with innovation strategy.

Technology-Environment

Technologies vary in their adaptive capacity (Perrow, 1967; Daft and Lengel, 1986; Scott, 1992). In Chapter 7, we categorized organizational technologies into routine and nonroutine. A routine technology is well-known and contains few exceptions with easy-to-solve problems; a nonroutine technology has many exceptions and more difficult problems (Perrow, 1967, p. 196). A routine technology is not adaptable and costly to change; a nonroutine technology is more adaptive to the requirements. Equivocality is confusion and lack of understanding about the environment. We propose two misfits between the organization's technology and the environment.

The routine technology will not adapt quickly or easily, where the high equivocal environment demands rapid change; it is a misfit. A routine technology fits well with a low equivocality environment as this environment is predictable, and a routine technology is sufficient to deal with the very limited change. Further, it is more likely to be efficient and meet low cost requirements of competition (Daft et al., 1986; Burton et al., 2002).

For this next proposition, the misfit is the opposite situation - nonroutine technology and low equivocality. Here, the nonroutine technology fits better with a higher equivocality, where adaptation is required (Daft et al., 1986; Burton et al., 2002).

Propositions on Technology - Environment Misfits

A routine technology is a misfit with a high equivocality environment.

A nonroutine technology is a misfit with a low equivocality environment.

Climate-Environment

Climate was categorized in Chapter 4 as internal process, group, rational goal and developmental (Zammuto and Krakower, 1991; Burton et al., 2002). The internal process climate is a formalized and a structured work place; the group climate is a friendly and familiar place; the rational goal is a hard driving and results oriented climate; and developmental climate is entrepreneurial and creative. The organizational climate is relatively enduring, influencing individual behavior and organizational outcomes (Glick, 1985, p. 607). We include four climate-environment misfits.

A group climate is tightly focused towards its own members, but is not focused on the environment, and has high resistance to change; thus, it cannot change quickly as required by the high equivocality environment (Hooijberg and Petrock, 1993; Burton et al., 2002).

An internal process climate is focused on its rules of action and behavior, and not on the organization's environment - greater focus on the inside of the organization than its outside environment. It has a high resistance to change and fits well with a low equivocality environment where requisite change is much less likely (Hooijberg et al., 1993; Burton et al., 2002). It is not likely to either see the shift, understand the need for change and does not have a capacity, which supports adaptation to such needed change. It is a misfit with the high equivocality where change is demanded.

A developmental climate encourages experimentation, adaptability, and risk taking; it has a broad range of potential behaviors. However, it is not likely to be very efficient, nor meet the competitive requirements of a low equivocality environment (Hooijberg et al., 1993; Burton et al., 2002).

Similarly, the developmental climate is a misfit with a low uncertainty environment that does not require adaptation (Hooijberg et al., 1993; Burton et al., 2002). The inefficiencies of the developmental climate cannot be incurred in a low uncertainty environment.

Propositions on Climate - Environment Misfits

Group climate is a misfit with a highly equivocality environment.

Internal process climate is a misfit with a high equivocality environment.

Developmental climate is a misfit with a low equivocality environment.

Developmental climate is a misfit with a low uncertainty environment.

Climate-Strategy

The strategy of an organization has to be aligned with views and feelings of those who implement it.

The internal process climate is a rules-oriented inwardly focused climate; it is well suited for a defender strategy where the focus on process is important. An internal process climate does not fit an analyzer with innovation or a prospector strategy. An internal process climate is focused on its own processes of managing with non-equitable rewards, high resistance to change, high conflict and low leader credibility, among other characteristics which lead to an emphasis on process control according to the procedures and rules. Innovative strategies require new products on a continuing basis with sense of experimentation and exploration. In most situations, the internal process will compromise the innovative strategy. To realize either strategy, the climate must change to one that will have a more external orientation. (Miles et al., 1978; Bluedorn and Lundgren, 1993)

The group climate reinforces its own values and ways of doing things; it is not very adaptive to external pressures. In contrast, the prospector strategy is an exploratory strategy that requires an organization to adjust quickly, thus, there is a misfit (Miles et al., 1978; Bluedorn et al., 1993).

A developmental climate explores and has low resistance to change (Bluedorn et al., 1993; Burton et al., 2002). Here, there is a misfit between a climate, which has low resistance to change, and a strategy that has a low need for change, but a high need for efficiency (Miles et al., 1978; Burton et al., 2002).

Propositions on Climate - Strategy Misfits

Internal process climate is a misfit with an analyzer with innovation strategy.

Internal process climate is a misfit with a prospector strategy.

Group climate is a misfit with a prospector strategy.

Developmental climate is a misfit with a defender strategy.

Technology-Strategy

Strategy and technology do not fit together when one is adaptive and the other is not (Miles et al., 1978). A routine technology is not adaptive; a prospector strategy is both experimental and adaptive (Miles et al., 1978, p. 56-57). The two do not fit (Miles et al., 1978; Burton et al., 2002). Similarly, a nonroutine technology and a defender strategy do not fit in the opposite way - the technology is adaptive, but the strategy is not (Miles et al., 1978; Burton et al., 2002).

Propositions on Technology - Strategy Misfits

Routine technology is a misfit with a prospector strategy.

Nonroutine technology is a misfit with a defender strategy.

Climate -Technology

Similar to previous argument the climate and technology do not fit when one is adaptive and the other is not. An internal process climate is internally and control oriented. It is rule-oriented and has high resistance to change; a nonroutine technology requires adaptation. A routine technology with a similar focus on implementing standard ways of doing things is a better fit for an internal process climate. An internal process climate is a misfit with a nonroutine technology (Zammuto and O'Connor, 1992; Burton et al., 2002).

The nonroutine technology requires adaptation and change; it fits well with a developmental climate. For a nonroutine technology, a climate with more flexibility would be better. The developmental climate can adapt and change, but the routine technology has a need for stability of process; hence, there is a misfit (Zammuto et al., 1992; Burton et al., 2002).

Propositions on Climate - Technology Misfits

An internal process climate is a misfit with a nonroutine technology.

A developmental climate is a misfit with a routine technology.

Technology-Leadership Style

The leadership style was in Chapter 3 categorized as leader, producer, entrepreneur, and manager. This categorization reflects the management's preference for being involved and taking risks in decision-making

A nonroutine technology is adaptive and requires managerial involvement to realize change (Miller et al., 1982). With limited time to process all the information in a nonroutine technology, the control-oriented manager will become overloaded and not be able to make the needed decisions in a timely manner for the non-small organization, thus creating a bottleneck. The manager's leadership style can work for a small organization where there are few decisions.

A routine technology is stable and does not readily support change and adaptation. The entrepreneur thrives on change, including new technologies and ways of doing things. With a routine technology, the entrepreneur will be overly restricted to deal with the uncertainty.

Propositions on Technology - Leadership Style Misfits

Nonroutine technology is a misfit with a manager leadership style, except in small organizations.

A routine technology is a misfit with an entrepreneur.

Climate - Leadership Style

The leadership style can also be a misfit with the climate of the organization (Hunt, 1991, p. 163). The leadership style should support the organizational climate.

A group climate utilizes its own intra-group relations to manage itself and maintain its stability. A group climate has a high degree of trust, low conflict, and medium to high morale, among others. It is internally oriented and has high resistance to change. An individual who wants to make decisions may threaten the trust and morale of the organization and actually increase the level of conflict. He/she may threaten the climate with his/her involvement and interference in the organization. Management does not need to be involved in details. A manager with his/her desire for decision-making involvement and need to avoid uncertainty can destroy the high morale, create conflict, and undermine management's credibility. Hence, there is a misfit between group climate and a manager (DiPadova, L. N. & Faerman, S. R. 1993). The group climate is also a misfit with an entrepreneur and leader style as both want to make decisions and a group climate, which permits the organization to set its own direction and make its own decisions.

A developmental climate has its own internal momentum and decisions are made widely and quickly. It is relatively flexible and externally oriented, characterized by low conflict, low resistance to change, and high leader credibility, among others. The uncertainty-avoiding individual generally also minimizes change. A manager will not be comfortable in a developmental climate and may introduce more control to reduce the uncertainty. Similarly, an entrepreneur will want to make decisions and change directions, which may be difficult.

An internal process climate is a rule-oriented climate that has low trust, a high degree of conflict and a high resistance to change. The leader and the entrepreneur will find the inertia and resistance to change difficult to deal with, and the low trust and high conflict are likely to become worse. This creates a misfit (DiPadova and Faerman, 1993; Burton et al., 2002).

Propositions on Climate - Leadership Style Misfits

Group climate is a misfit with a manager, leader, and entrepreneur.

Developmental climate is a misfit with a manager and entrepreneur.

Internal process climate is a misfit with a leader and entrepreneur.

Leadership Style - Environment

A manager is limited in his/her information processing capacity. The high equivocality environment generates a very large number of decisions and innumerable contingencies to consider (Galbraith, 1973). The manager will become overloaded and backlogged with decisions such that the contingencies of the high equivocal environment will not be considered in the relevant time frame. The information processing demands of the high equivocal environment will exceed the information processing capacity of the manager, and thus create the misfit.

Propositions on Leadership Style - Environment Misfits

A manager is a misfit with a high equivocality environment.

Leadership Style-Size Skill

The size-skill strategic factor was introduced in Chapter 5: size is the number of employees, which is modified by the skill level of the employees in the organization.

A manager and an entrepreneur must process a large amount of information, as each will be involved in detailed decision-making and control. The information processing demands are positively correlated with the size of the organization as the number of decisions and contingencies increase with the size. The information processing demands of the large organization are very large. It requires some degree of decentralization as the top managers cannot process the requisite information for a high degree of involvement. The leadership styles of a manager or an entrepreneur with low delegation is a misfit for a large organization where there are many decisions requiring great information processing demands.

Propositions on Leadership Style - Size Misfits

A manager or an entrepreneur is a misfit with a large organization.

CONTINGENCY MISFITS

Contingency fit is at the heart of contingency theory (Burns and Stalker, 1961; Thompson, 1967; Lawrence and Lorsch, 1967; Galbraith, 1973; Miles et al., 1978; Galbraith, 1995) and is fundamental to the multi-contingency theory as developed in Chapters 3 - 8. In Figure 1.3, the organizational relations include the strategic factors of environment, strategy, technology, size, leadership, and climate. These factors are related to the appropriate organizational configuration, such as functional and divisional as well as properties such as centralization, formalization, organizational complexity, and incentives. For example, a highly uncertain environment fits with an organization with a low degree of formalization, but this environment is a misfit with a highly formalized organization. The formalized organization is likely to be too rigid without the ability to adapt to changes in the environment. Below, we examine the contingency misfits for the multi-contingency model.

Environment Misfits

The core of the research of the fit between the environment and the organization, or the environmental imperative that the environment helps determine a good choice for the organization's structure, is a basic tenet in contingency theory (Burns et al., 1961; Lawrence et al., 1967; Duncan, 1972). As discussed above, the environment is measured along three dimensions: equivocality, uncertainty, and complexity, where the description is multidimensional. The organizational characteristics are centralization, formalization, complexity, and so on as shown in Figure 1.3. A particular organizational structure may fit well with a particular environment, but may be a misfit with others.

An environment that is low on all three dimensions of equivocality, complexity, and uncertainty is well understood; there are no unknown environmental properties. There are only a small number of important environmental parameters, and they are predictable. This environment fits well with an organization that operates smoothly and efficiently and where everyone has a known, well-defined job, and does it. It is a misfit with an organization, which has the capacity to deal with complex multidimensional issues in an informal and flexible manner. An informal organization is a misfit in a stable and simple environment; its lack of procedure leads to lack of direction and costly inefficiency in this environment (Burns et al., 1961; Lawrence et al., 1967; Lewin and Volberda, 1999). In addition, a matrix configuration with its multidimensional foci is a misfit as is low formalization.

Changing the environment from low to high complexity with low equivocality and low uncertainty requires a bureaucracy, albeit a complex one with many and complex rules. Formalization leads to efficiency. Here again, a matrix configuration and low formalization are misfits. Low organizational complexity is also a misfit; from the law of requisite variety, the internal complexity must be at least as large as the environmental complexity (Ashby, 1956). Thus, a low internal organizational complexity is a misfit for the high external environmental complexity (Burns et al., 1961; Lawrence et al., 1967).

Next, changing the environmental uncertainty from low to high, suggests decentralization and high organizational complexity. A simple organization and/or an organization with high centralization have limited information-processing capacity at the top and cannot deal well with high uncertainty and high complexity (Burns et al., 1961; Lawrence et al., 1967).

High equivocality requires an organization, which can adapt. Functional organizations and high formalization are efficient, but cannot change easily, as is required for high equivocality. The high equivocality is a misfit with a functional organization and high formalization (Burns et al., 1961; Lawrence et al., 1967).

Building on the argument above, the higher uncertainty makes the centralized decision-making inefficient, as information bottlenecks will be created (Burns et al., 1961; Lawrence et al., 1967; Galbraith, 1973).

If the environmental uncertainty and complexity are at their highest, a complex, informal or organic organization is desired. A mechanical organization such as a functional configuration, high formalization and high centralization is a misfit (Burns et al., 1961; Burton et al., 2002).

Propositions on Environment Misfits

An environment with a low equivocality, low complexity, and low uncertainty is a misfit with a matrix configuration and low formalization.

An environment with a low equivocality, high complexity, and low uncertainty is a misfit with a matrix configuration, low formalization, and low organizational complexity.

An environment with a low equivocality, high complexity, and high uncertainty is a misfit with a simple configuration, low organizational complexity, and high centralization.

An environment with a high equivocality, low complexity, and low uncertainty is a misfit with a functional configuration, and a high formalization.

An environment with a high equivocality, low complexity, and high uncertainty is a misfit with a functional configuration, high formalization, and high centralization.

An environment with a high equivocality, high complexity, and high uncertainty is a misfit with a functional configuration, high formalization, and high centralization.

Strategy Misfits

The fit between structure and strategy builds upon Chandler's (1962) proposition "structure follows strategy" (Naman and Slevin, 1993). When the structure and the strategy are misaligned, poor performance obtains. Here, the strategy misfits are grouped by the strategy:

prospector, analyzer with innovation and analyzer without innovation and defender (Miles et al., 1978).

A prospector strategy involves exploring new products and market opportunities. The very large information-processing demands on the organization can be met by a large number of individuals who can and will take action. Low formalization and low centralization are needed to move quickly (Miles et al., 1978, p. 61-64). A functional organization with its well-defined functions can be too narrow and limiting. High formalization narrows the possible responses of the organization and makes it not very adaptable. And, high centralization will lead to a bottleneck for the requisite information-processing (Galbraith, 1973). The prospector is then a misfit with the inflexible, centralized organization.

An analyzer with innovation both searches for new opportunities and then moves on to acceptable ones, and also keeps and maintains its existing business; it is a combined prospector and defender (Miles et al., 1978, p 68). This two dimensional focus, which is somewhat contradictory, requires organizational complexity to cope with this complex strategy (Ashby, 1956). An organization with low complexity cannot cope with the possibilities.

An analyzer without innovation is similar to a defender as it defends its existing business, but it also looks for new opportunities after others have demonstrated the business opportunity (Miles et al., 1978, p. 68). It follows others, but without innovation. There is little need for decentralization as the required coordination can suffer with multiple decision makers and it can be costly. There is a focus on the efficiency of specialization and associated rules of formalization. It is a misfit with low formalization, low organizational complexity, and low centralization.

A defender strategy focuses on keeping what the organization has (Miles et al., 1978) (p. 31). It has a limited number of products for a narrow market with an emphasis on efficiency. There is little need for variety in action and organizations, which are more complex and have higher information-processing capacity by introducing unneeded and unwanted potential variation. A matrix organization has excess information processing for the defender, and is a misfit. Low formalization and low centralization introduce unwanted variety. And a low organizational complexity is inefficient; greater efficiency can be obtained with a horizontal differentiation of tasks and the efficiency of specialization (Burton et al., 2002).

Propositions on Strategy Misfits

A prospector strategy is a misfit with a functional configuration, high formalization, high organizational complexity, and a high centralization.

An analyzer with innovation is a misfit with a low organizational complexity.

An analyzer without innovation is a misfit with a low formalization, low organizational complexity, and low centralization.

A defender strategy is a misfit with a matrix configuration, low formalization, low organizational complexity, and low centralization.

Technology Misfits

The technology imperative that the organizational structure must fit with the technology grew from Woodward (1965) and Perrow (1967). Routineness and divisibility are technology dimensions. Routineness is the degree to which today's activities are like yesterday's; and, divisibility is the degree to which the production process can be broken down into separable parts.

For a highly routine technology, the organization does not require large information-processing capacity; recent history creates the plan. High formalization and high complexity are desired (Woodward, 1965). The high information-processing capacities of a matrix configuration are not needed; they are costly and can lead to confusion. Low formalization and low complexity or low specialization can diminish the efficiency of routineness.

The complementary argument is that a nonroutine technology creates misfits for organizations that are limited in their information-processing capacity and are not adaptive to change. The functional organization, high formalization, high complexity, and high centralization limit the adaptive capacity of the organization and are misfits.

High divisibility requires relatively small information-processing capacity from the organization. A functional organization is a good fit (Woodward, 1965; Burton et al., 2002). A matrix organization has excess capacity to process information and coordinate activities that do not require coordination.

Low divisibility of technology requires integration and high information-processing capacity. A divisional configuration creates autonomous subunits or divisions, which cannot deal with technologies that are integrated across divisions, thus, creating a misfit.

Propositions on Technology Misfits

A highly routine technology is a misfit with a matrix configuration, low formalization, and low complexity.

A nonroutine technology is a misfit with a functional configuration, high formalization, high complexity, and high centralization.

A highly divisible technology is a misfit with a matrix configuration.

A non-divisible technology is a misfit with a divisional configuration.

Size Skill Misfits

Size misfits are grouped by large and small organizations. Size is measured by the number of employees, moderated by the skill level of the employees; higher skilled employees make the organization "larger" (Baligh et al., 1996p. 1658-1660).

A large organization needs to be compartmentalized into smaller administrative units (Whetten and Cameron, 2002) otherwise, the top of the organization becomes overloaded and cannot make decisions, direct activities and control the firm (Blau, 1970). It is a misfit with a simple configuration, or centralized decision-making. Further, the large organization benefits from specialization of task and formalized rules (Miller, 1987). We suggest that a large organization creates misfits when management attempts to run the organization informally without specialization of tasks. Briefly, there is too much for management to do - information overload and decision gridlock can occur (Blau, 1970; Galbraith, 1995; Burton et al., 2002).

In contrast, a small organization does not need the specialization and horizontal differentiation of high complexity, nor the formalization as the top management can make the decisions and coordination can be informal, rather than by rules (Whetten et al., 2002; Burton et al., 2002).

Propositions on Size Misfits

A large sized organization is a misfit with a simple configuration, low formalization, low complexity, and high centralization.

A small sized organization is a misfit with a high formalization and a high complexity.

Climate Misfits

An organization's internal climate is a "relatively enduring quality" that is experienced by its members and affects their behavior, and can be described as characteristics (Tagiuri and Litwin, 1968) (p. 27). Zammuto and Krackower found seven dimensions of climate that yielded four organizational climate types using a competing values framework: group, developmental, internal process and rational goal (Zammuto et al., 1991).

For a group climate, which is characterized as a "friendly place to work where people share much of themselves," low formalization, low centralization, non-high complexity and group incentives work well (Burton et al., 2002). The group itself wants to "run its own show" as determined by its own preferences and does not want to be constrained by internal rules and procedures, nor told what to do. It would also like to be independent of its environment to the extent possible. A group climate is a misfit with constraints on its choices and behavior, which are imposed by the organization and its assignment of tasks and procedures. It is a misfit with an organizational structure that impinges upon the members' scope of action through high formalization, complexity, centralization and the functional assignment of tasks (Zammuto et al., 1991, p. 95).

A developmental climate is characterized as a "dynamic, entrepreneurial, and creative place to work" with high trust, low conflict and a drive towards change. It works well for people who want to take responsibility and who can change as new opportunities emerge. It is a good fit with a matrix structure, low formalization, low-medium complexity, and low-medium centralization. The development climate is a misfit with a very structured, specialized, and centralized organization tasks (Zammuto et al., 1991, p. 95).

The internal process climate is described as "formalized and structured place to work" with low trust and high degree of conflict as well as a high degree of resistance to change. An internal process climate requires a very structured organization where each person knows his/her job - a bureaucracy with high formalization and complexity. It is a misfit with a simple organization or matrix structure, low formalization and lack of procedural rules and general tasks of low specialization and low complexity, (Zammuto et al., 1991, p. 95).

The rational goal climate is described as "results-oriented" with a drive for change, a low degree of trust, and a high degree of conflict. A rational goal climate requires the flexibility for change of low to medium formalization. It is a misfit with formalization that makes change more difficult and time-consuming, (Zammuto et al., 1991, p. 95).

Propositions on Climate Misfits

A group climate is a misfit with a functional configuration, high formalization, high complexity, and high centralization.

A developmental climate is a misfit with a machine bureaucracy, functional configuration, high formalization, high complexity, and high centralization.

An internal process climate is a misfit with a simple configuration, matrix configuration, low formalization, and low complexity.

A rational goal climate is a misfit with high formalization.

Leadership Style Misfits

Leadership style is measured along two dimensions: preference for delegation and uncertainty avoidance, as discussed in Chapter 3. Leadership styles can then be categorized as leader, producer, entrepreneur, and manager. Each leadership style has its own misfits.

For a leader, who is characterized as an individual with a high preference for delegation and a low level of uncertainty avoidance, misfits restrict the organization and narrow the possibilities of action. A functional configuration and a machine bureaucracy do not permit the organization the range of activities that the leader desires.

Similarly, high formalization with rules restricts broad action and quick adaptation. Leaders are results oriented and want the organizational members to focus on results, not process. Finally, leaders need new and relevant information from many sources and do not want limited information in either scope or source. Leaders are misfits for organizations, which are limited in their focus and operation.

For the producer, who is characterized as an individual with a high preference for delegation, but a high level of uncertainty avoidance, i.e., an individual who wants to know, but does not need to make decisions, the misfits are ones which do not permit the organizational members a good deal of freedom of action, but action that is monitored. A functional configuration and a machine bureaucracy require more top-level decision-making than the producer desires. The producer wants the organization to act and deal with the uncertainty; this is not compatible with process incentives and narrow information. The producer assumes the uncertainty of letting others act on his/her behalf, i.e., let the organization be the agent.

For an entrepreneur, who is characterized as an individual with a low preference for delegation and low uncertainty avoidance, misfits restrict direct intervention and limit the range of action as found in a divisional configuration, a machine bureaucracy, and an organization with high formalization, low centralization and an emphasis on process within narrow limits. The entrepreneur demands a range of options, which can be undertaken quickly upon his/her demand. An organization that limits these options is viewed as obstructive in the face of uncertainty.

For a manager, who is characterized as an individual with a low preference for delegation and high uncertainty avoidance, the misfits are ones that yield a loss of control beyond narrow ranges. Ad hoc, matrix, divisional and virtual network configurations - all require decision-making and action by others without the direct involvement of the manager at the top. The manager loses control with low formalization, a lack of standard procedures, low centralization and results based incentives. The manager can neither predict nor be able to control what will happen. Generally, the manager would feel disconnected with the organization and its operations, which would be a misfit for the manager.

Propositions on Leadership Style Misfits

A leader is a misfit with a functional configuration, a machine bureaucracy, high formalization, process based incentives, and non-rich information.

A producer is a misfit with a functional configuration, a machine bureaucracy, process based incentives, and non-rich information.

An entrepreneur is a misfit with a divisional configuration, a machine bureaucracy, high formalization, low centralization, process based incentives, and non-rich information.

A manager is a misfit with an ad hoc configuration, a matrix configuration, a divisional configuration, a virtual network, low formalization, low centralization, results based incentives, and a lack of standards.

MORE SPECIFIC MISFITS

In this section, we develop misfits that are more specific. For example, we examine product innovation, national culture, and leadership attitudes toward risk as well as misfits that add to those given above.

Environment - Strategy Misfits

This set of specific misfits includes misfits related to product innovation as well as more general environment - strategy misfits. With a low product innovation and an uncertain environment, the situation calls for a review and suggests that the organization should consider greater product innovation. Low product innovation means the same products are available for an extended period. In a certain environment with little change in customer demands and preferences, there is little need for new products. However, with increasing uncertainty in customer demand, new competitor strategies, possible governmental actions, shifting customer tastes, etc., current products are likely to be mismatched with this changed environment. New products and innovation will likely be required to adapt and meet the emerging needs and opportunities of the new environment.

With many markets and a low product innovation, the organization may want to consider whether the product innovation is high enough. Unless the many markets are very stable, it is likely that new products should be introduced; thus, some product innovation is needed to meet these variations in market demand.

A defender strategy is easier to manage for a few products or markets as management can focus attention on a few issues well. Defender strategies are difficult to sustain for a large number of products; there is a high probability that some products and markets will require innovation and new developments. The recognition of when to give up a defensive strategy for a given product or market requires a good deal of management attention and the need for change is frequently missed or realized late; needed adaptation is then even more difficult. For at least some of the many products or markets, a defender strategy is likely to be a mismatch.

A defender strategy is not innovative or adaptive. In an ill-defined environment, adaptation will be required to survive as new situations and issues will emerge. Here, the organization should change its strategy to an analyzer or prospector to adapt to the evolving and changing environment. With a highly uncertain environment, the environmental variable values change in large amounts and can vary in the extreme. The defender is trying to maintain the status quo and defend its market share, product lines, production processes, etc. This is risky as the market or other environmental values change outside the range of the defender's position. Then there is a serious mismatch between what the defender is doing and what the market will support. Here, the organization must prepare to shift quantities and perhaps products to meet the uncertainty in the environment. An analyzer strategy is probably more appropriate.

With the complexity of many factors in the environment, it may make it difficult for a defender to protect what it does and difficult to protect its established market position. A high uncertainty environment may deem an analyzer without innovation problematic as it cannot adapt to new circumstances. The analyzer without innovation works best with a medium environmental complexity - neither too much nor too little. When only few factors in the environment affect the organization, the analyzer with innovation strategy may not be a suitable one, as it will generate unneeded and costly innovations.

The prospector strategy requires an exploration of the unknown and the creation of the organization's own future. There is a high uncertainty about what will happen and how successful the organization will be, hence there is a misfit with national culture, which has high uncertainty avoidance.

Propositions on Environment-Strategy Misfits

Low product innovation is a misfit with a non-low environmental uncertainty.

High market diversity is a misfit with low product innovation.

A defender strategy is a misfit with high market or product diversity.

High environmental uncertainty is a misfit with a defender strategy.

High environmental complexity is a misfit with a defender strategy.

High environmental uncertainty is a misfit with an analyzer without innovation strategy.

A high or low environmental complexity is a misfit with an analyzer without innovation strategy.

A simple environmental complexity is a misfit with an analyzer with innovation strategy.

High uncertainty avoidance culture is a misfit with a prospector strategy.

Technology - Environment Misfits

In an uncertain environment, it is very likely that the customers will prefer variation in products and services. Competitors are likely to vary their strategies in products, prices, advertising, etc. New innovative strategies may be called for. A nonroutine technology will likely be required to adapt to an uncertain environment.

Most mass production operations are very limited in capacity to adapt and make different products. Mass production optimizes on the economies of specialization and standardization. A highly equivocal environment requires adjustment to the unknown as that environment becomes clearer. The possibility for mismatch of what the existing mass production can do and what will be required in the new environment is very high and further the economic consequences are likely to be great with low return. A highly equivocal environment calls for a more nonroutine production capability than most mass production operations have.

Propositions on Technology - Environment Misfits

A routine technology is a misfit with a non-low environmental uncertainty.

A high environmental equivocality is a misfit with a mass production technology.

Climate - Environment Misfits

The rational goal climate with its low leader credibility and low trust requires that the management must drive the activities to assure implementation. This is difficult in a highly equivocal environment as there are simply too many issues to consider in a very short time. A rational goal climate may not fit an environment that is highly equivocal, except when the situation becomes a crisis.

A developmental climate works well when there is a sense of collectivity and equality among the members. High power distance and high individualism make it difficult for a developmental climate to emerge, and thus create a misfit.

Propositions on Climate - Environment Misfits

A rational goal climate is a misfit with a high equivocal environment.

High power distance and high individualism is a misfit with developmental climate.

Climate - Strategy Misfits

A rational goal climate requires a good deal of intervention by the management to deal with the low trust, high conflict, low morale, etc. The prospector strategy is focused more on product and market innovation than internal issues and is thus a misfit with a rational goal climate.

Propositions on Climate - Strategy Misfits

A rational goal climate is a misfit with a prospector strategy.

Strategy - Technology Misfits

With a routine technology, developing new products or services will also be very difficult. With limited product and market opportunity, the range of prospector possibilities may exceed the environmental possibilities. The prospector needs to seek new markets as well as new products. When the markets do not exist or markets cannot be created, high costs of innovation without return will be incurred by the prospector.

Generally, more products are required for an analyzer. A few products may be reasonable in the short run, but an analyzer should be in constant consideration of new possibilities. When a few, unchanging products become the norm, the analyzer should broaden its scope of new opportunities. High uncertainty avoidance is a misfit with a prospector.

Similarly, a routine technology will not support high product innovation. A routine technology yields standard products with low variation. The need for product innovation creates a mismatch. Product innovation will be difficult to manage, expensive and inefficient. For product innovation, a more nonroutine and adaptable technology is required. Of course, the organization may also shift to markets and products where less product innovation is required and a routine technology is suitable.

Additionally, with many products and a low product innovation, the organization may want to consider increasing the product innovation. Low product innovation will likely create a mismatch with the product demands of the marketplace. New products and new product innovation will be required to meet the changing product needs.

A prospector strategy and low product innovation are a serious mismatch of strategy and capability. Normally a prospector strategy does not fit well with a mass production technology. A prospector explores with new products and services with innovative technology. A mass production will not be able to adjust quickly and support the prospector. The prospector requires a nonroutine technology that most mass production operations can support.

Propositions on Strategy - Technology Misfits

A prospector strategy is a misfit with a low market or product diversity.

An analyzer strategy is a misfit with a low market or product diversity.

A highly routine technology is a misfit with a high product innovation.

A high product diversity is a misfit with a low product innovation.

A prospector strategy is a misfit with low product innovation.

A mass production technology is a misfit with a prospector strategy.

Climate - Technology Misfits

A rational goal climate has an external orientation with a focus on control. Usually, there is high conflict, low trust, and low leader credibility, among others. A routine technology is a better match. A nonroutine technology is difficult to support in this climate. A nonroutine technology is better supported by an organizational climate with high flexibility, low conflict, high trust, and high leader credibility.

A group climate can support a nonroutine technology with its capacity to handle and process information that is more complex.

Propositions on Climate - Technology Misfits

A rational goal climate is a misfit with a nonroutine technology.

A group climate is a misfit with routine technology.

Technology - Size Skill Misfits

When the organization has a nonroutine technology, and profession-alization is low, that is, the workforce has a low level of education and training, the situation can create production and service difficul-ties. Such an organization usually requires an investment in the education and training. A nonroutine technology usually requires that individuals adapt work methods to the particular task. The indi-viduals must have a sufficiently high level of skill to make these ad-aptations. Low levels of education and training do better at routine tasks and technologies. With a nonroutine technology and low level of education and training, new training will be required for the work-force. This training should emphasize individual responsibility and decision-making for the quality of the service. It should provide new skills, which permit the individual to take the initiative for actions, which meets the customers' requirements. A routine technology pro-duces goods and services efficiently which are standard and without variation.

With a high capital requirement and an organization that is not large, the organization can be vulnerable. An organization with a high capital requirement and a few employees usually makes a few standardized products. Further, the technology is likely to be very limited in adaptitiveness. The organization is then vulnerable to changes in the environment, market and products changes. Smaller organizations with small capital requirements are frequently more adaptive. To reduce this vulnerability, the organization should con-sider creating a greater capability for adaptation, which will usually require more employees of higher skill, education and training.

Low capital requirements are usually associated with low barriers to entry. Small competitors can often enter the market with more advanced technology and lower costs. Large organizations are frequently slow to adjust and adapt. The smaller competitor will have the advantage and thus, be a threat. One alternative is to break up the large organization into a number of smaller ones.

Propositions on Technology - Size Skill Misfits

Low professionalization is a misfit with a nonroutine technology.

If the capital requirement is high and size is small then there is a potential strategic misfit.

A high capital requirement is a misfit with a small size.

A low capital requirement is a misfit with a large size.

Climate - Leadership Style Misfits

The developmental climate is adaptive and open to change which may be broad in the range of possibilities. The risk-averse manager will find this position uncomfortable and may take actions, which will threaten the developmental climate.

Propositions on Climate - Leadership Style Misfits

A developmental climate is a misfit with a risk-averse top management.

Leadership Style - Environment - Size Misfits

A large organization with a complex and dynamic environment may not fit with a leadership style as a manager. With a complex and dynamic environment, there are a very large number of changing situations to which to adjust. The management cannot access all the situations, analyze what needs to be done, and oversee the implementation. There is simply too much to do; there is too much information to deal with. A leadership style as a manager will usually lead to an information overload at the top and a delay in action when it is most needed. Despite a tendency for management to become even more involved in details, the situation requires more leadership and alternative approaches, such as greater decentralization.

Propositions on Leadership Style - Environment - Size Misfits

A leadership style as a manager is a likely misfit with a high environmental complexity a high environmental uncertainty, and large size.

Leadership Style - Strategy Misfits

A prospector strategy demands a projection into the unknown with new and innovative products and services, where the returns are uncertain. A risk-averse management will be very uncomfortable with this high level of risk. Risk-averse managers prefer situations with less uncertainty. It is possible to either change the prospector strategy or hire more risk assuming leaders. Usually a risk-averse management will control expenditures to reduce or eliminate the prospector projects. If the environment and markets call for a prospector strategy, a new leader may be preferable. Some risk-averse managers can adapt, but it is very difficult.

Conflict and confusion are likely results when the organization both has an analyzer strategy and a management with a short time horizon. An analyzer is searching for opportunities, which may not be within the current activities of the organization. Frequently, the organization will incur investments and startup costs, which will decrease short-term returns. Management should then develop a longer-term outlook for the organization.

Propositions on Leadership Style - Strategy Misfits

A risk-averse top management is a misfit with a prospector strategy.

A short-term top management is a misfit with an analyzer strategy.

Environment Misfits

The environment includes the national culture of the organization as discussed in Chapter 6. National culture has four dimensions individualism - collectivism, power distance, uncertainty avoidance, and masculinity-femininity. National culture differences indicate that individuals will behave differently and thus, misfits can emerge with structural characteristics.

A culture with a high power distance accepts that some individuals have the right to set the course for others and the organization; and formalization is one means to determine what will occur. A high power distance is then a misfit with a low formalization. Similarly, individuals do not want the decision-making authority of the low centralization.

Uncertainty avoidance is an attitude of members of the society about the future and its predictability. Formalization in an organization yields predictability of future events. When this formalization is lacking, there is a misfit with a culture with high uncertainty avoidance. High uncertainty avoidance is a misfit with an ad hoc or matrix configuration as both configurations are loosely structured to adjust quickly to uncertain events as they become evident. Individuals will find it difficult and uncomfortable to work in these organizations, and thus the misfit.

Low individualism or high collectivism indicates that individuals are strongly integrated and are cohesive. High horizontal differentiation indicates that the work tasks are divisible and small, which is a misfit for individuals who are close together.

When the environment is either highly equivocal or highly uncertain, the focus should be on results or results based incentives as we discussed in Chapter 6. For these environments, procedural based incentives are difficult to specify a priori, as procedures must emerge as events dictate and thus create a misfit.

Propositions on Environment Misfits

A national culture environment with a high power distance is a misfit with a low formalization.

A national culture environment with a high power distance is a misfit with a low centralization.

A national culture environment with a strong uncertainty avoidance is a misfit with a low formalization.

A national culture environment with a strong uncertainty avoidance and high power distance is a misfit with an ad hoc configuration or matrix configuration.

A national culture environment with a low individualism is a misfit with a high horizontal differentiation.

An environment with a high equivocality is a misfit with procedural based incentives.

An environment with a high uncertainty is a misfit with procedural based incentives.

Strategy Misfits

The prospector strategy focuses on developing new technologies and products/services; results based incentives put a focus on and reward the success for new ideas and their realization. Procedures are difficult to specify a priori in any detail and even so are not likely to direct effort for an emerging direction, and thus procedural based incentives are a misfit for a prospector strategy.

A reactor strategy is inherently risky and it will ultimately lead to failure. It is time to reconsider the strategy.

Propositions on Strategy Misfits

A prospector strategy is a misfit with procedural based incentives.

A reactor strategy is a misfit situation.

Climate Misfits

A developmental climate is emergent in its activities and procedures are therefore difficult to specify a priori. A developmental climate is thus a misfit with procedural based incentives. A rational goal climate focuses on the goals and is thus a misfit with procedural based incentives.

An internal process climate does not yield a high trust among the individuals in the organization. A virtual configuration, which is loosely connected across time and space, requires that the individuals have a trust of expectation concerning agreed activities. Without this level of trust, prediction of other activities and coordination breaks down. The virtual network configuration is then a misfit with an internal process climate.

Propositions on Climate Misfits

A developmental climate is a misfit with procedural based incentives.

A rational goal climate is a misfit with procedural based incentives.

An internal process climate is a misfit with a virtual network configuration.

Leadership Style Misfits

A leader and an entrepreneur both have low uncertainty avoidance, but want the organization to succeed regardless of the emerging situation. It is not possible to specify procedures; both activities and means emerge with the events, and thus a leader or an entrepreneur is a misfit with procedural based incentives.

A manager desires control, determining what will occur and monitoring activities. In a virtual network configuration, leadership is shared and temporary among the members; the needed coordination is developing and changing. A manager would become frustrated with the situation and overloaded in an attempt to control the organization; hence, it is a misfit.

Propositions on Leadership Style Misfits

A leader is a misfit with procedural based incentives.

An entrepreneur is a misfit with procedural based incentives.

A manager is a misfit with a virtual network configuration.

SUMMARY

In this chapter, we have developed a number of misfit statements congruent with the model depicted in Figure 1.3 in Chapter 1, and the theoretical developments from Chapter 3 through Chapter 8. These misfit statements resemble the managerial situation to look for things out of kilter and then fix the problem. This is the diagnosis and design process shown in Figure 9.1.

Figure 9.1. The Diagnosis and Design Process

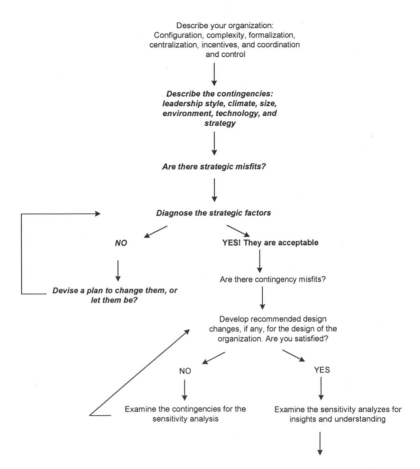

CHAPTER 10

Organizational Design Fit

INTRODUCTION

The objective of strategic organizational design is to obtain total design fit: strategic fit, contingency fit, and design parameter fit. These criteria were developed in Chapter 1. Briefly, strategic fit is fit among the input factors: leadership style, climate, size, environment, technology, and strategy. Contingency fit ensures that the contingency relations - if-then statements - have been followed and are compatible with each other. These relationships have been developed in Chapters 3-8. Design parameter fit considers the compatibility of the design recommendations on configuration, formalization, centralization, and so on. Total design fit is a consideration of all three fit criteria simultaneously. To illustrate we examine the case, Charles Jones.

Charles Jones operates The Jones Company, a small computer-based design company that provides individualized custom orders. Each order is an individual job, and procedures require considerable variation and adaptation to meet customer specifications. The environment is not complex: there are only a few customers, who focus on price and quality. Yet the environment is uncertain, as the price competition makes prices difficult to predict. The nature of the quality is continually evolving, and it, too, is difficult to predict. The equivocality is high because Mr. Jones is convinced that he does not understand what will be important to customers next year or even who his customers might be next year: this year his customers are advertising boutiques, last year they were architects and next year, may be some others.

He thinks that his employees respect him, and there is a great deal of trust with little conflict in The Jones Company. The resistance to change is moderate, so the climate could be categorized as a developmental climate.

Mr. Jones runs his business and is involved in every detail - at least, he would like to be. Despite a predilection to be involved in the designs themselves, he knows that he must search widely for new business in order to survive. He experiments with new customers and new design procedures and spends a good deal of his time developing technical proposals for new customers. He is an entrepreneur. Can we recommend an organizational design for the Jones Company?

For Jones, there are no strategic misfits. We can therefore concentrate on contingency fit propositions.

The issue is to balance the effect of the various contingency factors. We introduce the certainty factor approach (presented in Chapter 1) to balance and weigh the various factors. The effect of management style has been argued to be strong, and we have assigned a certainty factor (cf.) of 40. Individually, the other contingencies are less important and have been assigned a certainty factor of 20. Although we have assigned numerical values for the certainty factors, the literature gives only qualitative guidance on the relative importance. The certainty factors are qualitatively similar to those utilized in OrgCon. A certainty factor of 10 is a weak factor. The factors in our example are summarized in Table 10.1. The summary in the row labeled Total design has been calculated on the basis of the Mycin principle discussed in Chapter 1.

With respect to configuration, the situation fits both a simple, matrix and an ad hoc configuration, except Mr. Jones's leadership style favors a simple configuration. His leadership style has a strong effect on the choice, and thus a simple configuration is recommended with a high certainty factor.

The organizational complexity is clearly low. The uncertain environment, unit production, and nonroutine technology do not favor specialization or high vertical differentiation. There are contingency conflicts on the centralization issue. Most contingency parameters lead toward a high centralization. Only the prospector strategy recommends a low centralization. Therefore, centralization is recommended to be high but with a slightly lower certainty factor than for either configuration or complexity.

Formalization presents an interesting conflicting situation. All contingency factors, except leadership style, favor low formalization. Despite the fact that leadership style individually has a higher weight than the other contingency factors, the case for low formalization is very strong. Additionally, the design parameter fit rule suggests that centralization and formalization should not both be high. The recommendation is that formalization should be low. There is a unanimous recommendation, that direct supervision, group meetings, and a result-oriented incentive system should be the coordinating devices. For coordination and control, there is a general agreement that direct supervision and group meetings are appropriate. High media richness is required for the environment. And incentives should be results-based

This analysis is confirmed by OrgCon. The propositions presented in Chapters 3 through 8 have been the basis for the OrgCon expert system. In this system, each rule has been assigned a certainty factor according to the validation procedure described in Chapter 1. Thus, OrgCon represents a particular weighting of the various rules. The input is stored in the file "Jones.oc8" on the accompanying CD.

To obtain a design parameter fit the totality of the design parameters must also fit together to obtain a total design parameter fit.

Now assume that Mr. Jones leaves and the new person has a leadership style of the leader type. This means that management favors an ad hoc configuration, high organizational complexity, low centralization, and low formalization. The total design recommendation then changes to an ad hoc configuration. Organizational complexity will probably still be low, primarily due to the size. Centralization will now change to a decentralized structure, primarily due to management's preference. Formalization remains low, and the incentive and coordination devices remain the same. However, this design still represents a total design fit but for a very different organization. The simple structure was centered on a strong leadership style. The ad hoc structure is much more centered on a strong incentive scheme. This example shows the process of diagnosing and designing. In this particular organization there were no strategic misfits. When there are, they must be addressed first.

Table 10.1. Total Design Fit: The Jones Company. Note: cf = certainty factor

	Configuration	Organizational Complexity	Centralization	Formalization	Coordination and Control	Incentives
Entrepreneurial Leadership Style	Simple cf 40	Low cf 40	High cf 40	High cf 40		
Developmental Climate	Matrix cf 20	Medium cf 20	Low cf 20	Low cf 20	Planning, integrators, meetings, high media richness	Results cf 20
Size small	Ad hoc cf 20 Simple cf 20	Low cf 20	High/medium cf 20	Low cf 20		
Environmental equivocality, high, uncertainty high, complexity low	Simple cf 20 Ad hoc cf 20	Low cf 20	High cf 20	Low cf 20	Direct supervision, group meetings, high media richness	Results cf 50
Service organization, unit production divisible, non-routine technology	Ad hoc cf 10	-		Low cf 20	Group meetings high media richness	Results cf 50
Prospector strategy	Simple cf 20 Ad hoc cf 20	High or low cf 20	Low cf 20	Low cf 20		
Total design	Simple cf 69 Ad hoc cf 54 Matrix cf 20	Low cf 69 High cf 20 Medium cf 20	High cf 62 Low cf 36	Low cf 67 High cf 40	Planning and integrators, direct supervision, group meetings, high media richness	Results cf 75

SAS

Since 1950, SAS has experienced four distinct phases in its history: the relatively stable 1950-1975 period, the environmental threat of the 1975-1981 period, the initial Carlzon period of the 1980s, and today's new challenges. Each phase has its own special characteristics, and we want to examine each as well as the transitions. Strategic, contingency, and design parameter fit will be examined to illustrate the influence of misfits of the organization and how one misfit ripples throughout to create problems elsewhere. We begin with the first phase-1950-1975.

The 1950-1975 period for SAS has been described at some length. Briefly, it was a relatively stable period. To review, SAS was a large organization with a few thousand employees. The environment was described as one with low equivocality, low complexity, and low uncertainty. The technology was highly routine and nondivisible. SAS had a defender strategy and a group climate. The components of this situation fit well, and no misfits are identified.

During this period, SAS had a dominant functional configuration which overlay a geographical configuration. Organizational complexity, centralization, and formalization were high. Abundant rules were utilized to achieve coordination and control. Incentives were procedure based. This organizational design is consistent with propositions of design. There is wide latitude on the choice of a configuration, which includes a functional configuration. The stable environment further suggests a functional configuration. And finally, a defender strategy uniquely requires a functional configuration. The group climate suggests a less rigid structure. The organizational complexity should be high because of SAS's large size and its status as a nonpublic organization. This recommendation is augmented for a routine technology, and this is consistent with a defender strategy. Clearly, organizational complexity should be high.

A large organization should have high formalization. The stable environment calls for high formalization, medium complexity, and medium centralization. Routine technology, modern information technology, and a defender strategy all require a high formalization.

Considering the centralization, a large organization should be decentralized with an even higher centralization for this environment. These two recommendations are in conflict. Routine technology indicates a medium centralization, and the defender strategy indicates high centralization. Finally, coordination and control should utilize rules, planning, and procedure-based incentives.

The SAS organizational design is largely consistent with the recommended design. There is an excellent fit for the functional configuration, high formalization, high organizational complexity, and coordination rules. Centralization has conflicting recommendations of low, medium, and high.

This seeming conflict could suggest a centralized strategic decision-making and more decentralized decision-making on operative issues. This would be compatible with high formalization - using parameter fit rules. We note that high centralization and high formalization are probably overkill and would lead to a very inflexible organization. However, SAS was well designed for its situation during the 1950-1970 time period.

The second period between 1975 and 1981 created numerous contingency misfits. In general, the organizational situation changed with the oil crisis of 1975, but the organization did not change. In 1975, suddenly the environment changed from low equivocality, low uncertainty, and low complexity to high on all three dimensions. The crises also affected the organizational climate. The implications were great, and yet the adjustments were minimal. The functional configuration was now questionable, media rich meetings and integrators needed to replace rules, and incentives should be results based rather than procedural based.

The centralization, organizational complexity, and formalization should now be unambiguously lower. Suddenly, contingency misfits abound between the situation and the organizational design. Configuration, formalization, complexity, centralization, and coordination and control were out of balance. It is probable that SAS's management had a leadership style of the management type. The environmental change by itself created a number of contingency misfits. Besides the contingency misfits, the new environment created strategic misfits as well. High equivocality, high uncertainty, and high complexity are not compatible with a defender strategy. For the most part, SAS maintained a defender strategy, perhaps becoming more cost conscious than previously.

Therefore, in the 1975-1981 period, SAS had a serious contingency misfit problem, and probable strategic misfits, but the design parameter fit was fine. Within itself, SAS was a consistent organization, but was a misfit with its environment.

Jan Carlzon's appointment began the third period. Early in 1981, Carlzon began to change a number of things. He brought a leadership style that resembled that of a leader or an entrepreneur; he recognized the changed environment; he changed the strategy from a defender to an analyzer with innovation. This new situation now was consistent, and there were no strategic misfits.

The organizational configuration is recommended to be matrix or divisional for analyzer with innovation, but not a divisional configuration for a nondivisible technology. The new environment further supports a matrix configuration. As mentioned above, the new environment calls for low formalization, low complexity, and low complexity. Yet SAS remained a large organization, which suggests a high organizational complexity as does the leadership style. The routine technology similarly calls for a high complexity like the new strategy. Overall, complexity should remain high. All of the above is also consistent with the developmental climate.

Formalization is suggested to be high by the routine technology and the large size but is driven by the leadership style and developmental climate. Formalization is pulled in opposite directions.

In contrast, centralization should be low, due to the leadership style and developmental climate, large size, new environment, and to a lesser degree the new strategy.

The environment calls for coordination and control by group meetings, integrators with high media richness, together with results-based incentives. This again fits the developmental climate. However, the routine technology pulls in the opposite direction, with rules and procedure-based incentives.

Overall, the new SAS of 1981 fit the new situation well. The leadership style, the large size, the dynamic environment, the routine nondivisible technology, and the analyzer with innovation strategy fit together; there were no strategic misfits. The routine, nondivisible technology would require special attention in this environment. The contingency fit was good but not perfect. The organizational configuration of the recommended matrix is difficult to realize. SAS experimented with various divisional forms based on region, customer, and so on, but none were entirely satisfactory. Formalization remained high, due to the technology, as did complexity. Centralization was made low, and many rules were displaced. In brief, the contrasting dictates of the technology and large size versus the dynamic environment pull SAS in opposite directions. It is a delicate balance to obtain a well-performing organization.

The fourth period-the late 1980s and until the middle of the 1990s-was a difficult one for SAS. In the late eighties, Carlzon's measures to renew SAS's offerings and find new ways for exploration turned SAS's strategy into more of a prospector strategy. This however was not consistent with SAS's design parameters, and SAS lost control of costs and finally, the company's competitiveness deteriorated. Additionally, the problems affected the climate negatively by creating a higher level of mistrust and numerous conflicts.

From 1994 to 2001, Stenberg's attempt to change the company's focus on its core business allowing SAS to pursue an analyzer with innovation strategy seems more appropriate. There was strategic as well as contingency fit. And the company's configuration was consistent with the design parameters. The environment remained uncertain, complex, and equivocal. Technology in terms of routineness and divisibility were largely unchanged, despite enormous changes in technical details, particularly computers. In many ways, the airlines pioneered with integrated computer operations, and they continue to innovate. An analyzer strategy fits this new world of continuing change, where mergers and alliances have created even larger airlines but where also the small-niche airline has emerged. Jørgen Lindegaard continued to further develop the core focus of the airline activities. In his first years he has had to focus on crisis management as the environment in 2001 changed from highly competitive to almost hostile. This created a conflict in activities for survival and the more daily operation of the services. SAS can be formally analyzed by OrgCon for the four periods informally discussed in this chapter. Input is stored in the files "SAS.oc8."

    1.   Are there design fit issues in the periods (94-01) described above?

DESIGN PARAMETER FIT

Khandwalla (1973), in a study of seventy-nine manufacturing firms, did not find a significant difference on average scores on organizational variables between two groups of high and low performers. However, he did find a significant difference on the correlation of organizational variables. He investigated differentiation and integration along the lines of Lawrence and Lorsch (1967). Those organizations that had both a high integration and high differentiation or that had a low score on both did better than those that had a high-low combination. This is consistent with Lawrence and Lorsch's (1967) argument, but it also shows that design parameter fit is important. Only those organizations that have the right combination of design parameters will do well. This is also related to the discussion about configurations. Should we consider only a limited set of prototype configurations, as Mintzberg (1983) did? It may be easier to consider only such a limited set of configurations, but the applicability may be equally low.

Khandwalla's research showed that interaction effects are important. Interaction effects can come from either the contingency factors or from the design parameter factors. One approach to this issue into account would be for each possible situation to find the proper organization form - the particular combination of the structural dimensions from Figure 1.2. This was the approach used by Duncan (1979) using decision trees. However, even with few dimensions there are thousands of possible situations, which makes it is impossible to do so.

Therefore, some compound rules are appropriate as we have discussed previously with respect to several factors, especially size. In Chapters 3 through 8, we have presented simple rules which can be combined using the certainty factor principle presented in Chapter 1.

This can be supplemented with explicit design parameter propositions as specific rules. We will present some such rules.

A low centralization coupled with low formalization requires results-based incentives, but low centralization with high formalization calls for procedures-based incentives - that is, high formalization requires procedures-based incentives. If organizational complexity is high, it is less likely that a centralized organization can process the required information. There needs to be a very strong argument here if high centralization should be recommended. This view is in the knowledge base represented with a negative certainty factor, as described in Chapter 1. There is a tradeoff between formalization and centralization as well as between complexity and centralization.

Similarly both a high formalization and a high centralization may be too costly and even produce conflicts. A centralized management may not want to obey the rules given and then ask the employees to break them too.

Propositions on Design Parameter Fit

If centralization is low and formalization is low, then incentives could be based on results.

If centralization is low and formalization is high, then incentives could be based on procedures.

If organizational complexity is high, then it is less likely that centralization should be high.

If formalization is high, then it is less likely that centralization should be high.

If organizational complexity is low, then it is less likely that centralization should be low.

New Forms and Flexible Organizational Design

In earlier capsules new organizational forms were discussed. In Chapter 2, we argued that the traditional vocabulary and concepts can be utilized for new forms. In Chapter 6, we examined the relation between hyper competition and flexibility. In Chapter 8, we examined strategy and new forms. Here, we want to extend that analysis to suggest the new organizational designs for flexibility can fit into the contingency framework.

Flexibility usually means to be able to change quickly. First, we begin with two organizational designs which meet the notion of being "flexible." Second, we look at the situation or contingencies for which flexibility fits. In Figure 10.1 the two designs and the associated contingencies are given. The contingencies are not unique, i.e., for any resulting design there are a number of contingency sets which could yield that design. We have only included one contingency set for each design and included some supporting propositions for the fit.

The first flexible organization is an ad hoc configuration with low organizational complexity, low formalization, high centralization, meetings for coordination, high media richness and results based incentives. Can this organization adjust quickly? We argue it can. There are no rules to restrict change and tasks are general. Centralization is high, yet decisions can be made by many individuals. Success is measured directly by achieving the organizational goals and the incentives are matched to that goal. There is little organizational inertia here and it can change quickly. The next issue is the contingencies which give rise to this organizational design. These contingencies are not the only ones which could yield this particular design, but these particular contingencies fit and are possible. The entrepreneurial leadership style fits with a rational goal climate for a small organization. The environment is highly uncertain, complex, and equivocal. The technology is nondivisible and nonroutine. A prospector strategy then fits. In brief, this flexible organization is a small, goal driven organization which has a strategy of exploring the possible. It seems to meet the common notion of flexibility.

The second flexible organization is a divisional configuration with high complexity, low formalization, low centralization, coordination by planning and meetings, high media richness and results based incentives. Can it adjust quickly? Again, we suggest that it can. The divisions are only loosely coupled and can adjust independently. Here again, the low formalization and low centralization facilitate change. The results based incentives support the change as needed. The contingencies here are somewhat different. The environment is highly uncertain, complex and equivocal. The organization can be large and the technology needs to be divisible across the divisions. Indeed, the technology can be nonroutine and the strategy can also be a prospector, but could also be an analyzer. It is a large organization which can adjust as needed.

What is a flexible organization? From these two examples, we find low formalization, low organizational complexity, high media richness and results based incentives for both. For the contingencies, the environment is highly uncertain, complex and equivocal. The strategies are change oriented: prospector or analyzer with a nonroutine technology. Do they meet the fit criteria? We conclude that new organizational forms can be described and designed in sensible ways using an information processing framework and also utilizing the knowledge base from contingency theory. Yet, there is much to be learned about these new forms of organization to augment and supplement our understanding of organization.

Figure 10.1. New Organizational Forms

Contingency factors:	Flexible	Flexible
Leadership style	Entrepreneur	Leader
Climate	Rational goal	Rational goal
Size	Small	Large
Environment	High uncertainty, complexity, equivocality	High uncertainty, complexity, equivocality
Technology	Nondivisible, nonroutine	Divisible, nonroutine
Strategy	Prospector	Prospector, analyzer
Design:		
Configuration	Ad hoc	Divisional
Organizational complexity	Low	High
Formalization	Low	Low
Centralization	High	Low
Coordination	Meetings, integrators	Meetings, integrators
Media richness	High	High
Incentives	Results based	Results based

DESIGN SYNTHESIS

Design must be a synthesis of knowledge to obtain a good recommendation. As we argued earlier, both the environment (Chapter 6) and the technology (Chapter 7) are important determinants of a good design. So the question is now: how does one put the two together and what is the proper weighting and balance of the environment and the technology in determining a good design.

A design recommendation is a set of statements about the organization's properties and structural configuration. For example, the configuration is functional; centralization is high, and so on. How do we arrive at this recommendation? Let us examine again the recommended "centralization is high." Many contingencies could influence this result. The environmental uncertainty is low, the environmental equivocality is low, and the management has management type leadership style, and so on. Of course, there can be influences which suggest a high decentralization, such as a large organization. How does one combine in a reasonable way these numerous influences; some of which are mutually supportive and some opposing? As discussed in the Jones case above we use the Mycin rule explained in Chapter 1. The idea behind the rule is straightforward; for a given recommendation, more supporting information strengthens the recommendation, but at a decreasing marginal rate. For example, a low environmental uncertainty suggests a high centralization as does management's leadership style of the management type. The two influences are mutually supportive and the two together yield a stronger recommendation than either one by itself. However, the two supporting influences do not add linearly, but at a decreasing rate.

We now examine the design synthesis, the major contingencies that support a given recommendation or design. The idea is to examine when a given design is appropriate. For example, a simple configuration is appropriate when the management has en entrepreneurial leadership style. Other contingent propositions, such as a small organization, lead to the same recommendation. Neither is sufficient as both contingencies can be compatible with other configurations. In design, there is always a possible non-uniqueness or equifinality in the recommendations. Nonetheless, the two supporting contingencies for a simple configuration strengthen the recommendation and make that recommendation more likely than alternative recommendations.

In the next section, in each table we put the rules together that relate to the various structural design properties. We only incorporate the major factors. Refer to chapters 3 - 8 to get the full details of the support. All the propositions are also incorporated in OrgCon.

The Simple Configuration

The simple configuration consists of a top manager and individuals. There may be little functional specialization and no well-defined departmental structure with departmental heads. Decision-making, coordination, and control are usually done by the top manager. For the simple configuration the organization relies on the top manager (Table 10.2). An important contingency for the simple configuration to be viable is that the top manager is ready to assume that role. Thus, the simple organization requires a top manager who will assume responsibility and authority to get things done.

Table 10.2. Recommendations for a Simple Configuration

A simple configuration may be recommended when there is at least one of the following:

Contingency Factor	The Values
Leadership style	Entrepreneur
Size	Small
Environment	Max two of Complexity Uncertainty or Equivocality are high High hostility
Strategy	Prospector Analyzer and small

The strength of the recommended simple configuration is greater as the number of operative contingencies increases.

A small organization may be a simple configuration. Similarly a simple configuration is possible with an entrepreneurial leadership style.

When both of these conditions are met, then the simple configuration is more likely to be recommended. Taken together they provide strong support for a simple configuration. If the manager is not ready to take the responsibility, an ad hoc configuration may be recommended. But that will depend on other contingency factor values.

The simple configuration may be recommended when only one of the environment factors - complexity, uncertainty, or equivocality -is high. If more than one is high, the information processing capacity of the manager may be exceeded. Both equivocality and uncertainty may be high, but environmental complexity must be low for a simple configuration to be recommended. A hostile environment also suggests a simple configuration. A prospector strategy is compatible with a simple organization. The size of the organization is important for the choice of the simple configuration in a different way. If it is small and has an analyzer strategy, it strengthens the recommendation for a simple configuration.

When all the premises are true, it is very likely that the simple configuration will be appropriate. For example, when the top manager has an entrepreneurial leadership style and the organization is small, has an equivocal but simple and less certain environment, and has an analyzer strategy, then the simple configuration is a very highly recommended choice. Many different situations may lead to a recommendation that a simple configuration is appropriate. The more propositions that support such a recommendation the stronger the recommendation is. The Mycin certainty factor principle captures this idea; more supportive information strengthens the recommendation.

The Functional Configuration

The functional configuration has more levels and more horizontal specialization than a simple configuration. There is a well-defined departmental structure. The departments are created based on the functional specialization in the organization.

Generally, a medium or large organization is compatible with a functional configuration - but with other configurations as well. With medium or large size the functional specialization is usually desirable (Table 10.3). The functional organization is limited in its capacity to process large amounts of information and particularly information which is unusual and non standard in form. This is particularly true, when the climate is an internal process climate. High equivocality creates difficulties and excessive information processing demands. The functional configuration does not deal well with ambiguity and the unknown. If the equivocality of the environment is high, then the uncertainty and complexity have to be low for the functional organization to be appropriate.

Table 10.3. Recommendations for a Functional Configuration

A functional configuration may be recommended when there is at least one of the following:

Contingency Factor	The Values
Leadership style	Entrepreneur, Manager
Climate	Internal process
Size	Medium or large
Environment	Low equivocality, Low uncertainty, Low complexity
	Low equivocality, Low uncertainty, High complexity
	Low equivocality, High uncertainty, High complexity
	High equivocality, Low uncertainty, Low complexity
Technology	Not nonroutine
Strategy	Analyzer without innovation Defender

The strength of the recommended functional configuration is greater as the number of operative contingencies increases.

A functional organization operates efficiently and will continue within narrow environmental variations. It can deal with environmental uncertainty and complexity provided these parameters are well defined and value variation is small. Even so, high complexity and high uncertainty are likely to require additional information processing capacity in liaison activities and integrators. The functional configuration deals well with routine technology that does not have large information processing demands. Finally, innovation is difficult to obtain with a functional configuration. The defender and analyzer without innovation are compatible, but innovative strategies should use the functional configuration with caution. Innovation requires large amounts of information processing to effect the requisite change. In brief, the functional configuration is oriented to efficiency of specialization and does not support large information processing demands.

Divisional Configuration

The divisional configuration is characterized by organizational subunits based on a grouping of products, markets, or customers. The units are relatively autonomous unlike the units in the functional configuration.

A medium or large organization is compatible with a divisional configuration (Table 10.4). In this example, there are multiple design choices for a particular contingency variable value. For low equivocality the uncertainty and complexity can be either high or low, but not at the same time for a divisional configuration. This medium to low information processing situation is compatible with a divisional organization, provided that the non-divisible technology is not present. In brief, the divisional configuration deals with some information processing demands of a medium to large organization and a somewhat complicated environment by creating relatively self-contained divisions that employ divisible technologies. The divisional configuration is compatible with an analyzer strategy with or without innovation. This also fits a rational goal climate. The divisional organization can deal with a number of products and markets, provided they are relatively independent. A divisional carbon copy can be repeated many times in different locations.

Table 10.4. Recommendations for Divisional Configuration

A divisional configuration may be recommended when there is at least one of the following:

Contingency Factor	The Values
Leadership style	Leader, Producer
Climate	Rational goal
Size	Medium or large
Environment	Low equivocality, High complexity, Low uncertainty
	Low equivocality, Low complexity, High uncertainty
	Low equivocality, High complexity, High uncertainty
Technology	Divisible
Strategy	Analyzer

The strength of the recommended divisional configuration is greater as the number of operative contingencies increases.

Matrix Configuration

The matrix configuration introduces a dual hierarchy; it incorporates the essential functional and divisional configuration of an organization simultaneously.

There are a large number of matrix configuration propositions to match its complicated dual focus and hierarchy (Table 10.5). The leadership style has to be compatible with delegation. To handle the dual hierarchy a sharing or trustful climate is appropriate. Top management involvement in details will likely create an information overload at the top and delay activities as well as potentially create motivational issues for the matrix managers.

A matrix is compatible with a medium or large organization, but it is less likely to be appropriate for a small organization. The environment is a very important contingency for a matrix recommendation. The matrix is not necessary unless the environment is not well understood and requires a large information processing capacity, which a matrix organization can supply. By omission there is an implication that a matrix configuration may not be needed for a better defined environment.

A unit technology, where there are a number of customers or projects, is well suited for a matrix configuration. However, if the technology is divisible, the matrix is not needed - probably a divisional configuration is more appropriate, as argued above.

Table 10.5. Recommendations for a Matrix Configuration

A matrix configuration may be recommended when there is at least one of the following:

Contingency Factor	The Values
Leadership style	Leader, Producer
Climate	Group Developmental
Size	Medium or large
Environment	High equivocality, high complexity, high uncertainty
	High equivocality, high complexity, low uncertainty
Technology	Unit Not divisible
Strategy	Prospector Analyzer with innovation

The strength of the recommended matrix configuration is greater as the number of operative contingencies increases.

Finally for a prospector and an analyzer with innovation, a matrix configuration is a recommended configuration. These strategies with their demand for innovation have large information processing demands along multiple dimensions, and the matrix can be effective as well as efficient.

The matrix configuration is too costly to utilize in a small or simple situation. It can handle large information demands along multiple dimensions where both effectiveness and efficiency are important. The efficiency can be realized only if very scarce resources are utilized without opportunity losses.

Ad Hoc Configuration

The ad hoc configuration is a very loose configuration that employs Theory Y people with a high capacity for information processing.

Table 10.6. Recommendations for an Ad Hoc Configuration

An ad hoc configuration may be recommended when there is at least one of the following:

Contingency Factor	The Values
Leadership style	Leader
Climate	Group Developmental
Size	Small Not important
Environment	High equivocality, low complexity, high uncertainty
	High equivocality, high complexity, high uncertainty
Technology	Nonroutine
Strategy	Prospector

The strength of the recommended matrix configuration is greater as the number of operative contingencies increases.

The ad hoc configuration has much in common with the matrix configuration; it is recommended in similar situations (Table 10.6).

Virtual Network

The virtual configuration has individuals spatially separated, who operate asynchronously. It uses modern IT for communications and information processing. Face to face communication is minimal. The virtual configuration requires a high level of trust where individuals can expect others to perform tasks in a timely manner to permit co-ordination among the many individuals. Yet, the demands for a virtual network are flexibility and adaptation to new work demands as they develop. Wong and Burton (2000) found that shared goals and shared norms of behavior are very important for a successful virtual organization.

The virtual network requires the best of IT technology, but that is not sufficient for success. The climate for a virtual network should be a group or developmental climate. Here the individuals can create their own processes and make many of the decisions with a sense of overall purpose. The virtual network does not work well when the degree of tension is high. The virtual network can support a prospector strategy, perhaps with a moderate degree of innovation.

Table 10.7. Recommendations for a Virtual Network

A Virtual Network may be recommended when there is at least one of the following	
Contingency Factor	The Values
Climate	Group Developmental
Strategy	Prospector

The strength of the recommended matrix configuration is greater as the number of operative contingencies increases.

There is a fine balance between the virtual network's need for norms of behavior and trust in order to obtain coordination across time and space, and usual strategy requiring some innovation. It is easy for an imbalance to evolve.

As we discussed in Chapter 2, the virtual network can be found together with other configurations. As IT becomes more available and less costly, and the demands for coordination increase across time and space, we can think of the virtual network as a growing property of all organizations. Overtime, organizations will become more virtual, which has significant implications for the organization.

The ad hoc configuration is advisable only for very large or complex information processing situations where the level of the unknown is very high. Top management cannot be involved in details; there are too many new decisions to be made too often. People have to be able to work together in this loose organizational structure, so a group climate is appropriate. Size does not play an important role, but an ad hoc configuration functions best when the organization is small. An ad hoc configuration can cope with the ill-defined environment of high equivocality and complexity, which require a large information processing capacity. The ad hoc configuration is not suited for a low information "routine" technology. Whatever the source, the ad hoc makes sense only for a large information processing situation. A prospector strategy fits this configuration well.

The Professional Bureaucracy

A bureaucracy can be defined as an organization with the following characteristics (Weber, 1946):

- Division of labor
- Well-defined authority hierarchy
- High formalization
- Impersonal in nature
- Employment based on merits
- Career tracks for employees
- Distinct separation of members' organizational and personal lives.

The above is also often called a machine bureaucracy. If the members in a "bureaucracy" are highly skilled professionals and if some of the standardization is obtained via the professionalization, giving the professionals some decision authority, it is called a professional bureaucracy (Table 10.8).

The professional bureaucracy configuration is recommended for medium and large organizations. If the manager has a leadership style, then the professional bureaucracy is less likely. A nonroutine technology is not likely to be compatible with a bureaucracy. Even the professional bureaucracy relies heavily on routine procedures, albeit at professional levels, and nonroutine procedures require too much nonstandard information processing for the bureaucracy. Therefore, a prospector strategy will not be compatible with a professional bureaucracy. This is very compatible with an internal process climate.

Table 10.8. Recommendations for a Professional Bureaucracy

A professional bureaucracy may be recommended when there is at least one of the following:

Contingency Factor	The Values
Leadership style	Not a manager
Climate	Internal process
Size	Medium or large
Technology	Routine
Strategy	Not a prospector

The strength of the recommended professional bureaucracy is greater as the number of operative contingencies increases.

The Machine Bureaucracy

The machine bureaucracy focuses on the adherence to rules. In a bureaucracy, the information processing is very dependent upon the bureaucratic characteristics above; they define what information is required and how it is to be processed; the information technology can then be well specified. A bureaucracy can process large amounts of standardized information.

A manager who wants to be involved in detail is compatible with a machine bureaucracy. Medium and large organizations also can support a machine bureaucracy. As with the professional bureaucracy, the machine bureaucracy cannot support a nonroutine technology. At the heart of the machine bureaucracy are programmed procedures that are not compatible with the variations demanded with a non routine technology. This also precludes a prospector strategy.

Table 10.9. Recommendations for a Machine Bureaucracy

A machine bureaucracy may be recommended when there is at least one of the following:

Contingency Factor	The Values
Management style	Manager
Climate	Internal process
Size	Medium or large
Technology	Routine
Strategy	Not a prospector

The strength of the recommended machine bureaucracy is greater as the number of operative contingencies increases.

Organizational Complexity

The configuration is a general description of the organizational structure. Other characteristics are important to give a more complete design specification, which can be stated in numerous ways.

Organizational complexity is defined in Chapter 2; it is the degree of horizontal, vertical, and spatial differentiation. Horizontal differentiation is greater when there are several small tasks and specialization by experience, education, and training. Vertical differentiation is the number of hierarchical levels between top management and the bottom of the hierarchy. Spatial differentiation is greater when there are many locations of facilities and personnel. As the degree of organizational complexity increases, the difficulty of coordination issues and the requirements for information processing increase as well.

Organizational complexity depends on the leadership style, size, environment, technology, and strategy (Table 10.10). A high management involvement in details is likely to create a management overload for high complexity; the information processing demands are too large. However, the information overload may be decreased by an information processing system. Climate is related to the information processing capacity of the individuals. Thus, the internal process and rational goal climates with their requirements for a more structured workplace and less emphasis on the group can allow for a high complexity with specialization. In the group and developmental climates the complexity may be lower.

Table 10.10. Organizational Complexity Recommendations

A HIGH organizational complexity may be recommended when there is at least one of the following:

Contingency Factor	The Values
Leadership style	Producer, Entrepreneur, Manager
Climate	Internal process Rational goal
Size	Large and not public
Environment	Low equivocality, low complexity, high uncertainty
	Low equivocality, high complexity, high uncertainty
Technology	Routine Process
Strategy	Prospector Analyzer

The strength of the recommended high organizational complexity is greater as the number of operative contingencies increases.

Size is related to organizational complexity; increased size calls for increased organizational complexity - both vertical differentiation and horizontal differentiation. These propositions are well supported and widely accepted.

Table 10.10. Continued.

A MEDIUM organizational complexity may be recommended when there is at least one of the following:

Contingency Factor	The Values
Leadership style	Leader, Producer, Entrepreneur
Climate	Group Developmental
Size	Large
Environment	Low equivocality, low complexity, low uncertainty
	Low equivocality, High complexity, Low uncertainty
	High equivocality, Low complexity, Low uncertainty
	High equivocality, High complexity, Low uncertainty
Technology	Routine and small organization

The strength of the recommended medium organizational complexity is greater as the number of operative contingencies increases.

The environmental propositions are quite involved. Two arguments underlie these propositions. First, greater environmental equivocality calls for less organizational complexity. With high equivocality, it is not clear what will need to be done to survive, and thus, general skills will be more appropriate than specialized ones. Second, greater uncertainty calls for greater complexity. Here, small adjustments will need to be made to understand environmental forces. Specialists can react quickly and effectively in the small. However, if the environment is very hostile, the organizational complexity should be low to be able to react quickly.

A more routine technology supports greater horizontal differentiation and a greater span of control. A less routine technology suggests a greater vertical differentiation. The routine technology permits greater specialization without greater information processing demands. The nonroutine technology requires greater information processing demands that can be supported with greater vertical differentiation. However, modern information systems mitigate this effect.

Table 10.10. Continued.

A LOW organizational complexity may be recommended when there is at least one of the following:

Contingency Factor	The Values
Climate	Group
Size	Small
Environment	High equivocality, Low complexity, High uncertainty
	High equivocality, High complexity, High uncertainty, Hostile
Technology	Nonroutine
Strategy	Prospector

The strength of the recommended low organizational complexity is greater as the number of operative contingencies increases

A prospector needs low or high complexity and either generalists or specialists but not hybrids. The defender needs the high internal efficiency of high organizational complexity with its specialization. Finally, the analyzer with or without innovation relies on high complexity to focus on many areas in an efficient manner.

In summary, a low organizational complexity recommendation has the greatest support when the leadership style has a high preference for details and involvement in decision making, the organization is small, the environment is somewhat unsettled or hostile, and the strategy is prospector (Table 10.10). Notice here that the situation fit is realized only for some of the possible situations. A prospector strategy is appropriate in an unsettled environment and with a management that participates and makes quick decisions. Referring back to the configuration discussions, such a situation will also call for a simple configuration - for a design parameter fit. However, the simple structure is not recommended when all three dimensions of the environment score high.

Suppose the organization is large and private, the environment has high uncertainty but scores low on the other environmental dimensions and the technology is a routine process technology. All strategy types can fit a high organizational complexity. Again, multiple situations can lead to high organizational complexity.

Formalization

For many organizations it is efficient to obtain a standardized behavior of the members of the organization. This standardization can lead to low cost high product quality, and generally efficient operations.

Formalization is one way to obtain such standardized behavior and, as such, is a means to obtain coordination and control. Formalization represents the rules in an organization.

Formalization is related to management preference, size, environment, technology, and strategy (Table 10.11). A leadership style that either wants to manage or produce calls for high formalization. Generally, formalization gives the detailed control desired by the management. When trust is low, and the level of conflict is high, the rules and regulations must be in place. Formalization should increase with size to capture the economics of specialization. There is a caveat and clarification for professionals: namely, those professionals should have formalization between their professional units. The professionalization itself serves the same purpose.

Generally, as the environment becomes less well defined, formalization should decrease. Low equivocal and low uncertainty environments call for high formalization. Rules and procedures can

ments call for high formalization. Rules and procedures can be effective for this relatively certain situation, and efficiencies can be realized. For the other extreme of high equivocality and high uncertainty, formalization should be low. Appropriate actions are not known, adaptation is mandatory, and rules will only impede the needed flexibility. The intermediate environments call for medium complexity. As before, if environmental hostility is high, then the organization has to react fast and low formalization is needed.

Table 10.11. Formalization Recommendations.

A HIGH formalization may be recommended when there is at least one of the following:

Contingency Factor	The Values
Leadership style	Producer, Manager
Climate	Internal Process
Size	Large
Environment	Low equivocality, Low complexity, Low uncertainty
	Low equivocality, High complexity, Low uncertainty
Technology	Routine Process
	With modern information technology
Strategy	Analyzer without innovation Defender

The strength of the recommended high formalization is greater as the number of operative contingencies increases.

A routine technology calls for high formalization; nonroutine technology, low formalization: moderated for a professional organization and strengthened for a service organization. Service organizations do not have machines to regulate work, so formalized rules take on increased importance and relevance. Technology affects the formalization: increased formalization for process technology, as the technology itself does not drive the workplace. And information technology itself leads to greater formalization.

Generally, increased innovation calls for less formalization. A defender needs the economies of formalization to survive; an innovator needs the flexibility of less formalization.

Table 10.11. Continued.

A MEDIUM formalization may be recommended when there is at least one of the following:

Contingency Factor	The Values
Climate	Rational goal
Size	Medium
Environment	Low equivocality, Low complexity, High uncertainty
	Low equivocality, High complexity, High uncertainty
	High equivocality, Low complexity, Low uncertainty
	High equivocality, High complexity, Low uncertainty
Strategy	Analyzer

The strength of the recommended medium formalization is greater as the number of operative contingencies increases.

Again, high and low formalization can fit with a number of situations. Low formalization is strongly recommended when the leadership style is of the leader type, the environment is unsettled, and the size of the organization is small with a nonroutine technology. A prospector strategy would be appropriate.

A leadership style of the manager or producer type in a large organization in a stable environment using a routine technology following a defender strategy will have a perfect contingency fit with high formalization.

Table 10.11. Continued.

A LOW formalization may be recommended when there is at least one of the following:

Contingency Factor	The Values
Leadership style	Leader, Entrepreneur
Climate	Group Developmental
Size	Small
Environment	High equivocality, Low complexity, High uncertainty
	High equivocality, High complexity, High uncertainty Hostile
Technology	Nonroutine
Strategy	Prospector

The strength of the recommended low formalization is greater as the number of operative contingencies increases.

Centralization

Centralization is the degree to which formal authority to make discretionary choices is concentrated in an individual, unit, or level (high in the organization). Decentralization is low centralization. We measure centralization by how much direct involvement top management has in gathering and interpreting the information it uses in decision-making and the degree to which top management directly controls the execution of a decision. The above issues are important to determine who has authority to influence a decision aside from actually making the decision.

The organization's centralization is influenced by a number of contingency factors (Table 10.12). Centralization is quite dependent on the preference of the management. The management's preference for high involvement requires a centralized organization; a low preference, a decentralized organization. This is at variance with the maxim of universal greater decentralization. Rather, these propositions indicate that the organization should match and follow the style of the management. With a high level of trust, high work morale, and low level of conflict, the centralization can be lower leading to a higher capacity for information processing than otherwise. When the opposite is the case more centralization may be needed.

Table 10.12. Centralization Recommendations

A HIGH centralization may be recommended when there is at least one of the following:

Contingency Factor	The Values
Management style	Entrepreneur, Manager
Climate	Internal process
Size	Small
Environment	Low equivocality, Low complexity, Low uncertainty
	High equivocality, Low complexity, High uncertainty
	High equivocality, Low complexity, Low uncertainty Hostile
Technology	Routine
Strategy	Analyzer without innovation Defender

The strength of the recommended high centralization is greater as the number of operative contingencies increases

Large organizations should be decentralized; small organizations need not be so. These propositions are strongly supported in the literature and managerial experience and widely recommended. Large organizations simply cannot be run only by those at the top; information demands are too great, and motivational incentives for lower management are wanting. Smaller organizations can be centrally managed.

Table 10.12. Continued

A MEDIUM centralization may be recommended when there is at least one of the following:

Contingency Factor	The Values
Climate	Developmental Internal process Rational goal
Size	Large and private Medium
Environment	Low equivocality, High complexity, Low uncertainty
	Low equivocality, Low complexity, High uncertainty
Technology	Routine
Strategy	Analyzer

The strength of the recommended medium centralization is greater as the number of operative contingencies increases.

The environment affects the desired centralization. The well-defined environment calls for high centralization; the information processing demands are low, permitting a high degree of centralization. High centralization is warranted in a very equivocal and uncertain environment but low complexity. Here centralization permits a coordination unity and quick reaction when the environment does not have many factors in the face of the unknown. In contrast, the high-complexity environment with high equivocality and high uncertainty calls for low centralization. Here the environment has many factors that are unknown, and decentralization is the only hope as a centralized organization cannot handle the information demands in a timely fashion. Environmental complexity is quite important, as it affects the amount of the information processing demands. Medium centralization is proposed for intermediate situations of moderate information demands. High hostility requires a centralized decision-making in order to react quickly to the threats. Technology routineness suggests a centralized organization; there are few decisions to make, and top management can make them.

Table 10.12. Continued

A LOW centralization may be recommended when there is at least one of the following:

Contingency Factor	The Values
Leadership style	Leader, Producer
Climate	Developmental Group
Size	Large and not private
Environment	Low equivocality, High complexity, High uncertainty
	High equivocality, High complexity, High uncertainty
	High equivocality, High complexity, Low uncertainty
Strategy	Prospector

The strength of the recommended low centralization is greater as the number of operative contingencies increases.

Finally, strategy and centralization are related. A prospector should be decentralized. Both information demands and motivational issues support this proposition. In contrast, the defender should be centralized to obtain a coordinated response to the environment. The analyzers suggest a medium centralization. The analyzer without innovation also suggests a high centralization, as does the defender. The best level of centralization can be assessed by the organization's need for timely activity but, more importantly, by the information processing demands. High information processing demands indicate decentralization; low information processing demands permit centralization.

A perfect fit with a recommendation for a high centralization is a situation with a manager with a high preference for involvement in a small organization. An environment that is stable or at least not complex, a routine technology, and a defender strategy fit as well. A decentralized organization provides a perfect contingency fit when the organization is large, management has a low preference for involvement, the environment is unsettled, and the technology is nonroutine. A prospector strategy is appropriate as well.

Coordination and Control

Coordination and control can be obtained in many ways. As discussed earlier, when both centralization and formalization are low, coordination and control have to be obtained using other means. This is also one of many rules that address the situation fit problem. Additionally, there may be interaction effects: the effect of technology, for example, depends on the size of the organization. In this section, propositions on coordination and control are summarized.

Size, by itself, does not tell us much about coordination and control requirements. It must be coupled with another contingency. The environment is a major influence on coordination and control requirements, where the less well defined the environment, the greater the demand for coordination mechanisms. Finally, technology routineness is important.

Consider leadership style as a manager on the requirement for coordination and control. The situation of a small organization that is not centralized or formalized, with a management with a leadership style as a manager, suggests that control will be required and liaison mechanisms will be appropriate. Quite simply, management, despite its preferences, cannot be involved in everything and will require some means of coordination. Other mechanisms have been eliminated, and thus, liaison activities are to be utilized. A functional configuration with a management that has a high preference for control over information and decisions is consistent. In contrast low formalization and a leadership style as a leader suggests that other coordination approaches such as meetings and liaison activities will be required to obtain the needed coordination among the functional units. Otherwise, the functional units will drift off into non-related activities.

The environment affects coordination and control requirements. A low equivocality, low complexity, and low uncertainty environment can be coordinated by rules, procedures, planning, and direct supervision. It is a low information processing situation. As the need for information processing increases due to a less simple environment, planning becomes important and liaison activities, integrators and group meetings will be required. The distinction between environmental equivocality, complexity, and uncertainty yields more precise recommendations for particular situations. Finally, consider technology. A more routine technology can be coordinated with direct supervision and planning; as it becomes nonroutine, the information needs increase and group meetings will be required. Thus, if the technology has no dominant technology, other factors must be relied on.

Media Richness and Incentives

The appropriate media richness and incentives depend upon the environment and the technology among other variables. Generally, the well-defined environment calls for procedure-based incentives; the increasingly ill-defined environment calls for incentives based on results. As the environment becomes ill defined, it is less clear what to do *a priori*, and thus, the incentives should reward adaptability and quick response by rewarding results. To follow incorrect procedures precisely is the height of folly. Similarly, a well-defined environment requires a less media richness than does a less defined environment.

Finally, consider the effect of technology routineness on incentives. For a routine technology, procedures are well defined, and it is appropriate to perform them well. With a nonroutine technology, the goal is to obtain the derived outcome or results; it is less important which procedures are utilized. A routine technology requires low media richness whereas nonroutine technology requires high media richness.

In general, incentives should be procedure based when the procedures are well known and will yield appropriate results; if not, the incentives should focus on the results and encourage adaptation to obtain the desired results.

SUMMARY

In this chapter, we put the pieces together for organizational design. The design propositions have been grouped for design implications; when would a particular design recommendation be made? A functional organization, for example, is likely to be recommended when the organization is medium or large, the environment is relatively certain, noncomplex, and unequivocal, the technology is routine, and the strategy defender or analyzer without innovation. Each antecedent makes a functional recommendation more likely but not certain. Even though some of these conditions are met, the particular organization may have other contingencies that are more important and lead to a different recommendation. As you can see there are several situations that will lead to the same recommendation. Secondly, a particular situation may lead to more than one recommendation. The combination of the knowledge into a recommendation is done by using the certainty factor approach, compound rules and specific design parameter propositions.

To find the right organizational structure an iterative design process is often appropriate. Figure 10.2 shows such an iterative process. The organizational design part of the process is in bold and italics. OrgCon facilitates sensitivity analyses.

Figure 10.2. The diagnosis and Design Process

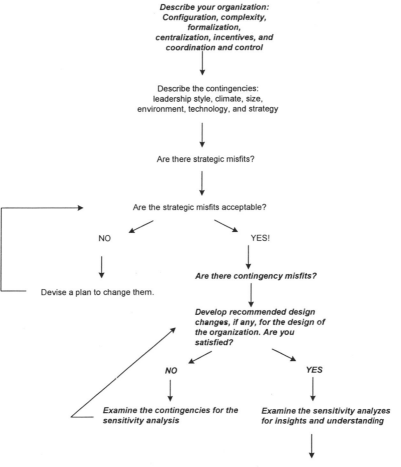

CHAPTER 11

The Dynamics of the Change Process

INTRODUCTION

Growth at Applied Computer Science, Inc., created a conflict for the design of the organization. As a small organization, a simple configuration would be appropriate. However, management's preference for delegation, low uncertainty avoidance, and a producer leadership style suggested more decentralization. Further, the company's increased size indicated the organization needed more specialization and decentralization. Generally, for a decision on a suitable organizational structure. Applied Computer Science had to assess the effect of all contingency factors.

In this chapter, we further develop the neo information-processing model for organizational diagnosis and design. We examine the change process of diagnosis using misfits and design using the total fit; it is a dynamic and continuing process of design.

SAS and the Lifecycle

SAS illustrates organizational change. The 1950–1975 period was a relatively calm time when SAS was well organized to meet its challenge. In the mid-1970s, the environment changed rather dramatically, and SAS did not adapt quickly. Jan Carlzon became the organizational change agent; he introduced a new management style, manipulated the climate, and changed the company's strategy. SAS was changed to a new organization. Our informal analysis and Organizational Consultant results provide recommendations for change as well as a useful guide for a retrospective view of what happened. For the most part, SAS adjusted appropriately and well; however, SAS continues to face difficult issues. Our analysis suggests that major organizational changes are not required.

SAS is an interesting illustration, but is it general to the change process? Most organizations face the challenge of a new environment, new technology, or even a new leadership. The requisite change is usually difficult. SAS illustrates both well. However, SAS is not a good illustration of the rate of change that many organizations are experiencing today. SAS had five or six years to begin to change. Many organizations, particularly smaller organizations, have around five or six months. Consequently, organizational change is now a continuing process, where previously it was more intermittent. Configurations and organizational characteristics are in a constant state of flux.

DIAGNOSIS AND DESIGN

The description of the organization and its activities together with an assessment whether it meets its needs is diagnosis. In Chapter 9, we developed the concept of misfit as a diagnostic. Design is devising structures to attain goals. In Figure 1.3, the goals are effectiveness, efficiency, and viability of the design. In Chapter 10 the "devised structures" are the properties (complexity, formalization, centralization, and so on) and the structural configuration of the organization. The "ought to be" or normative aspects of design are the recommendations themselves - that is, a recommended design is one that will be effective, efficient and viable. As we have argued throughout this book, choosing a good design recommendation is a difficult task. There are a number of issues. First, most of the knowledge base was originally developed in the positive science tradition to explain the world, not directly to recommend good practice and aid managers; some translation is necessary. Second, the knowledge is diverse and fragmented as there are numerous models using different variables, which are only loosely related. Third, design must be a synthesis or putting together of knowledge to obtain a recommendation

Strategic misfits occur when the situation in which the organization has to operate is not internally consistent. As discussed in previous chapters, Bon Goût experienced a change in environmental uncertainty and environmental equivocality that might not fit with its strategy and technology. Similarly, the airline industry became more dynamic and unpredictable, making SAS's environment more uncertain and not compatible with the leadership style. Most strategic misfits occur when a balanced situation is changed. The environment may change, new technologies may be adopted or new management or new strategies may create an unbalanced situation. It is unlikely that a proper organizational design exists in an unbalanced situation. To obtain a balanced situation, strategic choices may be necessary.

When strategic or contingency misfits are observed, a natural reaction is to fix them - to change the organization so it matches the situation. However, two problems have to be dealt with. First, with strategic misfits present, the various contingency propositions may lead to conflicting recommendations for the design parameters, making it difficult or even impossible to obtain design parameter fit. Second, the organizational design theory that provides the basis for the contingency fit does not provide a one-to-one correspondence between strategic factors and design parameters. There may be other elements of organizational design that may fit a particular situation– a nonunique recommendation or equifinality in the terms of the goals. For example, both a matrix configuration and a functional configuration with liaison devices could be appropriate for the same situation. Similarly, there may be more recommendations for other design parameters. However, not all possible combinations may be appropriate. The situation may recommend both high and low centralization and high and low formalization. This gives four possible combinations. Usually, only high centralization with low formalization and low centralization with high formalization are appropriate designs. High centralization with high formalization may be an overkill while low centralization with low formalization is appropriate only with a proper incentive structure.

From the discussion it follows that one cannot infer from a design parameter misfit to either strategic or contingency misfit. A strategic misfit may lead both to contingency and design parameter misfits. There may be no strategic misfit and no contingency misfit and yet, a design parameter misfit may exist. Strategic organizational diagnosis is the analysis of strategic, contingency, and design parameter misfits. Strategic organizational design is creating total design fit as presented in Chapter 1.

Organizational design is both process and product. The product is the statement of the organizational design in terms of the configuration and the properties. The process is the set of organizational activities we undertake to realize the design product. Organizational design as process and product is sequential and dynamic and occurs over time. Organizational design is usually re-design.

Design as process involves organizational change; it is time dependent and dynamic. The process of design is embedded in the reality of the current organization; we seldom if ever begin with a blank piece of paper and consider the organization at the beginning. The lifecycle of the organization has only one new beginning, but many changes in its evolution. The management of the organization's lifecycle requires a number of different organizational designs over time and for different circumstances. The understanding for the need to change and the realization of the change are part of the process of organizational design. The OrgCon incorporates both diagnosis and design. It aids the design process by asking the designer questions about the current organization, the contingency factors and then offers recommendations on the design - the configuration and properties as shown in Figure 1.3.

LIFECYCLE MANAGEMENT: DIAGNOSIS AND DESIGN

Design is managing the lifecycle. The lifecycle includes an evolutionary phase of moderate stability, which is interrupted by short periods of revolutionary change between the evolutionary changes. The main idea in lifecycle management is that over time, contingencies and circumstances change, thus, the organizational configuration and properties should then change. The change is not random; we should maintain fit in response to new size or growth, new environments, new leadership, new technology, a different strategy and a modified climate. As any one of the contingencies is likely to change within a short time, design is an ongoing dynamic process, where management diagnoses and continues to search for the correct organizational design - adjusting the configuration and particularly the properties on an ongoing basis. This is the life of a viable organization.

The lifecycle model is a metaphor, which views the organization like the biological lifecycle of birth, growth, maturity, and death. There are a number of variations on the general theme. The lifecycle models are positive models, which describe the organization over time and find that organizations tend to follow a lifecycle. Here we want to examine the lifecycle as a management problem. The organization's design is then a matter of choice at each point and time, and management must choose to re-design the organization over time if it is to survive.

Throughout this book, we have examined SAS and its evolution. It has a lifecycle, which follows the general pattern. The early days were very focused on getting the job done of flying airplanes. The next phase involved a consolidation of function and management where it both found its place in the air service business and began to rationalize its organization and processes. Given its regulatory environment and need for control, it quickly moved to a more formalized organization, which did serve it well when competition heightened. Since the early 80's, SAS has become a more complex organization to address its more complex environment. SAS has a lifecycle. At this point, we can see more clearly its beginning and early development. Now in the maturity, it continues to evolve and adjust. Will it decline, as suggested by the lifecycle? It may at some time in the future, but it continues to strive and address its challenges. It is in a continuing maturity phase.

There are a number of variations on the lifecycle model. Greiner (1972) presented a five stage lifecycle: creativity, direction, delega-tion, coordination, and collaboration. He suggested different foci, structures, management styles, control systems, and rewards for the various phases. Cameron and Whetton (1981) elaborated and con-firmed the stages as entrepreneurial, collectivity, formalization and control, elaboration of structure and decline. Each stage has its own requirements.

The entrepreneurial stage is a small organization, dominated by the founder who does what is necessary with a prospector strategy and non-routine technology in an uncertain environment. He or she is involved in almost everything, yet has a developmental climate. As shown in Figure 11.1, a simple configuration is recommended with low organizational complexity, low formalization, high centralization where coordination is realized through direct supervision and meet-ings. The incentives should be results based. This organization fits together around the entrepreneur. He or she directs what is to be done; decision-making is centralized. There is no need for formalized rules or elaborate job complexity. Rules would only get in the way of the entrepreneur and restrict the flexibility that is needed to pursue the prospector strategy and adjust to the uncertain environment.

Table 11.1. Lifecycle Designs, Cameron & Whetton Stages, (1981) Supporting proposition

Strategic factors:	Entrepreneurial	Collectivity	Formalization and control	Elaboration of structure
Leadership style delegation preference	Entrepreneur	Producer	Manager	Leader
Climate	Developmental	Developmental	Rational goal or internal process	Rational goal
Size	Small	Medium	Large	Large
Environment Equivocality Uncertainty Complexity Hostility	Medium High Medium Medium	Low Medium High Low	Low Medium High Low	Medium Medium High Low
Technology	Nonroutine	Medium routine, nondivisible	Routine, nondivisible	Medium routine divisible
Strategy	Prospector	Analyzer with innovation	Analyzer with innovation	Analyzer with innovation
Misfits	Nonroutine technology and low training level, low delegation and developmental climate	Numerous situational and design misfits	Medium uncertainty and routine technology, low equivocality and analyzer with innovation, Internal process	Rational goal climate and leader
Recommendations Configuration Complexity Formalization Centralization Coordination	Simple Low Low High Meetings, planning, direct supervision	Matrix Medium Medium Medium Meetings, planning	Functional High High Medium Rules	Divisional High Medium Medium Professionalization
Incentives	Results	Results and procedural(	Procedural	Results

On the accompanying CD, you can find the file Lifecycle.ocd which contains the input data for OrgCon for the entrepreneurial stage. You can review in more detail the data on the leadership style, climate, size, environment, technology, and strategy. Using the OrgCon, you can then run the program and review the recommendations for the Entrepreneur and the strategic misfits. These files contain more information than in Table 11.1. Further, you can do sensitivity analysis on the input data to investigate the nature of the entrepreneurial stage of the organization.

For the entrepreneurial stage, the OrgCon found three strategic misfits: a non-routine technology and a low level of training for the workforce, a prospector and a few products, and low delegation and a development climate. These misfits are then the signals that a more revolutionary change may be needed. There are many ways to deal with these misfits. First, they can be observed as management problems, but not dealt with. Second, they may point to larger problems, which cannot be ignored. The entrepreneur may consider increasing the workforce skill level for the non-routine technology and the prospector strategy in order to develop more products. At the same time, he or she will have to back off from the low delegation and high centralization to develop the climate. Frequently, the need for revolutionary change is observed that the size becomes too large for the highly involved entrepreneur and he or she cannot process the information. Here we can see other possible misfits, which can also lead to the need for more revolutionary change. A sensitivity analysis on the environment indicates the effects and limits of the environment on this stage. We suggest that the need for change goes beyond size alone.

The second stage of the lifecycle is the collectivity stage. In Table 11.1, the contingencies and recommendations are given. It is a medium size organization with a medium level of delegation by management and a developmental climate. The environment has high complexity, medium uncertainty, low equivocality, and low hostility. The technology is medium routine and the strategy is an analyzer with innovation. The recommended configuration is a matrix or functional, with medium levels of organizational complexity, formalization, and centralization. Coordination should be with meetings and planning and the incentives should be results based.

This is a larger organization where the entrepreneur is giving way to less direct involvement; there are the beginnings of more defined organizational properties. There is a need for coordination, which is obtained with planning and meetings, rather than direct supervision. The misfits are of two kinds: strategic and design. The strategic misfit suggests that the environment may not require an innovative strategy. The design misfits suggest that the level of rules, procedures, etc., may not bc sufficient, which hints at a need for a more revolutionary change. For more detail, examine the file Lifecycle.ocd on the CD.

The third stage of the lifecycle is the formalization and control stage in Table 11.1. Here the organization has become even larger, and the management is concerned with quality and routinization and the climate is now rational goal or internal process. The environment is complex with medium uncertainty and low equivocality and low hostility. The technology is routine and non-divisible and the strategy is an analyzer with innovation. The recommended configuration is functional with high organizational complexity, high formalization, and medium centralization. Coordination is achieved through rules and the incentives are then procedural; we want the rules and procedures implemented. The strategic misfits suggest that a routine technology and an uncertain environment as well as a low equivocality with an analyzer with innovation strategy may not fit. The organization may be too well defined to adjust to the changing environment. The formalization organization is on the CD under the file Lifecycle.ocd.

The fourth stage of the lifecycle is the elaboration of structure stage in Table 11.1. The organization has grown even larger, the climate is a rational goal, and the management has to direct as a leader. Uncertainty remains high with medium uncertainty and medium equivocality and low hostility. The technology is divisible and medium routine, and the strategy is an analyzer with innovation. The recommended configuration is now divisional with high complexity, medium formalization, and low centralization. Coordination involves professionalism and the incentives should be results based. It is a classic divisional organization with design fit and good total fit. The organization can process the requisite information to deal with its challenges. You can review the elaboration on the CD in file Lifecycle.ocd. The fifth stage is the decline stage, which is not included in Table 11.1. Here the organization usually adopts a reactor strategy which is not viable.

The lifecycle model is both evolutionary and revolutionary. The organization must adjust continually within each stage; it must change more fundamentally between stages. Yet, the identification of stages and stage changes is problematic and easier to observe ex post than in the management process. The time duration of each stage is not obvious to the manager. Triggering mechanisms for stage change are more involved than size alone. From a managerial view, the design of the organization is ongoing. Some changes are larger than others are, but the managers themselves may not know the distinction between evolution and revolution in design. Our examination of SAS can be mapped onto a lifecycle and it is interesting to identify the various stages. The lifecycle is an interesting metaphor, but at the same time, an organization is not an animal. It is an artifact of our creation, which can and should change in fashion that is more complicated. There are many contingencies and many responses; the organization will evolve in a very complicated manner. The traditional organizational lifecycle is one path of many, which the organization may take over time. The design question is to choose from many possible paths on an ongoing basis - to maneuver in the face of this multivariable and complex problem. This is the challenge to keep the organization viable.

THE DYNAMICS OF DIAGNOSIS AND DESIGN

A dynamic approach introduces time explicitly (Stacey, 1996). A dynamic feedback model involves the observation of ongoing events, and then making adjustments or corrections. For organizational design, we monitor misfits on a continuing basis and then take corrective action as misfits occur. This is an event-driven feedback, dynamic approach. The misfit mechanism is not one observation and one correction, but involves a sequence of misfit corrections that we will consider later in this chapter. Assume that there is total fit - no misfits. Then the following questions become important:

- What might trigger a misfit?
- What happens over time when misfits are fixed?
- Are there different approaches to fixing misfits?

- Does it matter in which sequence the misfits are fixed?
- What happens if new shocks occur before all misfits are fixed?
- Are there lead and lag effects and are the variables in the model interrelated over time?

Next, we consider these issues.

Misfits: Where Do They Come From?

Misfits can arise from two fundamental sources: external shocks or changes outside the organization and managerially initiated changes. (See Table 11.1.) The organization's environment can change and seems to be changing at an increasing pace: new competitors, new products, new markets, etc. The technology can also be given and changed from the outside as new technologies are introduced. The exogenous factors are frequently beyond management's control, but can be assessed and adapted to. Donaldson (2001p. 247) suggests, "there is no theory of why the organization moves out of fit into misfit in the first place." Nevertheless, he (p. 249) posits, "the crisis of poor performance was required to trigger needed adaptive change." Poor performance may drive change, but there can be other anticipatory actions as well. We suggest that management itself can create misfits: the strategy can be modified, the size can be changed through merger or downsizing, the leadership can be changed in terms of actions or personnel, new technology can be introduced. These endogenous changes are the choice of management. The organizational climate can be changed by management's actions; or, an external event, such as a rumor of merger, can also change the climate. Similarly, the organization does have discretion over some technology changes - particularly those it creates itself. Even among new technologies from outside, it may have some choice in timing of adoption. The climate can be either exogenous or endogenous, depending upon the source of the change. Management both reacts to the exogenous changes and creates the endogenous changes that create misfits. Therefore, management's responsibility is dual: management should fix misfits as they occur, but management must also create misfits for higher-level organizational goals. For example, a new strategy may be needed even though it creates short-term misfits for the organization, which must be fixed later on. The management may disrupt the current situation for higher level and longer-term goals. The management is then to fix misfits and to create them, i.e., either seeks equilibrium in the organization or also creates disequilibrium for the organization. It is a constant cycle.

Table 11.2. External and Internal Misfits

External Misfits	External and Internal Misfits	Internal Misfits
Exogenous External Shocks Spontaneous Reactive	Exogenous and endogenous	Endogenous Created Proactive
Sources:	Sources:	Source: Management
Environment. Changes are not controlled or selected by the organization.	Technology. Can be given from the outside or developed internally.	Strategy. Management can change in the short term with long term implications.
	Climate. Cannot be changed much in short term, but can be for the longer term.	Size. through mergers, management can change in short term, can grow internally in the longer term
		Leadership. management is not likely to re create itself, but it can be changed, i.e., new management

The dynamic equilibrium state is likely to be one of constant existence of misfits and constant fixing through fit recommendations. For long run viability, it is not likely that the first would even find itself, or even want to find itself without misfits. Misfits come from many sources, including management itself. If the environment and technology are in constant change, then misfits are very likely, i.e., to have high probability of existence. If the organization does not change, then it can be very efficient in the short run, but not likely for the long run. This balance of misfit and fit is similar to the balance of exploration and exploitation (March, 1991). An organization that exploits well is likely to be in design fit, but is unlikely to be in contingency fit. An organization that explores well is unlikely to be in strategic fit and contingency fit.

External Changes: Evolutionary and Revolutionary

Meyer, Goes and Brooks (1993, p. 72) categorize external changes into evolutionary and revolutionary, which require incremental adaptation and frame-breaking metamorphosis, respectively. The exact breakpoint is difficult to assess precisely, but it is a significant qualitative difference resulting in two phases. The revolutionary phase has related concepts in punctuated equilibrium (Tushman and Romanelli, 1985; Gersick, 1991), and hypercompetition (D'Aveni, 1994; Volberda, 1996) , among others.

Punctuated equilibrium begins with a state of equilibrium that is shocked into disequilibrium and then seeks a new and perhaps different equilibrium. Tushman and Romanelli (1985, p. 173) define punctuated equilibrium as a process where "organizations progress through convergent periods punctuated by reorientations which demark and set bearings for the next convergent period." Convergent periods are relatively long periods of incremental adaptation; convergent periods are relatively short periods of discontinuous change in strategy and organization. They posit that executive leadership is the mediating force for change. To begin, the organization is in a state of equilibrium; it is shocked by an external force; and, it then seeks a new equilibrium. There are a number of issues, which arise: what do we mean by the equilibrium. What is a shock, how big is a shock, and where does it come from. What is the time interval between shocks, can the organization find a new equilibrium between shocks, i.e., does it have sufficient time; should the organization try to adjust to the external shocks; can an organization exist in a state of permanent disequilibrium. The external shock is one source of disequilibrium or misfit for an organization.

In contrast, hypercompetition suggests that the environment is changing quickly, and there is little time for management to adjust. That is, there is not the long period of convergence, but environments are more likely to be a more permanent state of punctuations.

Hypercompetition (D'Aveni, 1994, p. 154) is defined as "an environment characterized by intense and rapid competitive moves," and the "behavior is the process of continuously generating new competitive advantages and destroying, or neutralizing the opponent's competitive advantage, thereby creating disequilibrium..." Hypercompetition, or Schumpeterian competition, frequently involves shorter product life cycles, say 18 months rather than seven years, new competitors from outside the normal set, e.g., typewriters have been eliminated by PC's, discontinuities, and general unpredictability about what will occur in the environment, and its importance for the organization. Volberda (1996p. 367) suggests that hypercompetition is facilitated by disruptive activities - creating activities of firms that are capable of breaking some new ground, pioneering new fields, promoting radical innovation, and, in the process, partially or completely transforming the environment and the needed organizations.

The environment is ever increasingly changing for the organization. There are many descriptions of this new environment: hypercompetition, high velocity, high variety, chaotic, complex, and co-evolutionary. In addition, there are descriptions in more traditional terms: unstable, unpredictable, highly complex, highly uncertain, and equivocal. There are also notions that the environment is not given to the organization, but enacted by the organization itself, or evolving jointly in a co-evolutionary process (Koza and Lewin, 1999). The new environmental imperative looks at the environment, both in new ways and with new values to more traditional measures. For both, there is a renewed emphasis on the relationship between the organization and the environment in order to obtain desired performance.

Internal Changes: Intended Misfit Creation

Management must initiate and create misfits for a larger purpose. These internal shocks create disequilibrium in the short term for a larger goal. For example, Chandler's (1962) famous dictum that "structure follows strategy" creates a misfit as it is implemented - the new strategy is a shock and creates a misfit between the new strategy and old structure. To reestablish equilibrium, the structure must now follow the strategy, i.e., a fit is now to be established. The difference is that the source of the shock is internal by management. Once the misfit has been created, there is managerial action to re establish the equilibrium, or fit.

Management has broad choice. Management can create misfits through:

- A new strategy for the firm.
- A new product or new product line.
- A new marketing strategy and customer approach.
- A new technology or means of satisfying customer needs.
- A new information system.
- A different approach to human resource management and the incentive system.
- A merger with another firm.
- A strategic alliance with other firms.
- A partnership with another firm.
- Hiring or promoting a new team with a different leadership style.
- A major expansion or growth.
- Generally, the introduction of continuing change.

Each of these situations creates a misfit for the organization and thus, requires some response and a fix to establish a new equilibrium for the organization.

Management then needs a higher level of goals and objectives for creating misfits than the efficiency goals embedded in fit requirements. Misfit creation goals result from longer term strategic goals for the firm. The current strategy is not viable for the long term; thus, a modified strategy must be adopted. This, too is a misfit argument, namely, the there is a misfit between the current strategy and a desired strategy. The introduction of desired strategy resolves the strategy misfit, but then creates an organizational design misfit that then must be fixed. It is a cascade process.

Long-term viability goals lead to misfits for the firm in strategy, size, technology, or leadership style. Fixing these strategic misfits then creates organizational design misfits that must be addressed for efficiency. This fit condition is only temporary; it will be knocked out of kilter by external events, or by internal management created changes. The more quickly the environment or other factors change, the shorter will be the fit condition. For some industries in today's environment, it is likely that the fit is never obtained, but remains an elusive ideal. However, it does provide a framework for managerial action. Further, it provides a direction for desired change by the management.

THE MECHANISMS OF MISFITS

Fit involves a number of elements to be in balance as indicated in Figure 11.1. It is the simultaneous realization of fit on the multiple dimensions. Therefore, when an organization is in an equilibrium condition of total fit, and then is shocked by either internal or external forces; it is not a simple process of identifying a misfit and then fixing it to establish again equilibrium for total fit. Here, we want to examine the fit, misfit, and fit process in more detail.

Consider an organization that is in total fit, and then it is shocked; a misfit is created. What is the process back to total fit? We take a particular situation to illustrate the process and the intermediate states. Referring to Figure 11.2, assume the organization is in the lower left hand quadrant. This is a strategic fit condition for the organization. For this illustration, we will not include the contingency fit; we will address total fit later in this section. Now, assume there is an external environmental shock; the environment is changed from low equivocality, low uncertainty, and high complexity to high equivocality, high uncertainty, and high complexity. That is, the environment has changed from the lower left hand quadrant to the upper left hand quadrant. Now, the challenge is to bring it back to strategic fit situation. The shortest route is to change the environment back to its previous state; but usually this is not possible. So, how do we change other strategic elements to fit the new environment? Here, we examine the one by one step process in detail.

With the change of the environment from the low equivocality, low uncertainty, and high complexity to the high equivocality, high uncertainty, and high complexity, five misfits now have been created. This new environment is a misfit with the other five strategic states of: leadership, manager; climate, internal process; size, large; technology, mass; and strategy, defender.

Now, let us assume we fix the strategy to fit the new environment; that is, change the strategy from defender to analyzer. What are the misfits now? We have fixed one by aligning the strategy with the environment. Nevertheless, four misfits remain between the environment and the leadership, climate, size and technology. In addition, four new misfits have been created between the new analyzer strategy and leadership, climate, size and technology. Now, there are eight misfits - four plus four. In fixing a misfit, we have created more.

Continuing with the one by one process of fixing misfits, let us assume that we now change the leadership to fit the new environment and strategy. Thus, the number of misfits is nine: three each for the new environment, the new strategy, and the new leadership. This is the maximum number of misfits in the process. It is intuitive that the maximum number of misfits will occur at the midpoint of the change over, when half of the elements have been changed.

Figure 11.1. Strategic Fit

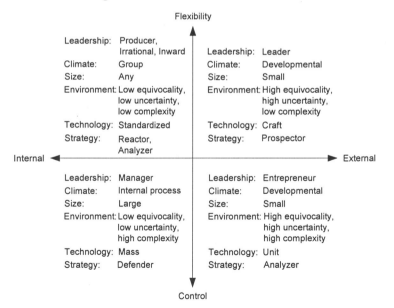

Let us continue by changing the climate to the new environment. We have a fit among the environment, strategy, leadership, and climate, but each does not fit with the size and technology. This yields eight misfits: two each for size and technology with environment, strategy, leadership, and climate.

Technology is brought into fit with the new situation. This gives five misfits as each of environment, strategy, leadership, climate and size does not fit with the size.

The organization is in strategic fit as it has moved from the lower left hand quadrant to the lower right hand quadrant with the change of size.

This one by one process of fixing misfits has a pattern of initially creating more misfits and then finally fixing all of them. Here, for six strategic factors, and the one by one process, the pattern is 0, 5, 8, 9, 8, 5 and 0 misfits.

In the above, we only considered one initial change and then one path to achieve strategic fit for the new environment. However, there are many possible paths, sequences, or orderings to follow in the one by one process. We want to develop the number of possible number of paths, say from the lower left quadrant to the upper left quadrant. Recall that we began with a shock to the environment, but there are five other possibilities: strategy, leadership, climate, size, and technology. Each could have been changed from external forces, or changed by management. Then, there is the choice of five elements that could now be changed. Then, four remain and so. There are 6! possible sequences, or 720 possible paths or orderings to change the strategic factors from the lower left to the lower right quadrant. In general, there are n! paths. Of these 720 possible paths, some are be better than others.

This one by one process for fixing misfits generalizes in a straightforward fashion. The model in Figure 11.1 for strategic and organization fit includes twelve basic variables: environment, strategy, management style, climate, size, technology, configuration, centralization, complexity, formalization, communication, and incentives. Beginning with any one element and making one by one changes to fix it, the number of misfits then follows the same pattern: 11, 20, 27, 32, 35, 36, 35, 32, 27, 20, and 11.

Figure 11.3 shows the results for this one by one process, where the x-axis represents the number of misfits fixed and the y-axis represents the number of misfits present. This simple model illustrates what many firms have experienced, namely that if you start to fix one problem, you may create more problems than you solve. If the changes take place over time, the performance may deteriorate dramatically. As above, the graph in Figure 11.2 assumes that the dimensions are related in a way that getting out of balance on one dimension will create imbalances with all other dimension.

The number of possible sequences has grown dramatically. For six elements, the number of possible orderings is 720. Now, for twelve elements, the number is 12!, or 4.79×10^8. The number of orderings goes up nonlinearly and quickly as n increases.

Figure 11.2. Total Number of misfits as misfits are fixed one by one

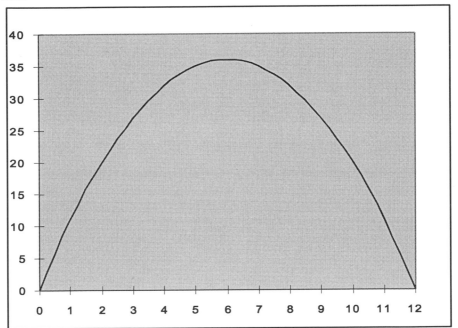

In the illustration above, we changed the strategy from defender to analyzer in light of the new environment. There is considerable evidence that it is very important to have the organization's strategy fit its environment (Porter, 1985). But we could have also chosen to change the climate first. Of course, the climate should be changed, but it takes a longer time and opportunity loss of the misfit of the environment with the strategy is likely to be quite costly. It is likely to minimize the opportunity loss of misfits by changing the strategy first (Burton, Obel et al., 2003). Nonetheless, we do not have much evidence on which are the better paths in the one by one process.

There are some important observations here. First, when an organization is shocked out of fit by a single change, it is normally not possible to bring it back into fit by adjusting to that change with a single change. In following the shock, the initial adjustments will create even more misfits. In the situation above, if the environment changes, and the organization adjusts its strategy only, it will have created even more misfits. Second, the maximum number of misfits occurs when the half of the elements fit, and the other half fit, but the two halves do not fit - when three strategic elements are in the one quadrant, and three in another. Third, there are n! possible paths in the one by one process; and further, we do not have much evidence on which of these paths are reasonable and which are not. These and other issues about a dynamic treatment of misfits will be dealt with in the next section.

HOW TO FIX MISFITS

In a recent paper, Zajac et al (2000) develop and test a model of dynamic strategic fit. Their position is that a fit model has to be dynamic, multivariate and include organizational strategy and environmental factors. They present four scenarios of strategic fit and misfit as shown in Table 11.3.

Table 11.3. Four Possible Scenarios of Dynamic Fit

		Does Strategic Change occur	
		Yes	No
Is Strategic Change needed to establish dynamic strategic fit	Yes	Dynamic fit	Dynamic misfit
	No	Dynamic misfit	Dynamic fit

The four scenarios reflect that fit and misfit occur related to the need for change and the actual change. Misfits may arise if a change is needed and the firm does not change. Misfits may also occur if no change is needed, and the firm makes changes, e.g., if other firms in the industry are making changes. Table 11.3 simplifies the dynamic fit issue as it assumes that appropriate changes take place. In many cases, the firm may recognize the need for change, but does not make the correct change. This will not provide the appropriate fit. Another important issue is the timing of the change and a company may change, but too late. Miles and Snow (1978) reactor category represents a type of company that reacts too late and often in an incorrect way.

This shows that there are a large number of approaches to fixing misfits. Total fit can be regained by bringing the misfit back to its original state. For example, in the above, bring the environment back to its initial state. This may only be possible if an unnecessary change caused the misfit. If that is not the case the organization must adapt to the new situation, e.g., adapting to the new environment.

The adaptation to a new situation can be done in various ways. Here, we want to consider and compare two approaches: the incremental one by one approach discussed above, and compare it with the one time total fix approach, i.e., making all the changes to move the organization at the same time. For example, in Figure 11.1, move from the lower left quadrant to the lower right quadrant in one move by making simultaneous changes in the strategic elements. Next, we want to compare the benefits and costs of the two approaches.

We compare the two approaches to fixing misfits: the incremental one by one approach and the total one fix approach. For the incremental one by one approach, we know that there are n! sequences possible, each of which could be different in their benefits and costs, but we will assume all of them are the same.

The benefit to changing to the new situation is to avoid the opportunity losses that are incurred by the misfits. So, if we assume that the opportunity losses due to misfits are positively related to the number of misfits, then the total opportunity losses are related to the number of misfits as shown in Figure 11.2. In order to minimize these losses from misfits, the organization should move as quickly as possible to the right hand end of the curve to a state of total fit. It is quite clear that the total fit approach is preferred to any one by one approach, as the opportunity losses will be minimized.

The cost of organizational change suggests that it is not possible to fix all the misfits immediately. The cost to change can be considered on three dimensions:

- The scope of the possible changes, i.e., can the organization actually change to fix the misfit. For example, an organization may not be able to change its climate and strategy in either order.

* The resistance to change in the organization.
* The time to change, i.e., how much time is required to make the changes.

First, the scope of the possible changes is greater when the organization can make the changes involved to fix the misfits, i.e., any of the n! paths is possible. Gresov (1989) found that not all misfits can be changed and it is not unusual for an organization to continue in an extended misfit situation. Volberda (1996) has discussed the concept of flexibility as a measure of an organization's capability to change to a new organization. He then makes an argument that an organization with greater flexibility is more likely to survive and prosper. His argument follows closely the concept of requisite variety: only variety can destroy variety (Ashby, 1956; Kuhn, 1986). That is, the scope of change must be as great as the scope of the changes imposed by external changes. For the one by one approach, the scope of possible changes must be sufficient to realize the sequence chosen; and further, it is expected that some paths are not feasible. For the total fit approach, the scope must be sufficiently large that the organization can move to the new situation. However, it does not require that all possible paths are feasible.

Second, resistance to change means the organization's capacity or lack of capacity to change. An organization with a high resistance to change has high inertia and it will remain with its current design. An organization's climate is frequently difficult to change. The leadership may be difficult as well; but it may be relatively easy to change top management. In general, an organization with many established routines will have higher resistance to change than an organization with few established routines.

Thirdly, the time to change is measured as the time it takes the organization to fix a misfit. In Figure 11.2, it is the time it takes to move the organization along the x-axis, either incremental, or in one total move. These three costs together make up the cost of change.

On balance, an organization should try to move to the new fit condition as quickly as possible to realize maximum benefits, and realizing that the move may be impeded by the costs that are realized in terms of the feasible scope, the resistance to change and the time to change.

One pragmatic approach is to the compare the benefit of a change with its cost on a one by one basis. Using this approach, the changes can be ordered by their benefit, and by their costs. Then, choose changes that have high benefit, or value and low cost. This is a myopic approach. It is reasonable, but we do not know whether it has local peaks that are difficult to go beyond. That is, we do not know how rugged the response surface is. Nonetheless, myopic search is reasonable and is relatively easy to operationalize.

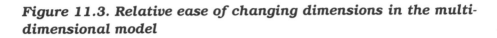

Figure 11.3. Relative ease of changing dimensions in the multi-dimensional model

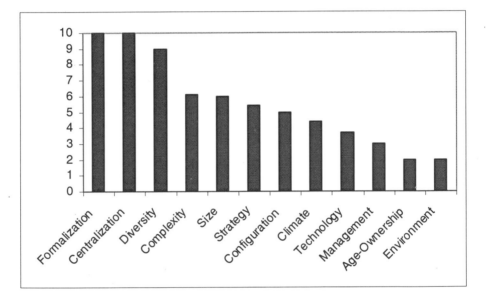

In many companies, the ease of changing the parameters in the model from Figure 1.3 is shown in Figure 11.3. Generally, the organizational dimensions are easier to change than the strategic dimensions.

Our executive MBA students developed a relative ease of change graph as shown in Figure 11.3. They argued that formalization is relatively easy to change; and, the environment is the most difficult. Of the n! paths, they suggest this is the least cost path. Even so, we do know of firms that continue to use formalization, or routines in the face of a strategy that calls for different formalization levels. The flexible organization calls for low formalization; and, many firms fail to realize it. Further, they posit that the organizational dimensions are easier than the strategic dimensions to change with the configuration being the exception. However, normally we argue that structure should follow strategy, but it may not be the easier to implement. It may be easier to change some organizational elements before changing the strategy. It is an interesting issue of what is a good ordering of implementation for fixing misfits or initiating managerial change.

HOW FAST IS FAST ENOUGH

What is a good approach for the fast changing world? We review the static approach, the punctuated equilibrium adaptation approach, and the time-paced organization. The later two are dynamic in that time is explicitly considered and managed.

For the static approach where changes are quick, but moderate enough for adjustment without changing the organization, Volberda (1996, p. 366-367) recommends a flexible organization. He defines flexibility as "the degree to which an organization has a variety of managerial capabilities and the speed at which they can be activated, to increase the control capacity of management and improve the controllability of the organization."

Flexibility is defined in terms of managerial capability and speed to obtain necessary control. For the requisite flexibility, he suggests a non-routine technology, organic structure and an innovative climate. The basic notion here follows Ashby's (1956) Law of Requisite Variety that only variety can destroy variety. That is, the managerial variety of the organization must exceed the variety in the environment (Kuhn, 1986). If the variety of the environment exceeds that of the organization then there will come a time when the organization cannot continue to be viable in its environment and it will not survive. This law is not new, but what is new is the variety we have in today's environment. Quite simply, the variety has increased in the environment; thus, the variety of possible responses by the organization must also increase in a commensurate fashion. Volberda (1996) then specifies that the commensurate response should be an organic structure, non-routine technology, and an innovative culture with lots of information gathering and processing. Tom Peters suggests: crazy times call for crazy organizations.

In Chapter 10, we developed two kinds of flexible organizations. The first is small, entrepreneurial, results oriented, informal, with high information utilization. The second is large, decentralized, results oriented, yet informal organization that requires lots of information gathering from the outside and information processing through meetings and other integrators. The organizational climate for both is a rational goal climate that has low resistance to change and high tension. Both have a non-routine technology, where the small organization does not require a divisible technology, but the large organization does. Each is a fit for an environment that is described as highly uncertain, highly complex, and highly equivocal. We argue that these organizations are capable of a wide scope of response of activities, low resistance and can change quickly.

A dynamic approach introduces time explicitly. The punctuated equilibrium is a shock to the organization that cannot be absorbed by normal adjustments by the organization, e.g., by increasing the production quantity without changing the production methods. The disequilibrium is observed as a misfit, either in the strategic or contingency factors. We demonstrated that the equilibrium restoration or obtaining total fit is not a one-step process, but involves many changes, which can be done either one by one, or in one big change. When the cost of change is sufficiently low to yield total fit, by either approach, before a new shock occurs then the adaptation can be realized in a fit, misfit, total fit adaptation process. But we have noted that the time between shocks may be short and it is not possible to realize total fit before a new shock occurs. Total fit may never be obtained. The question now is whether it is appropriate to fix as many misfits as possible, observe the new misfits and determine a new set of misfits to fix. The process is to fix misfits, knowing that total fit will never be realized. One question is how to choose which misfits to fix and which to ignore at any point in time. This is particularly important if not all misfits are fixed. Is this a good approach, or is there a better approach? Another approach is to monitor misfits intermittently, rather than on a continuing basis. Then, the misfit process would be to fix the observed set ignoring the misfits that had been created in the meantime. With approach, the time between observations becomes a critical issue.

The process of fixing misfits is complex. Earlier, we demonstrated that the one by one process involves n! possible paths to choose from. Here, we consider the issue of a good approach to fixing misfits when the interval between shocks is shorter than the time required to obtain total fit. The process of fixing misfits over time is also complicated by interrelationships between the variables in the multidimensional contingency model. The leadership behavior may affect the organizational climate. For market leader companies, choices of strategy may effect the environment. Additionally some misfits may have an immediate effect on performance while other misfits may affect the performance in the long run. Thus introducing time in the fit-misfit sequence complicates the analysis significantly.

The third approach is a time-paced organization. For a rapidly changing world, Eisenhardt and Brown (1998) propose time paced change for the organization. Based upon their research of high technology firms, they develop nine rules:

Rule 1. Advantage is Temporary.
Rule 2. Strategy is Diverse, Emergent, and Complicated.
Rule 3. Reinvention is the Goal.
Rule 4. Live in the Present.
Rule 5. Stretch Out the Past.
Rule 6. Reach into the Future.
Rule 7. Time Paced Change.
Rule 8. Grow the Strategy.
Rule 9. Drive Strategy from the Business Level.
Rule 10. Repatch Businesses to Markets and Articulate the Whole.

The first three rules state the required strategy that is diverse, emergent, and complicated and requires lots of information processing. There is an emphasis on time and that change is to be expected. Therefore, we can contrast this notion of strategy with traditional notions of long term, fixed and narrowing of what the organization will do. Here, the strategy is broad and opportunistic. Rather than aiming for a long-term objective, the goal for the organization is to look continually at present opportunities and adjust.

The next four rules present the nature of the organization, where the focus is on time and its management. The organization keeps both the past and future in mind to create change in the present. Change should be time paced, and not event paced. This is a central notion in their research. That is, the organization should introduce change according to its own clock, and not wait or be driven by external events, such as technology development or competition. The organization must make its own future by creating its own disruptions or misfits. Intel is an example; they build a new factory every 18 months and they make Moore's Law of doubling computational capacity every 18 months a reality. They make the future according to their own time schedule.

The last three rules on leadership suggest a kind of quilt work process that implies guided and rationalized decentralization, which makes up a consistent whole. Various metaphors emerge; here, leadership is tending an English garden according to the seasons rather than a designing a French one.

Eisenhardt and Brown (1998) state their design on complementary dimensions to organizational structure and characteristics. Their emphasis is on how time is viewed, managed, and used by the organization. Time paced means a regularity of activity and change by the organization. Change becomes usual and expected. Herein lies the paradox; change may become so usual, that it is no longer change. That is, time pacing makes change routine: it becomes ordinary, normal, regular, and expected. They (1998, p. 66) present three basics for the time paced organization:

- *Performance metrics*. Time to do something: product development, product launch, sales growth realization, etc. becomes the performance measure, rather than an emphasis on profits, sales, and financial measures.
- *Transitions*. These are organizational changes, such as product changes, new product introductions, absorbing acquisitions are the important management activities.
- *Rhythms*. The normal cycles, which drive the business, include the relations with customers, suppliers, and competitors.

The rhythms should be frequent enough to meet the business needs and make change normal and expected in the organization, but not so often as to create the motion of change without productivity. That is, the organization is in a state of flux that utilizes all of its resources and energy while realizing the output of products or services. There is an optimal rhythm.

There are many advantages to setting one's own destiny in a fast changing world. Expectations are set and clear; uncertainties within the organization are reduced; change is normal. Yet, the organization is opportunistic and adjusts broadly to its opportunities. The approach is consistent with creating managerial variety, consistent with the Law of Requisite Variety. But, are there risks? Yes! The organization may not create the variety demanded by the environment. Eisenhardt and Brown (1998) discuss Intel as an example; Intel creates its own future and drives the technology for computation. Is it possible that a new approach could out flank them? IBM in the 70's had a similar hold on mainframe computing, and almost missed the PC revolution. Earlier, Polaroid created its own destiny with the Polaroid camera and instant pictures, but has more recently switched its focus to a more analyzer strategy. Today, 3M demands 30% of its revenue must come from new products every year. These products can come from a broad range of products, technologies, and markets. It creates its own future within much broader parameters than Intel does. The time-paced organization does not eliminate risks, but shifts the nature of the risks from forecasting and adjusting to projection and exploration. Can the organization be time paced and adjust quickly. Population ecology questions whether the organization can adapt. Hannan and Freeman (1977) posit that it is the environment that optimizes. Nonetheless, the time-paced organization may be the best strategy when the environment is unpredictable and changing.

DESIGN IS EXPLORATION AND EXPLOITATION

March (1991; 1994) developed the concepts of exploration and exploitation as action frames for the future of an organization. Exploration means "adaptiveness" and includes "result, dynamic and long run" (1994, p.1) and "search, variation, risk taking, experimentation, play, flexibility, discovery, innovation" (1991, p.71); exploitation includes "efficiency, rules and routines, static, short run," (1994, p.1) and "refinement, choice, production, efficiency, selection, implementation, execution" (1991, p.71). For our design goals, exploitation is closely related to efficiency and effectiveness; exploration to effecness and viability. Either approach to the exclusion of the other is likely to lead to death of the organization. Exploitation alone leads to a trap of short run efficiency but in the long run a mismatch with its challenges and opportunities. Exploration alone is costly without the benefits of producing products or services efficiently. Management must find a balance of exploration and exploitation.

In our terms, exploitation is very close to finding a "fit" for the organizational design. We want to find an organization, which exploits the environment and is efficient and effective. Exploration is more concerned with the viability of the organization and the creation of selected "misfits," mostly strategic misfits. Of course, viability cannot be assured through the creation of misfits alone and hence, viability incorporates exploitation as well. In Figure 1.3, the concept of fit includes three aspects: strategic, contingency and design.

Exploration can generate misfits in at least two ways: managerial and natural. Managerial misfits are conscious and explicit creation of strategic misfits by the adoption of a new leadership style, strategy, or technology for the organization that in the long run will obtain a better performance. For example, management may adopt a new technology, which does not fit with the current strategy. In this book and in the OrgCon, we suggest what the strategic misfits are and where to explore for resolution, if management chooses. Strategic misfits can thus be purposeful and part of the organization's exploration.

Strategic misfits can also arise naturally outside the organization, without managerial choice. One frequent natural strategic misfit occurs when the environment for the organization changes and the organization's strategy does not. E.g., a defender strategy, which was appropriate for a low uncertainty environment, may not be appropriate for a highly uncertain environment. In short, nature may explore itself and create misfits.

Exploitation is misfit reduction and establishing the design fit for the organization. In Figure 1.3, the design fit criteria are strategic, contingency and design. Here, the goal is to find a design, which is efficient and effective. In the diagnosis and design of an organization, we could consider the design questions:

- If we want an organization which will explore well and exploit less well, what should the organization design be?

- If we want an organization which will exploit well and explore less well, what should the organization design be?

There are very large numbers of possibilities here. We propose a two-step process: first, we will suggest likely strategies, leadership styles, size, and technology for exploration and exploitation, second, we will utilize the design concepts to suggest a good design. The first phase is somewhat exploratory itself. The second phase exploits what we know. Exploration and exploitation are then intermediate design properties in this view.

- If the organization is an explorer, then the strategy should be prospector; the leadership style should be entrepreneur; the size should be small or medium; and the technology should be nonroutine.

This organization is similar to the entrepreneur stage of the lifecycle. It is one possible way to be an explorer.

- If the organization is an exploiter, then the strategy should be defender; the leadership style should be manager; the size should be medium or large; and, the technology should be routine.

This organization is similar to the later stages of the lifecycle.

These propositions are in part definitional. Exploration is adaptive. A prospector is adaptive in the extreme and creates its own environment. An analyzer is also adaptive to its environment. Exploration requires a lot of information to be processed from many sources. A manager will become overwhelmed. A large organization usually will not be able to explore and act quickly. In addition, a non-routine technology is required to explore. Exploitation focuses on efficiency and doing things well. It need not adapt well or quickly; it can defend. Routine technology is more efficient and management can be more involved and still manage the large organization.

The second phase, the design phase is then:

- If organization is an explorer, then the organization should be (have) a simple, matrix, or ad hoc configuration, low formalization, high decentralization, medium organizational complexity, coordination by meetings, liaison roles, committees, results based incentives, and high media richness.

- If organization is an exploiter, then the organization should be (have) a functional, machine bureaucracy configuration, high formalization, high centralization, high organizational complexity, coordination by rules, procedural based incentives, and low media richness.

In many ways, the exploring organization is the mirror image of an exploiting organization, and vice versa. The optimal balance between the two extremes depends upon a more detailed statement of contingencies for the organization as suggested in Figure 1.3. The exploration-exploitation pair provides a way to think about the realization of the efficiency, effectiveness, and viability goals of the organization.

The exploration-exploitation pair can also be considered in two phases: the diagnosis and design phase, and the operations phase. In the above, the diagnosis and design phase have been discussed. That is, given the situation, what is a good design for the organization. Here we want to turn to the second phase: the operations phase for the organization. For the above organizations, we posit that:

- If the organization has a simple, matrix, or ad hoc configuration, low formalization, high decentralization, low organizational complexity, coordination by meetings, etc., results based incentives and high media richness, then the organization can explore well.
- If the organization has a functional or machine bureaucracy configuration, high formalization, high centralization, high organizational complexity, coordination by rules, procedure based incentives and low media richness, then the organization can exploit well.

We can generalize that an organization which explores well will not exploit well and vice versa. There is a tradeoff and a choice. Although the discussion above states the propositions in contrasts and extremes, we also suggest the choice is more complicated; there are multiple intermediate designs which strike different balances between exploration and exploitation.

DESIGN IS LEARNING

March's exploration-exploitation is also a frame for learning - organizational learning. As discussed above, the exploration-exploitation requires a balance and either extreme is not viable. March (1991) focuses on two other aspects of the balance: risk and investment. Risk attitude is central to striking the balance - greater exploration requires a greater tolerance for risk and failure or preference for uncertainty. March posits that most new ideas or explorations are bad ideas, or will turn out to be bad ideas. Obviously, we do not know which ones *a priori*, or we could avoid them from the beginning. Exploration then requires activities which have unknown outcomes with a high probability of failure. If failure is severely punished, then individuals will avoid exploration and the organization will not learn, or learn very little and slowly. In order to explore and learn, we must not only tolerate, but also encourage risk taking which necessarily involves failure. Sitkin (1992) argues that small failures are preferred to large ones and that a learning organization should reward taking risk and encourage small failures, which permit the organization to learn for greater success. This approach creates a balance of exploration with risk. Whether an organization should encourage very large explorations and associated risks will depend upon the desired balance.

Levinthal and March (1993) argue that many, if not most organizations strike an improper balance, excessive emphasis on short run exploitation to the detriment of longer run exploration; it is too myopic. In the short run, individuals are rewarded for risk-averse behavior, which puts a premium on exploitation. Over the longer run, there is too little exploration and the performance of the organization suffers. Of course, exploration is not meant to be random and without purpose; and further, success should be rewarded more than a failing - even a valiant attempt. There is an underinvestment in exploration and an overinvestment in short run exploitation, which will persist until exploration is rewarded by the organization.

They offer two mechanisms for learning and exploration: simplification and specialization. Simplification involves the construction of buffers: specialized individuals who are charged to learn with targets, search, and slack. Decomposition of problems into smaller subproblems is the second approach. Decomposition and its relation to the organization structure is the most basic mechanism. Loosely coupled organizations are better for learning, where tightly coupled are better for error detection and control (p.92). As we have discussed in Chapter 2, divisional configurations are more loosely coupled than are functional configurations. The divisions are less connected than are the functions. Further, low formalization with fewer rules and high decentralization tend to spread decision-making and decouple the organization. Decomposition is one means to keep the risk of failure from exploration small. Clearly, the incentives for risk taking are important. Galunic and Eisenhardt (1996) found that divisional organizations can obtain new areas to explore and develop through the assignment of charters to the division. That is, a division will be chartered to initiate a new product or line of business. The large corporation can learn by assigning learning tasks to the smaller divisions, i.e., decomposing the problem into a smaller one.

Cohen and Levinthal (1989; 1990) develop the concept of absorptive capacity of an organization as a measure of the organization capacity to learn and innovate. They demonstrate that organizations, which have research capability, learn more quickly than organizations, which do not have a research capacity. This capacity may not lead to new and fundamentally new ideas, but it does facilitate the understanding and evaluation of what others are doing, and then leads to a reasoned and timely adoption of innovations and new technologies. The organization can exploit outside information effectively and timely. Research capacity keeps the organization at the edge of new technology and permits it to innovate, even if it does create the new edge of technology. Here, learning has two important elements: evaluation of what is important and secondly, knowing early on. Without the capacity, organizations tend to learn too late for the innovation to be of value. It helps explain why an organization should invest in basic research, even though most basic research is in, or will become part of the public domain.

To examine barriers to learning in organizations, and why organizations have difficulty taking on new challenges and ideas is another approach to learning. Probst and Buchel (1997, p. 71-72) argue that the hierarchal organization design itself can create barriers to information processing and learning. Greater organizational complexity means that the organizational work is partitioned and broken up into finer pieces and the individuals who are assigned to the separate jobs can be further apart. Thus, there is a greater need to process information for coordination. The focus is first on operations. Second, the "new" and different ideas needed for learning may be driven out. A higher vertical differentiation creates more levels between the top and bottom of the organization, where each level can facilitate information flow, but can also impede it. The greater the number of levels, the more likely that distortion and impediment will occur. The organization is then limited in its ability to learn from within itself, but also to share information for learning from the outside. One solution is to increase the information processing of the various layers. Another approach is simply to eliminate some of the layers, typically middle management in what we call "delayering." But will delayering increase organizational learning? Not necessarily, the fundamental problem remains. New ideas may not be generated, considered and adopted. We conclude that high organizational complexity can be a barrier to learning, but simple delayering may not generate learning by itself; it only makes it easier.

In a similar manner, the horizontal differentiation of a functional organization can be a barrier to learning. Learning can take place within the functions, but it is much more difficult to generate challenges and new ideas, which span the functions. Individuals within the functions observe problems and opportunities more readily within functions and further they can deal with them easier. Inter-functional problems, e.g., a production problem across marketing and operations, are more difficult to observe, define and mount the necessary actions to than issues within the function. A traditional approach is to create liaison activities as boundary spanners, committees, taskforces, etc. to solve problems, but also to learn in new ways. In brief, the argument is that the greater the partitioning of the work, the more difficult it is to learn. The challenge is to find mechanisms, which facilitate learning. Divisional configurations partition the organization into to loosely coupled subunits. However, the coupling should not be total such that each unit is independent. The organization must be coupled enough to observe and learn from others and from the interfaces, but at the same time, not coupled so tightly as to impede experimentation and learning.

Some organizational designs are more conducive to learning than other designs. There is a needed balance between being coupled just enough to learn and not too tightly connected to drive out experimentation and exploration. In terms of configurations, we suggest that simple, ad hoc and divisional configurations are more conducive to learning than functional and bureaucratic configurations. As discussed above, greater complexity is likely to impede learning. Clearly, greater formalization will minimize experimentation and learning. Greater decentralization is more likely to enhance learning. Higher media richness and general knowledge about the outside as Levinthal and March (1993) argue should support learning. Finally, results based incentives, which are tolerant of failure, should also help. Organizational learning remains a rich area for research, despite recent development and progress.

SUMMARY

In this chapter, we have developed a dynamic model for strategic and organizational diagnosis and design. In Figure 11.4, there is a summary flowchart of diagnosis and design, which describes the dynamic process for the organization.

We illustrated the process by examining the lifecycle of the organization. We then turned to a more detailed examination of the dynamics of diagnosis and design. A misfit can be triggered either by an external shock as a changing environment, or an internal managerial change which disrupts the organization in the short term for longer term performance. Poor performance is a frequent motivation. The fixing of misfits is not a simple response, but usually involves a long path. Beginning with a total fit situation, one misfit will create more misfits as strategic or organizational adaptations are made. Initially, more misfits are created, before the number decreases as the organization is brought into total fit. In general, for n fit elements, there are n! one by one paths to obtain fit that the organization may take. The one by one approach incurs the opportunity performance loss of misfits during the recovery time. A one time total change incurs less opportunity loss. Balancing the misfit performance loss is the cost of change. Three cost elements include: the scope of the strategic and organizational change and is it feasible; the resistance to change; and, the time it takes to implement the changes. Pragmatically, we would like to introduce these changes according to their ease of implementation. Even so, the external shocks may come rapidly so that it is not possible to realize a total fit before a new shock occurs. Diagnosis and design is an ongoing process.

Figure 11.4. Diagnosis and Design Process

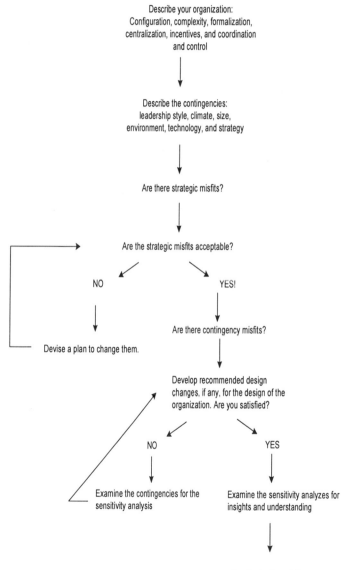

We then examined the concepts of exploration and exploitation and their implications for the appropriate design and also the diagnosis and design process. An exploring organization is like a prospector; an exploiting organization is like a defender. As process, exploration puts an emphasis on diagnosis and strategic issues; exploitation puts an emphasis on design and adjustment. Generally, exploration is creating misfits; exploitation is fixing them.

We then extended these notions to examine design as learning. Here the issue is to devise good experiments for organizational learning. Given the organizational goals and strategy, some organizations should devise exploring experiments, others exploiting experiments.

In brief, the diagnosis and design process is an ongoing dynamic process, which has a very large number of possible paths.

REFERENCES

Aldrich, H. E. 1972. Technology and Organization Structure: A Reexamination of the Findings of the Aston Group. *Administrative Science Quarterly*, 17(1): 26-43.

Alexander, C. 1964. *Notes on The Synthesis of Form.* Cambridge: Harvard University Press.

Amburgey, T. L. & Dacin T. 1994. As the Left Foot Follows the Right? The Dynamics of Strategic and Structureal Change. *Academy of Management Journal*, 37(6): 1427-1452.

Anderson, T. & Warkov, S. 1961. Organizational Size and Functional Complexity: A Study of Administration in Hospitals. *American Sociological Review*, 23-28.

Ansoff, H. I. 1965. *Corporate Strategy.* New York: McGraw-Hill.

Argote, L. 1999. Organizational Learning: Creating, Retaining and Transferring Knowledge. Boston: Kluwer.

Argote, L. & Ingram, P. 2000. Knowledge Transfer: A Basis for Competitive Advantage in Firms. *Organizational Behavior and Human Decision Processes*, 82(1): 150-169.

Arrow, K. 1974. *The Limits of Organization.* New York: W.W. Norton.

Ashby, W. R. 1956. *Introduction to Cybernetics.* London: Methuen.

Ashkanasy, N. M., Wilderom, C. P. M., & Peterson, M. F. 2000. *Handbook of Organizational Culture and Climate.* Thousand Oaks: Sage.

Baligh, H. H. 1994. Components of Culture: Nature, Interconnections, and Relevance to the Decisions on the Organization Structure. *Management Science*, 40(1): 14-27.

Baligh, H. H., Burton R.M, & Obel, B. 1996. Organizational Consultant: Creating a Useable Theory for Organizational Design. *Administrative Science Quarterly*, 42(12): 1648-1662.

Baligh, H. H. & Burton, R. M. 1982. The Movable Boundaries Between Organizations and Markets. *International Journal of Polity Analysis and Information Systems*, 6(4): 435-449.

Darlett, C. A. & Ghoshal, S. 1988. Organizing for Worldwide Effectiveness: The Transnational Solution. *California Management Review*, 54-74.

Barlett, C. A. & Ghoshal, S. 1989. *Managing Across Borders: The Transnational Solution.* Boston: Harvard Business School Press.

Birkinshaw, J., Nobel, R., & Ridderstråle, J. 2002. Knowledge as a Contingency Variable: Do the Characteristics of Knowledge Predict Organization Structure? *Organization Science*, 13(3): 274-289.

Blau, P. M. 1970. A Formal Theory of Differentiation in Organizations. *American Sociological Review*, 201-218.

Blau, P. M. & Schoenherr, R. A. 1971. *The Structure of Organizations.* New York: Basic Books.

Bluedorn, A. C. & Lundgren, E. F. 1993. A Culture-Match Perspective for Strategic Change. *Research in Organizational Change And Development*, 7: 137-179.

Boje, P., Madsen, T. K., Obel, B., and Sørensen, C. 1990. De Erhvervsmæssige Konsekvenser og Muligheder for Fyn set i Lyset af den Faste Storebæltsforbindelse og EF's Indre Marked,

Bourgeois, J. & Eisenhardt, K. 1988. Strategic Decision Processes in High Velocity Environments: Four Cases in the Microcomputer Industry. *Management Science*, 34: 816-835.

Brynjolfsson, E. 1994. Information Assets, Technology, and Organization. *Management Science*, 30(12): 1645-1662.

Brynjolfsson, E., Malone, T. W., Gurbaxani, V., & Kambil, A. 1994. Does information Technology Lead to Smaller Firms? *Management Science*, 40(12): 1628-1644.

Buchko, A. A. 1994. Conceptualization and Measurement of Environmental Uncertainty. An Assessment of the Miles and Snow Perceived Environmental Uncertainty Scale. *Academy of Management Journal*, 37(2): 410-425.

Buraas, A. 1972. *Fly over Fly*. Oslo: Gyldendal Norsk Forlag A/S.

Burns, T. & Stalker, G. M. 1961. *The Management of Innovation*. London: Tavistok.

Burton, R.M and Kuhn, A. J. 1979. Strategy Follows Structure! The Missing Link of Their Intertwined Relation..

Burton, R. M., Lauridsen, J., & Obel, B. 2002. Return on Assets Loss from Situational and Contingency Misfits. *Management Science*, 48(11): 1461-1485.

Burton, R. M. & Obel, B. 1984. Designing Efficient Organizations: Modelling and Experimentation. Amsterdam: North Holland.

Burton, R. M. & Obel, B. 1986. Environmental-Organizational Relations: the Effects of Deregulation. *Technovation*, 5(1): 23-34.

Burton, R. M. & Obel, B. 1988. Opportunism, Incentives, and the M-form Hypothesis: A Laboratory Study. *Journal of Economic Behavior and Organization*, 10: 99-119.

Burton, R. M., Obel, B., & Lauridsen, J. 2003. How the CEO View of Organizational Climate and Strategy affects Performance! *Human Resource Management (Submitted)*.

Byles, C. M. & Labig, C. E. Jr. 1996. Hospital Strategy and its Relationship to Administrative Practice and Performance: A Partial Test of the Miles and Snow Typology. *Journal of Managerial Issues*, 8(3): 326-342.

Cameron, K. & Quinn, R. E. 1999. *Diagnosing and Changing Organizational Culture*. Addison-Wesley Pub. co.

Cameron, K. & Whetton, D. A. 1981. Perceptions of Organizational Effectiveness Over Life Cycles. *Administrative Science Quarterly*, 26.

Cameron, K. S. & Freeman S.J. 1991. Cultural Congruence, Strength, and Type: Relationships to Effectiveness. *Research in Organizational Change And Development*, 5: 23-58.

Campbell, J., Dunnette, M. D., Lawler III, E. E., & Weick, K. E. 1970. *Managerial Behavior, Performance, and Effectiveness*. McGraw-Hill.

Caplow, T. 1957. Organizational Size. *Administrative Science Quarterly*, 484-505.

Carillo, P. M. & Kopelman, R. E. 1991. Organization Structure and Productivity. *Group and Organization Studies*, 16(1): 44-59.

Carley, K. M. 1995. Computational and Mathematical Organizational Theory: Perspective and Directions. *Computational and Mathematical Organization Theory*, 1(1): 39-56.

Carlzon, J. 1985. *Riv Pyramiderne Ned!* København: Gyldendal.

Chandler, A. D. 1962. Strategy and Structure: Chapters in the History of the Industrial Enterprises. Cambridge, MA: MIT Press.

Chenhall, R. H. 2003. Management Control Systems Design Within its Organizational Context: Findings from Contingency-based Research and Directions for the Future. *Accounting, Organizations and Society*, 28(2-3): 127-168.

Child J. 1982. Discussion Note: Divisionalization and Size: A Comment on the Donaldson/Grinyer Debate. *Organization Studies*, 251-353.

Child, J. 1972. Organization, Structure, Environment and Performance: The Role of Strategic Choice. *Academy of Management Review*, 561-575.

Child, J. 1973a. Parkinson's Progress: Accounting for the Number of Specialists in Organizations. *Administrative Science Quarterly*, 328-346.

Child, J. 1973b. Predicting and Understanding Organization Structure. *Administrative Science Quarterly*, 18: 168-185.

Cohen, W. & Levinthal, D. 1990. Absorptive Capacity: A New Perspective on Learning and Innovation. *Administrative Science Quarterly*, 28(223): 244.

Cohen, W. M. & Levinthal, D. A. 1989. Innovation and Learning: The Two Faces of R & D. *The Economic Journal*, 99: 569-596.

Conference Board Chart Collection. 1991. New York: The Conference Board.

Cooper, R. B. & Quinn, R. E. 1993. Implications of the Competing Values Framework for Management Informations Systems. *Human Resource Management*, 32(1): 175-201.

Covin, J. G. & Slevin, D. P. 1989. Strategic Management of Small Firms in Hostile and Benigh Environments. *Strategic Management Journal*, 10: 75-87.

Cullen, J. B., Victor, B., & Bronson, J. W. 1993. The Ethical Climate Questionnaire: An Assessment of its Development and Validity. *Psychological Reports*, 73: 667-674.

Cyert, R. M. & March, J. G. 1963. *A Behavioral Theory of the Firm.* Englewood Cliffs, NJ: Prentice Hall.

D'Aveni, R. A. 1994. Hypercompetitive Rivalries: Competing in Highly Dynamic Environments. New York: The Free Press.

Daft, R. 1992. *Organization Theory and Design.* St. Paul, MN: West.

Daft, R. & Lengel, R. H. 1986. Organizational Information Requirements, Media Richness and Structural Design. *Management Science*, 32(5): 554.

Daft, R. & Lewin, A. 1993. Where are the Theories for the 'New' Organizational Form? An Editorial Essay. *Organization Science*, 4(2): i-vi.

Daft, R. L. & Weick, K. E. 1984. Toward a Model of Organizations an Interpretation Systems. *Academy of Management Review*, 9: 254-295.

Davidov, W. H. & Malone, M. S. 1993. *The Virtual Corporation.* New York: Harper.

Delery, J. E. & Doty, H. D. 1996. Modes of Theorizing in Strategic Human Resource Management: Tests of Universalistic, Contingency, and Configurational Performance Predictions. *Academy of Management Journal*, 36(4): 802-835.

Demski, J. S. & Feltham, G. 1978. Economic Incentives in Budgetary Control systems. *The Accounting Review*, LIII(2).

Denison, D. R. 1990. Corporate Culture and Organizational Effectiveness. Wiley.

Denison, D. R. 1996. What is the Difference Between Organizational Culture and Organizational Climate? A Native's Point of View on a Decade of Paradigm Wars. *Academy of Management Review*, 21(3): 619-654.

Denison, D. R., Hooijberg, R., & Quinn R.E. 1995. Paradox and Performance: Toward a Theory of Behavioral Complexity in Managerial Leadership. *Organization Science*, 6(5): 524-540.

DeSanctis, G. & Poole, M. S. 1994. Capturing the Complexity in Advanced Technology Use: Adaptive Structuration Theory. *Organization Science*, 5(2): 121-146.

DeSanctis, G., Staudenmayer, N., & Wong, S. S. 1999. Interdependence in Virtual Organizations. *Trends in Organizational Behavior*, 6: 81-103.

DiPadova, L. N. & Faerman, S. R. 1993. Using the Competing Values Framework to Facilitate Managerial Understanding Across Levels of Organizational Hierarchy. *Human Resource Management*, 32(1): 143-174.

Donaldson, L. 1982. Divisionalization and Size: A Theoretical and Empirical Critique. *Organization Studies*, 321-337.

Donaldson, L. 1987. Strategy and Structural Adjustment to Regain Fit and Performance: In Defense of Contingency Theory. *Journal of Management Studies*, 24(1): 1-24.

Donaldson, L. 2001. *The Contingency Theory of Organizations*. Thousand Oaks, CA: Sage Publications.

Doty, D., Glick, H. W. H., & Huber, G. P. 1993. Fit, Equifinality, and Organizational Effectiveness: A Test of Two Configurational Theories. *Academy of Management Journal*, 38(6): 1198-1250.

Downey, H. K., Hellriegel, D., & Slocum, J. W. Jr. 1975. Environmental Uncertainty: The Construct and its Application. *Administrative Science Quarterly*, 20: 613-629.

Drazin, R. & Van de Ven, A. H. 1985. Alternative Forms of Fit in Contingency Theory. *Administrative Science Quarterly*, 30(4): 514-539.

Duncan, R. B. 1972. Characteristics of Organizational Environments and Perceived Environmental Uncertainty. *Administrative Science Quarterly*, 17(3): 313-327.

Duncan, R. B. 1979. What is the Right Organization Structure? *Organizational Dynamics*, Winter: 59-79.

Døjbak, D., Burton, R. M., Lauridsen, J., and Obel, B. 2003. Misfit Losses Between Leadership and Strategy for the Firm.

Eisenhardt, K. M. 1975. Control: Organizational and Economic Approaches. *Management Science*, 31(1): 134-149.

Eisenhardt, K. M. 1989. Making Fast Strategic Decisions in High-Velocity Environments. *Academy of Management*, 32(3): 543-576.

Eisenhardt, K. M. & Brown, S. L. 1998. Time Pacing: Competing in Markets that won't stand still. *Harvard Business Review*, 59-69.

Ekvall, G. 1987. The Climate Metaphor in Organization Theory. In Bass, B. M. & Drenth, P. S. D. (Eds.), *An International Review*: Sage Publications.

Etzioni, A. 1964. *Modern Organizations*. Englewood Cliffs, NJ: Prentice-Hall.

Fiedler, F. E. 1977. What Triggers the Person-Situation Interaction in Leadership? In Magnusson, D. E. N. S. (Ed.), *Personality at the Crossroads: Current Issues in International Psychology*: Hillsdale NJ: Erlbaum.

Fink, S. L., Jenks, R. S., & Willits, R. D. 1983. *Designing and Managing Organizations*. Homewood, IL: Irwin.

Fredrickson, J. W. 1984. The Effect of Structure on the Strategic Decision Process. *Academy of Management Proceedings*, 12-16.

Fredrickson, J. W. 1986. The Strategic Decision Process and Organization Structure. *Academy of Management Review*, 11(2): 280-297.

Fry, L. W. 1982. Technology-Structure Research: Three Critical Issues. *Academy of Management Journal*, 25(3): 532-552.

Galbraith, J. R. 1973. *Designing Complex Organizations*. Reading, MA: Addison-Wesley.

Galbraith, J. R. 1974. Organization Design: An Information Processing View. *Interfaces*, 4(3): 28-36.

Galbraith, J. R. 1976. Organization Design: An Information Processing View. *Interfaces*, 28-36.

Galbraith, J. R. 1995. An Executive Briefing on Strategy Structure and Process. San Francisco: Jossey-Bass.

Galunic, D. C. & Eisenhardt, K. M. 1994. Reviewing the Strategy-Structure-Performance Paradigm. In Staw, B. & Cunnings, L. (Eds.), *Research on Organizational Research*: 215-255.

Galunic, D. C. & Eisenhardt, K.M. 1996. The Evolution of Intracorporate Domains: Divisional Charter Losses in High-technology, Multidivisional Corporations. *Organization Science,*(3): 255-282.

Geeraerts, G. 1984. The Effect of Ownership on the Organization Structure in Small Firms. *Administrative Science Quarterly,* 29(2): 232-237.

Gersick, C. J. G. 1991. Revolutionary Change Theories: A Multi-Level Exploration of The Punctuated Equilibrium Paradigm. *Academy of Management Review,* 16(1): 10-36.

Gerth, H. H. & Mills, W. 1946. *Essays in Sociology.* New York: Oxford University Press.

Giek, D. G. & Lees, P. L. 1993. On Massive Change: Using the Competing Values Framework to Organize the Educational Efforts of the Human Resource function in New York State Governemnt. *Human Resource Management,* 32(1): 9-11.

Giffi, C., Roth, A. V., & Seal, G. M. 1990. Competing in World-Class Manufacturing: America's Twentyfirst Century Challenge. Homewood, IL: Irwin, Inc.

Glick, W. H. 1985. Conceptualizing and Measuring Organizational and Psychological Climate: Pitfalls in Multilevel Research. *Academy of Management Review,* 10(3): 601-616.

Gouldner, A. W. 1957. Cosmopolitan and Locals: Toward an Analysis of Latent Social Roles. *Administrative Science Quarterly,* 2: 281-306.

Greiner, L. E. 1972. Evolution and Revolution as Organizations Grow. *Harvard Business Review,* 50.

Gresov, C. 1989. Exploring Fit and Misfit With Multiple Contingencies. *Administrative Science Quarterly,* 34(3): 431-454.

Gresov, C. & Drazin, R. 1997. Equifinity: Functional Equivalence in Organization Design. *Academy of Management Review,* 22(2): 403-428.

Grinyer, P. H. 1982. Discussion Note: Divisionalization and Size: A Rejoinder. *Organization Studies,* 339-350.

Grinyer, P. H. & Ardekani, M. 1980. Dimensions of Organizational Structure: a Critical Replication. *Academy of Management Journal,* 405-421.

Grønhaug, K. & Falkenberg, J. S. 1989. Exploring Strategy Perceptions in Changing Environments. *Journal of Management Studies,* 26(4): 249-359.

Habib, M. M. & Victor, B. 1991. Strategy, Structure and Performance of U.S. Manufacturing and Service MNCs: A Comparative Analysis. *Strategic Management Journal,* 12: 589-606.

Hage, J. 1965. An Axiomatic Theory of Organizations. *Administrative Science Quarterly,* 10: 366-376.

Hall, D. & Saias, M. 1983. Strategy Follows Structure! *Strategic Management Journal,* 1: 149-163.

Hall, R. M. 1991. *Organizations, Structures, Process, Outcomes.* Englewood Cliffs, NJ: Prentice Hall.

Hall, R. M., Hass, E. J., & Johnson, N. 1967. Organization Size, Complexity, and Formalization. *American Sociological Review,* 32(6): 903-912.

Hambrick, D. C. 1983. Some Tests of the Effectiveness and Functional Attributes of Miles nad Snow's Strategic Types. *Academy of Management Journal,* 25(1): 5-26.

Handy, C. 1995. Trust and the Virtual Organization. *Harvard Business Review,* 40-50.

Hannan, M. T. & Freeman, J. H. 1977. The Population Ecology of Organizations. *American Journal of Sociology,* 82: 929-964.

Harmon, P. & King, D. 1985. *Expert Systems: Artificial Intelligence in Business.* New York: Wiley & Sons.

Hawley, A. W., Boland, W., & Boland M. 1965. Populations Size and Administration in Institutions of Higher Education. *American Sociological Review*, 252-255.

Heneman, R. L. & Dixon, K. E. 2001. Reward and Organizational System Alignment: An Expert System. *Compensation and Benefits Review*.

Hickson, D. J., Pugh, D. S., & Pheysey, D. C. 1979. Operations Technology and Organization Structure: An Empirical Reappraisal. *Administrative Science Quarterly*, 24: 375-397.

Hofstede, G. 2001. *Culture's Consequences*. Sage publications.

Hofstede, G. 1997. Cultures and Organizations: Software of the Mind. McGraw-Hill.

Hooijberg, R. & Petrock, F. 1993. On Cultural Change: Using the competing Values Framework to Help Leaders Execute a Transformational Strategy. *Human Resource Management*, 32(1): 29-50.

Howard, R. 1992. The CEO as Organizational Architect: An Interview with Xerox's Paul Allaire. *Harvard Business Review*, 107-121.

Huber, G. P. 1990. A Theory of the Effects of Advanced Information Technologies on Organizational Design, Intelligence, and Decision Making. *Academy of Management Review*, 15(1): 47-71.

Hunt, J. G. 1991. *Leadership: A New Synthesis*. Beverly Hills, CA: Sage.

Hunter, S. 1998. Information Technology & New Organization Forms. Duke University.

Jackson, J. H. & Morgan, C. P. 1978. *Organization Theory: A Macro Perspective for Management*. Englewood Cliffs, NJ: Prentice-Hall.

James, L. R. & Jones, A. P. 1974. Organizational Climate: A review of Theory and Research. *Psychological Bulletin*, 81(12): 1096-1112.

James, W. L. & Hatten, K. J. 1994. Evaluating the Performance effects of Miles' and Snow's Strategic Archetypes in Banking, 1983 to 1987: Big or Small. *Journal of Business Research*, 31(1,2): 145-154.

Jensen, J. M. 1993. Strategic Framework for Analyzing Negative Rumors in the Market Place: The Case of Wash & Go in Denmark. In Sirgy, M. J. & Bahn, K. D. E. T. (Eds.), *Proceedings of the Sixth Bi-Annual International Conference of the Academy of Marketing Science.*: 559-563. Istanbul.

Jin, Y. & Levitt, R. E. 1996. the Virtual Design Team: A Domputational Model of Project Organizations. *Computational and Mathematical Organization Theory*, 2(3): 171-196.

Jurkovich, R. 1974. A Core Typology of Organizational Environments. *Administrative Science Quarterly*, 19(3): 380-394.

Kerr, S. 1975. On the Folly of Rewarding A, While Hoping for B. *Academy of Management Journal*, 18(4): 769-783.

Ketchen, D. J. J., Combs, J. G., Russell, C. J., Shook, C., Dean, M. A., Runge, J., Lohrke, F. T., Nauman.S.E., Haptonstahl, D. E., Baker, R., Becksterin, B. A., Handler, C., Honig, H., & Lamoureux, S. 1997. Organizational configurations and performance: A meta-analysis. *Academy of Management Journal*, 40(1): 223-240.

Khandwalla, P. 1973. Viable and Effective Organizational Designs of Firms. *Academy of Management Journal*, 16: 481-495.

Kimberly, J. R. 1976. Organization Size and the Structuralist Perspective: A Review, Critique and Proposal. *Administrative Science Quarterly*, 571-597.

Koberg, C. S. & Ungson, G. R. 1987. The Effects of Environmental Uncertainty and Dependence on Organizational Structure and Performance: A Comparative Study. *Journal of Management*, 13(4): 725-737.

Kohlberg, L. 1971. Stage and Sequence: The Cognitive-Developmental Approach to Socialization. In Goslin, D. A. (Ed.), *Handbook of Socialization: Theory and Research*: 347-480. Chicago: Rand McNally & Company.

Kotter, J. P. 1988. *The Leadership Factor.* New York: Free Press.

Kowtha, N. R. 1997. Skills, Incentives, and Control. An Integration of Agency and Transaction Cost Approaches. *Group and Organization Management,* 22(1): 53-86.

Koys, D. J. & DeCotiis, T. A. 1991. Inductive Measures of Psychological Climate. *Human Relations,* 44(3): 265-285.

Koza, M. P. & Lewin, A. Y. 1999. The Coevolution of Network Alliances: A Longitudinal Analysis of an International Professional Service Network. *Organization Science,* 10(5): 638-643.

Krokosz-Krynke, Z. 1998. Organizational Structure and Culture: Do Individualism/Collectivism and Power Distance Influence Organizational Structure. *Emerging Economies International Conference Proceedings*: Budapest: ABAS.

Kuhn, A. J. 1986. Organizational Cybernetics and Business Policy: System Design for Performance Control. University Park: Pennsylvania State University Press.

Kurke, L. B. & Aldrich, H. E. 1983. Mintzberg Was Right! A Replication and Extension of the Nature of Managerial Work. *Management Science,* 29(8): 975-984.

Lachman, R., Nedd, A., & Hinings, B. 1994. Analyzing Cross-National Management and Organizations: A theoretical Framework. *Management Science,* 40(1): 40-55.

Lawrence, P. R. 1981. Organization and Environment Perspective. In Van de Ven, A. H. & Joyce, W. F. (Eds.), *Perspectives on Organization Design and Behavior:* Boston: Harvard Business Press.

Lawrence, P. R. & Lorsch, J. W. 1967. *Organization and Environment.* Boston: Harvard Business Press.

Lemmergaard, J. 2003. Tolerance for Ambiguity. The Intersection Between Ethical Climate, Psychological Climate, and Ethnic Diversity. Ph.D dissertation, University of Southern Denmark.

Lennby, Peter, 1990, *Flyv mod Højere Højder,* Sweden

Lenox, M. 2002. Organizational Design, Information Transfer, and the Acquisition of Rent-Producing Resources. *Computational and Mathematical Organization Theory,* 8(2): 113-131.

Levinthal, D. A. & March, J. G. 1993. The Myopia of Learning. *Strategic Management Journal,* 14: 95-112.

Levitt, R. E., Thomson, J., Christiansen, T. R., Kunz, J. C., Yin, J., & Nass, C. I. 1999. Simulating Project Work Processes and Organizations: Towards a Micro-Contingency Theory of Organizational Design. *Management Science,* 45(11): 1479-1495.

Lewin, A. & Stephens, C. V. 1994. CEO Attitudes as Determinants of Organizational Design: An Integrated Model. *Organizational Studies,* 15(2): 183-212.

Lewin, A. & Volberda, H. W. 1999. Prolegomena on Coevolution: A Framework for Research on Strategy and New Organizational forms. *Organizational Science,* 10(5): 519-534.

Likert, R. 1967. *The Human Organizations.* New York: McGraw-Hill.

Litwin, G. H. & Stringer Jr., R. A. 1968. *Motivation and Organizational Climate.* Harvard University.

Luthans, F., Hodgetts, R. M., & Rosenkrantz, S. A. 1988. *Real Managers.* Cambridge, MA: Ballinger.

Madsen, T. K. & Jensen, J. M. 1992. Analyse Klassifikation og Behandling af Negative Rygter. *Ledelse Og Erhvervsøkonomi,* 56(1).

Malone, T. & Rockart, J. 1990. Computers, Networks, and the Corporation. *Scientific American*, 128-136.

March, J. G. 1991. Exploration and Exploitation in Organizational Learning. *Organization Science*, 2(1): 71-87.

March, J. G. 1994. Three lectures on Efficiency and Adaptiveness in Organizations,

March, J. G. & Simon, H. A. 1958. *Organizations*. New York: John Wiley & Sons.

Marsh, R. M. & Mannari, H. 1989. The Size Imperative? Longitudinal Tests. *Organizational Studies*, 10(1): 83-95.

Mayhew, B. H., Levinger, R. L., McPherson, J. M., & James, T. F. 1972. System Size and Structural Differentiation in Formal Organizations: A Baseline Generator for Two Major Theoretical Propositions. *American Sociological Review*, 629-633.

McGregor, D. 1969. *The Human Side of Enterprise*. New York: McGraw-Hill.

Merton, R. K. 1957. Patterns of Influence: Local and Cosmopolitan Influentials. *Social Theory and Social Structure. Revised and Enlarged Edition*: 387-420. New York: The Free Press.

Meyer, A. D., Goes, J. B., & Brooks, G. R. 1993. Organizations Reacting to Hyperturbulence. In Huber.G. & Glick, W. (Eds.), *Organizational Change and Redesign*: 66-111. New York: Oxford University Press.

Meyer, A. D., Tsui, A. S., & Binings, C. R. 1993. Configurational Approaches to Organizational Analysis. *Academy of Management Journal*, 36(6): 1175-1195.

Meyer, M. W. 1972. Size and the Structure of Organizations: A Casual Analysis. *American Sociological Review*, 434-440.

Miles, R. E. & Creed, W. E. D. 1995. Organizational Forms and Managerial Philosophies. In Cummings, L. L. & Staw, B. M. (Eds.), 333-372.

Miles, R. E. & Snow, C. C. 1978. *Organizational Strategy, Structure and Process*. New York: McGraw-Hill.

Milkovich, G. T. & Boudreau, J. W. 1988. *Personnel/Human Resource Management*. Plano, TX: Business Publications.

Miller, C., Glick, H. W. H., Wang, Y., & Huber, G. P. 1991. Understanding Technology-Structure Relationships: Theory Development and Meta-Analytic Theory Testing. *Academy of Management Journal*, 34(2): 370-399.

Miller, D. 1988a. Relating Porter's Business Strategies To Environment And Structure: Analysis and Performance Implications. *Academy of Management Journal*, 31(2): 280-308.

Miller, D. 1988c. Relating Porter's Business Strategies To Environment And Structure: Analysis and Performance Implications. *Academy of Management Journal*, 31(2): 280-308.

Miller, D. 1988b. Relating Porter's Business Strategies To Environment And Structure: Analysis and Performance Implications. *Academy of Management Journal*, 31(2): 280-308.

Miller, D. 1987a. Strategy Making and Structure: Analysis and Implications for Performance. *Academy of Management Journal*, 30(1): 7-32.

Miller, D. 1987b. The Structural and Environmental Correlates of Business Strategy. *Strategic Management Journal*, 8: 55-76.

Miller, D. 1989. Complexurations of Strategy and Structure: Towards a Syntheses. In Asch, D. & Bowman C. (Eds.), *Readings in Strategic Management*. New York: MacMillan.

Miller, D. 1991. Stale in the Saddle: CEO Tenure and the Match Between Organization and Environment. *Management Science*, 37(1): 34-52.

Miller, D. 1992. Environmental Fit Versus Internal Fit. *Organizational Science*, 3(2): 159-178.

Miller, D., de Vries, M. F. R. K., & Toulouse, J.-M. 1982. Top Executive Locus of Control and Its Relation to Strategy-Making, Structure and Environment. *Academy of Management Journal*, 25(2): 237-253.

Miller, D. & Dröge, C. 1986a. Psychological and Traditional Determinants of Structure. *Administrative Science Quarterly*, 31: 539-560.

Miller, D. & Friesen P. 1980. Archetypes of Organizational Transition. *Administrative Science Quarterly*, 25: 268-299.

Miller, D. & Toulouse, J.-M. 1986b. Chief Executive personality and Corporate Strategy and Structure in Small Firms. *Management Science*, 32(11): 1389-1409.

Miller, G. A. 1987c. Meta-analyses and the Culture Free Hypothesis. *Organizational Studies*, 8(4): 309-326.

Milliken, F. J. 1987. Three Types of Perceived Uncertainty About the Environment: State, Effect, and Response Uncertainty. *Academy of Management Review*, 12(1): 133-143.

Mills, P. K. & Moberg, D. J. 1982. Perspectives on the Technology of Service Operations. *Academy of Management Review*, 7(3): 367-478.

Mintzberg, H. 1980. *The Nature of Managerial Work*. Englewood Cliffs, NJ: Prentice-Hall.

Mintzberg, H. 1979. *The Structuring of Organizations*. Englewood Cliffs, NJ: Prentice-Hall.

Mintzberg, H. 1983. *Structures in Fives*. Englewood Cliffs, NJ: Prentice Hall.

Murphy, P. R. & Daley, J. M. 1996. A Preliminary Analysis of the Strategies of International Freight Forwarders. *Transportation Journal*, 35(4): 5-11.

Nadler, D. & Tushman, M. L. 1984. A Congruence Model for Diagnosing Organizational Behavior. In Kolb, D. A., Rubin, J. M., & McIntyre, J. M. (Eds.), *Organizational Psychology: Readings on Human Behavior in Organizations*: 587-603. Englewood Cliffs, NJ: Prentice-Hall.

Nadler, D. & Tushman, M. L. 1988. *Strategic Organization Design: Concepts, Tools and Processes*. Glenview, IL: Scott-Foresman.

Naman, J. L. & Slevin, D. P. 1993. Entrepreneurship and the Concept of Fit: A Model and Empirical Tests. *Strategic Management Journal*, 14(2): 137-153.

Nicholson, N., Rees, A., & Brooks-Rooney, A. 1990. Strategy, Innovation and Performance. *Journal of Management Studies*, 27(5): 511-534.

O'Leary, D. E. 1996. Verification of Uncertaind Knowledge-based Systems. *Management Science*, 42(12): 1663-1675.

Obel, B. 1986. SAS: changes in Competition, Strategy and Organization. In Burton R.M & Obel, B. (Eds.), *Innovation and Entrepreneurship in Organizations. Strategies for Competitiveness, Deregulation, and Privatization.*: Amsterdam: Elsevier.

Ouchi, W. G. 1979. Design of Organizational Control Mechanisms. *Management Science*, 25(9): 833-848.

Pennings, J. M. 1987. Structural Contingency Theory: A Multivariate Test. *Organization Studies*, 8(3): 223-240.

Perrow, C. 1967a. A Framework for Comparative Analysis of Organizations. *American Sociological Review*, 32: 194-208.

Pfeffer, J. 1982. Organizations and Organization Theory. Boston: Pittman.

Phatak, A. V. 1992. *International Dimensions of Management*. Boston: DWS-Kent.

Pondy, L. R. 1969. Effects of Size, Complexity, and Ownership on Administrative Intensity. *Administrative Science Quarterly*, 14: 47-60.

Poole, M. S. 1985. Communication and Organizational Climate: Review, Critique, and a New Perspective. In McPhee, R. D. & Tompkins, P. K. (Eds.), *Organizational Communications: Traditional Themes and New Directions*: Sage Publications.

Poole, M. S. & McPhee, R.D. 1983. A Structural Analysis of Organizational Climate. In Putman, L. L. & Pacahowsky, M. E. (Eds.), *Organizational Communication: An Interpretive Approach*:Sage Publications.

Porter, M. E. 1980. Competitive Strategy: Techniques for Analyzing Industries and Competitors. New York: The Free Press.

Porter, M. E. 1985. Competitive Advantage: Creating and Sustaining Superior Performance. New York: The Free Press.

Powell, W. W. 1990. Neither Market Nor Hierarchy: Network Forms of Organization. *Research in Organization Behavior*, 12: 295-336.

Probst, G. J. B. & Buchel, B. S. T. 1997. *Organizational Learning: The Cocmpetitive Advantage of the Future*. London: Prentice-Hall Europe.

Pugh, D. S., Hickson, D. J., Hinings, C. R., & Turner, C. 1969. The Context of Organization Structures. *Administrative Science Quarterly*, 14(91): 114.

Quinn, R.E. & Spreitzer, G. 1991. The Psychometrics of the Competing Values Culture Instrument and an Analysis of the Impact of Organizational Culture on Quality of Life. *Research in Organizational Change And Development*, 5: 115-142.

Quinn, J. B., Mintzberg, H., & James, R. M. 1988. *The Strategy Process: Concepts, Contexts, and Cases*. Englewood Cliffs, NJ: Prentice-Hall.

Quinn, J. B. & Rohrbaugh, J. 1983. A Spatial Model of Effectiveness Criteria: Towards a Competing Values Approach to Organizational Analysis. *Management Science*, 29(3): 363-377.

Quinn, R. E. & Kimberley, J. R. 1984. Paradox, Planning, and Perseverance: Guidelines for Managerial Practice. In Kimberly, J. R. & Quinn, R. E. (Eds.), *Managing Organizational Transitions*: 295-314. Dow-Jones-Irwin.

Rajaratnam, D. & Chonko, L. B. 1995. The effect of business strategy type on Marketing Organization Design, Product-market Growth Strategy, Relative Marketing Effort, and Organization Performance. *Journal of Marketing Theory and Practice*, 3(3): 60-75.

Rice, R. E. 1992. Task Analyzability, Use of New Media and Effectiveness: A Multi-site Exploration of Media Richness. *Organizational Science*, 3(4): 475-500.

Richardson, J. 1996. Vertical Integration and Rapid Response in Fashion Apparel. *Organization Science*, 7(4): 400-412.

Robbins, S. P. 1990. *Organization Theory: Structure, Design and Applications*. Englewood Cliffs, NJ: Prentice-Hall.

Robey, D. 1982. Designing Organizations: A Macro Perspective. IL: Irwin.

Roth, A. V. & Miller, J. G. 1990. Manufacturing Strategy, Manufacturing Strength, Managerial Success, and Economic Outcomes. In Ettlie, J. E., Burstein, M. C., & Fiegenbaum, A. (Eds.), *Manufacturing Strategy*: Norwell, MA: Kluwer.

Rousseau, D. M. 1988. The Construction of Climate in Organizational Research. *International Review of Industrial and Organizational Psychology*, 3: 139-158.

SAS 2002. SAS Annual Reports 1982-2002.

Schein.E.H. 1992. *Organizational Culture and Leadership*. San Francisco: Jossey-Bass.

Schermerhorn, J. R. JR., Hunt, J. G., & Osborn, R. N. 1991. *Managing Organizational Behavior*. New York: Wiley.

Schneider, B. Schneider, B. (Ed.) . 1990. *Organizational Climate and Culture*. San Francisco: Jossey-Bass.

Schneider, B. & Reichers, A. E. 1983. On the Etiology of Climates. *Personnel Psychology*, 36: 19-39.

Schneider, B. & Snyder, R. A. 1975. Some Relationships Between Job Satisfaction and Organizational Climate. *Journal of Applied Psychology*, 60(3): 318-328.

Schoonhoven, C. B. 1981. Problems with Contingency Theory: Testing Assumptions Hidden within the Language of Contingency Theory. *Administrative Science Quarterly*, 26: 349-377.

Scott, W. R. 1998. *Organizations, Rational, Natural and Open Systems*. Englewood Cliffs, NJ: Prentice-Hall.

Segev, E. & Gray, P. 1990. Business Success: Strategic Unit Comprehensive Computer-Based Expert Support System. Englewood Cliffs, NJ: Prentice-Hall.

Sharda, B. D. & Miller, G. 2001. Culture and Organization in the Middle East: A Comparative Analysis of Iran, Jordan, and the USA. *International Review of Sociology*, 11(3): 2001.

Siehl, C. & Martin, J. 1990. Organizational Culture: A Key to Financial Performance. In Schneider, B. (Ed.), *Organizational Climate and Culture*: Jossey-Bass.

Simon, H. A. 1981. *The Science of the Artificial*. Cambridge, MA: MIT Press.

Sims, R. & Keon, T. L. 1997. Ethical Work Climate as a Factor in the Development of Person-Organization Fit. *Journal of Business Ethics*, 16(11): 1095-1105.

Sitkin, S. B. 1992. Learning Through Failure: The Strategy of Small Losses. *Research in Organization Behavior*, 14: 231-266.

Slater, R. O. 1985. Organization Size and Differentiation. In Bacharach, S. B. & Mitchell, S. M. (Eds.), *Research in the Sociology of Organizations*: 127-180. Greenwich, CT: JAI Press.

Smith, K. G., Guthrie, J. P., & Chen, M.-J. 1989. Strategy, Size and Performance. *Organization Studies*, 10(1): 63-72.

Søndergaard, M. 1994. Hofstede's Consequences: A Study of Reviews, Citations, and Replications. *Organization Studies*, 15(447): 456.

Sorenson, O. 2003. Interdependence and Adaptability: Organizational Learning and the Long-Term Effect of Integration. *Management Science*, 49(4): 446-463.

Spencer, H. 1998. *Principles of Sociology*. New York: Appleton.

Staber, U. 2002. Organizational adaptive capacity: A structuration perspective. *Journal of Management Inquiry*, 11(4): 408.

Stacey, R. D. 1996. *Strategic Management and Organisational Dynamics*. London: Pitman Publishing.

Starbuck, W. H. 1965. Organizational Growth and Development. In March, J. G. (Ed.), *Handbook of Organizations*: 451-433. Chicago, Il: Rand McNally.

Tagiuri, R. & Litwin, G. H. 1968. *Organizational Climate*. Harvard University.

Thomas, A., Litschert, R. J., & Ramaswamy, K. 1991. The Performance Impact of Strategy-Manager Coalignment: An Empirical Examination. *Strategic Management Journal*, 12: 509-522.

Thompson, J. D. 1967. *Organizations in Action*. Oxford: Oxford University Press.

Thomsen, J. 1998. The Virtual team Alliance (VTA): Modeling the Effects of Goal Incongruence in Semi-routine, Fast-paced Project Organizations. Ph.D. Dissertation, Stanford University.

Tosi, H. 1992. The Environment/Organization/Person Contingency Model: A Meso Approach to the Study of Organization. JAI Press.

Tosi, H., Aldag, R., & Stoney, R. 1973. On the Measurement of the Environment: An Assessment of the Lawrence and Lorsch Environmental Uncertainty Scale. *Administrative Science Quarterly*, 18: 27-36.

Tung, R. L. 1979. Dimensions of Organizational Environments: An Exploratory study of Their Impact on Organizational Structure. *Academy of Management Journal,* 22(4): 672-693.

Tushman, M. L. & Romanelli, E. 1985. Organizational Evolution: A Metamorphosis Model of Convergence and Reorientation. In Cunnings, L. L. & Stacer, B. M. (Eds.), *Research in Organizational Behavior.* 171-222. San Francisco: SAI Press.

Venkatraman, N. & Prescott, J. E. 1990. Environment-Strategy Coalignment: An Empirical Test Of Its Performance Implications. *Strategic Management Journal,* 11(1): 1-23.

Victor, B. & Cullen, J. B. 1987. A Theory and Measure of Ethical Climate in Organization. In Frederick, W. C. (Ed.), *Research in Corporate Social Performance and Policy:* 51-71.

Victor, B. & Cullen, J. B. 1988. The Organizational Bases of Ethical Work Climates. *Administrative Science Quarterly,* 33: 101-125.

Victor, B. & Cullen, J. B. 1990. A Theory and Measure of Ethical Climate in Organizations. In Frederick, W. C. & Preston, L. E. (Eds.), *Business Ethics: Research Issues and Empirical Studies.* Greenwich, Connecticut: JAI Press Inc.

Volberda, H. W. 1996. Toward the Flexible Form: How to Remain Vital in Hypercompetitive Environments. *Organization Science,* 7(4): 359-374.

Vroom, V. H. & Yetton, P. H. 1973. *Leadership and Decision-Making.* Pittsburgh: University of Pittsburgh Press.

Wang, E. T. G. 2001. Linking Organizational Context with Structure: a Preliminary Investigation of the Information Processing View. *Omega.*

Wang, E. T. G. 2003. Effect of the Fit between Information Processing Requirements and Capacity on Organizational Performance. *International Journal of Information Management,* 23(3): 239-247.

Weber, M. 1946. From Max Weber: Essays in Sociology, Translated and Edited by H.H. Gerth and C. Wright Mills. New York: Oxford University Press.

Weick, K. E. 1969. *Ther Social Psychology of Organizing.* Reading, MA: Addison-Wesley.

Weisenfeld-Schenk, U. 1994. Technology Strategies and The Miles and Snow Typology: A Study of the Biotechnology Industry. *R & D Management,* 24(1): 57-64.

Weitzel, W. & Jonsson, E. 1989. Decline in Organizations: A Literature Integration and Extension. *Administrative Science Quarterly,* 34: 91-109.

Whetten, D. A. & Cameron, K. S. 2002. *Developing Management Skills.* Upper Saddle River, NJ: Prentice-Hall.

Williamson, O. E. 1975. Markets and Hierarchies: Analysis and Antitrust Implications. New York: Free Press.

Woodward, J. 1965. *Industrial Organization, Theory and Practice.* Oxford: Oxford University Press.

Yeung, A. K. O., Brockband, J. W., & Ulrich, D. O. 1991. Organizational Cultures and Human Resource Practices: An Empirical Assessment. *Research in Organizational Change And Development,* 5: 59-81.

Yukl, G. A. 1981. *Leadership in Organizations.* Englewood Cliffs, NJ: Prentice-Hall.

Zajac, E. J., Kraatz, M. S., & Bresser, R. K. F. 2000. Modeling the dynamics of strategic fit: A normative approach to strategic change. *Strategic Management Journal,* 21(4): 753-773.

Zalesnik, A. 1977. Managers and Leaders: Are They Different? *Harvard Business Review,* 15: 67-68.

Zammuto, R. F. & Krakower, J. Y. 1991. Quantitative and Qualitative Studies in Organizational Culture. *Research in Organizational Change And Development*, 5: 83-114.

Zammuto, R. F. & O'Connor, E. J. 1992. Gaining Advanced Manufacturing Technologies' Benefit: The Roles of Organizational Design and Culture. *Academy of Management Review*, 17(4): 701-728.

Zeffane, R. 1989. Computer Use and Structural Control: A Study of Australian Enterprises. *Journal of Management Studies*, 26(6): 621-648.

Index

Ad hoc...47, 70-74, 119, 121,
..................................128-134, 170, 176, 182, 221-229, 231, 233,
..........................239, 240, 244, 262, 263, 273, 296, 299, 300,
............................339, 349-353, 359, 362, 365, 374-376, 391,
....392, 395, 396, 419, 420, 423
Adhocracy....................6, 13, 52, 58, 90, 94, 151, 153, 155, 157, 163, 196
Alcazar.. 272
Alignment.............................21, 90, 91, 94, 131, 167, 280, 298, 314, 315
Analyzer.....................................36, 46, 272, 289-293, 296, 297, 304-312,
....316-318, 321, 330-332, 368, 340-345,
....350, 353, 359, 362, 365, 366, 369,
....370-373, 381, 384, 387, 388, 391-395,
....398, 403, 406, 415, 418
Autocratic 96, 102, 103, 106, 113, 123, 126, 130
Autonomy 141, 142, 146
Bargaining 205, 207, 279
Bilka 253
Bon Goût 185-187, 195, 196, 201, 202, 207,
....210, 212-219, 241, 271, 277, 315, 387
Boundary..3, 16, 19, 280, 423
Budgeting..92, 95, 303, 304, 308
Budgets... 125
Bureaucracy............................... 47, 48, 71, 72, 80, 87, 128, 129, 132-134,
....152, 160, 161, 182, 239, 240, 244, 262, 299,
....300, 328, 336-338, 377-379, 419, 420
Capital requirement 296, 305, 315, 347, 348
Centralization...................., 87-89, 92, 93, 98, 100, 116, 121-129,
.... 132-136, 138, 144, 152, 153, 156, 157,
....161-165, 168, 169, 176-179, 185, 187,
....199, 209-220, 223, 225-229, 233,
....237-246, 256-259, 266, 268, 270, 275,
....279, 288, 299-304, 307, 308, 311,
....312, 327-339, 347-353, 358-363,
....386, 387, 390-396, 405, 419, 420
Certainty factor 26, 35-43, 45, 135, 176, 348, 349,
....350, 351, 358, 359, 366, 399
Chain-of-command hierarchy 127
Climate 128, 131-139, 141-147, 149-169,
....313, 319-327, 334-336, 353, 359,
....362, 367-398, 403-406, 409, 411
Climate measure 131, 137, 142, 144
Collectivism..236, 350, 351
Collectivity stage 116, 393
Communication..................................... 6, 11, 43, 47, 54, 79, 85, 90, 96, 110,
....116, 169, 170, 201, 237, 244, 266, 270,
....300, 375, 405

Compensation 89, 94, 139, 172
Competing values approach......................98, 131, 135-138, 146, 149, 168
Competition 1, 2, 7, 24, 43, 58, 73, 163, 187, 193, 195,
....196, 200, 201, 207, 233, 241, 272, 273,
.... 292, 316-318, 348, 359, 390, 400, 414
Concern for quality.. 295, 298, 302, 306, 310, 313
Conflict 39, 42, 51, 70, 96, 128, 135, 137, 141,
....144-147, 150-155, 158-160, 163, 168
....321, 324, 325, 335, 336, 344, 346, 348,
....350, 353, 385, 391,
Contingency fit................................... 21-23, 27, 39, 43, 166, 312, 313, 327,
....347, 348, 353, 387, 389, 396, 399, 403
Contingency model..................................... 19-21, 24, 31, 98, 131, 138, 139,
.... 143, 149, 169, 311, 313-315, 328, 413
Contingency theory- 10, 11, 20-23, 28, 42, 47, 92, 131,
....186, 187, 190, 193, 194, 196, 313, 315,
....327, 328, 359
Control-flexibility.. 136
Cooperation.. 4, 24, 76, 88, 95, 145, 198
Coordination and control- 43, 46-48, 50, 81, 85, 88, 91-93, 117,
....120, 121, 125, 129, 133, 134, 153, 156,
.... 157, 161-166, 183, 187, 222, 225-227,
.... 230, 231, 244, 256, 257, 265, 274,
.... 277, 292, 300, 303, 304, 308, 311,
....349, 353, 385, 396, 397,
Coordinators 148
Creative...116, 118, 126, 135, 148, 233, 319, 335
Creativity 95, 142, 390
Culture 79, 128, 131-138, 142, 144, 167, 168,
....233, 235-241, 342, 350, 351, 412
Current organization 44, 388
Customization 55, 61
Databases 265, 270, 271
Decentralization 87-89, 98, 120, 124, 164,
.... 169, 173, 176-179, 182, 198, 199,
.... 216, 218, 238, 266-269, 287, 288,
.... 313, 327-331, 349, 363, 385, 390,
.... 391, 394, 396, 414, 419-421, 424
Decentralized 29, 268, 412
Decline stage.. 395
Decomposability...249, 250, 252, 255
Decomposable 249, 250, 255
Decouple.. 198, 421
Defender 272, 287-294, 296, 301-309, 313, 315-317, 320,
.... 322, 330-332, 340, 341, 353, 368, 369, 384,
.... 387-389, 392, 395-398, 403, 406, 417, 418, 427
Delegate 88, 102, 111, 116, 120, 124, 128, 130, 132, 137, 179
Delegation 1, 13, 88, 95, 98, 100-103, 106, 110, 111, 114,
.... 116, 118, 119, 122-127, 130-132, 138, 179,

....327, 336-338, 372, 385, 390-393
Democratic...............96, 102, 103, 106, 112, 113, 118, 122, 243
Department.. 15, 16, 50, 52, 53, 64, 70, 77,
.... 86, 178, 182, 212, 237, 274
Deregulation 24, 43, 185-187, 196, 201
Design parameter fit 347, 349, 350, 353, 357-359, 385, 387
Developmental climate 128, 146, 148, 154-158, 163, 144, 154,
.... 155, 157, 158, 319-323, 325, 335, 336,
.... 343, 348, 351, 353, 375, 380, 390-393
Developmental culture..135, 136, 137, 144
Diagnose.. 2, 28, 43
Diagnosis 1, 2, 10, 13, 16-19, 21, 23, 34, 39, 41, 44-47,
.... 91, 311-313, 355, 385-389, 396, 399, 417,
.... 419, 424-427
Differentiation 48, 79, 80-85, 91, 133, 168, 197, 199, 205,
....212, 215, 237, 238, 241, 260, 270, 277,
....278, 287, 292, 349, 358, 380, 382, 383
Diversification 24, 167, 181, 182, 277, 292
Diversified.. 1, 277
Diversity- 77, 166, 246, 249, 250, 288, 305, 307,
....309, 314, 315, 341, 345
Divisibility................................. 251, 252, 255, 256, 332, 333, 353
Divisional configuration...........................43, 56-65, 72-75, 82, 88, 92, 124
.... 164, 165, 227, 228, 238, 250, 262
....308, 311-314, 333, 337-339,
....353, 359, 369-372, 421, 423
Divisionalization..57, 166, 314
Divisions 8, 43, 48, 56-63, 66, 68, 77, 80, 86, 88
....120, 124, 166, 226, 262, 333, 359, 370, 421
Domestic..77, 186, 272, 315
Downsizing...164, 171, 185, 397
Dynamism... 191
Ecology 187, 196, 197, 416
Economies of scale 57, 69, 182, 205, 256, 307
Effective 2, 5, 6, 9, 11, 17-19, 23, 47, 63, 70, 93, 115,
....127, 131, 134, 149, 164, 167, 169, 185, 186,
....286, 299, 306, 313, 373, 386, 400, 416, 417
Effective performance 127
Effectiveness 3, 4, 5, 8, 11, 34, 43, 44, 46, 68, 70, 91,
.... 94, 116, 136, 153, 157, 161, 165, 170,
....183, 257, 294, 308, 373, 386, 416, 419
Efficiency 4, 5, 8, 11, 34, 43, 44, 46, 52, 63,
.... 65, 69, 70, 80, 84, 90, 91, 116, 148, 153,
....163, 165, 170, 183, 219, 250, 256, 285,
.... 291, 301-311,
....316, 317, 321, 329, 331, 333, 369, 373,
.... 384, 386, 402, 416, 418, 419
Elaboration of structure stage 395
Electronic information system 267

Electronic information technology................................. 269, 270
E-mail................................ 6, 170, 243, 265, 267, 268, 270, 271
Empower.. 273
Entrepreneurial stage.......................................116, 390, 393
Environmental complexity ....:...........28-30, 32, 34, 191, 192, 200,
.... 202-205, 210-218, 221,
....224-228, 233, 247, 329, 340,
....341, 349, 365, 394
Environmental equivocality 187, 215, 218, 342, 363, 383, 397
.... 351, 387
Environmental hostility 37, 195, 218, 219, 229, 386
Environmental uncertainty...37, 42, 191-194, 197,
.... 198, 199, 205, 207, 213, 217, 229,
....232, 329, 341, 342, 349, 363, 368, 387
Equifinality 27, 314, 315, 364, 387
Equivocal 166, 216, 232, 264, 316-318, 326, 342, 343,
....351, 353, 359, 366, 386, 393, 400, 412
Euroclass 238, 272
Expert systems.. 28, 35, 265
Exploitation...76, 297, 398, 416-421, 426
Exploration 67, 321, 341, 353, 398, 416-421, 423, 426
Export.. 77
Fairness..150, 142, 146
Femininity.. 236-239, 350
Five forces...205, 207, 233
Fixing misfits 396, 403-405, 408, 411, 413
Flexible form 73, 233
Flexible organization 73, 233, 277, 316, 359, 411, 412
Formalization 69, 72, 73, 85-88, 92-94, 98, 100, 101,
.... 116, 120, 121, 124, 125, 128, 129, 132-
....134, 136, 138, 144, 152, 153, 156, 157,
.... 160-165, 168, 169, 173, 176, 179-182, 185,
.... 187, 199, 209-221, 223-229, 233, 238-245,
....256-259, 266, 269-271, 274, 275, 279,
....287, 288, 299, 300, 303, 304, 307, 308,
....311-313327-339, 347, 349-353, 358, 359,
.... 362, 377, 385-397, 405, 410, 419, 421, 424
Formalization and control stage.. 394
Function 47, 52, 65, 67
Functional configuration 48-51, 53-57, 65, 69, 71, 72, 77,
.... 78, 80, 82, 86, 88, 92, 132, 181,
.... 222, 223, 229, 250, 262, 263, 303,
....307, 314, 315, 329-333, 336-338,
....353, 366-369, 387, 397, 421
Generalists..197, 299, 384
Global configuration..77, 78, 314, 315
Group climate 128, 144, 147, 150-156, 168, 169
.... 319-321, 324, 325, 335, 336, 346, 353, 376
Group culture 135, 137, 144

Hierarchy 8, 43, 46, 47, 51, 64, 68, 72, 79, 81-84, 86,
.... 92, 170, 198, 212, 213, 215, 217, 237,
....247, 269, 270, 280, 371, 372, 377, 380
High technology..87, 233, 413
Horizontal differentiation 38, 47, 78-83, 93, 133, 161, 177, 199,
.... 211, 215, 237, 238, 240, 241, 260-262,
....332, 334, 351, 352, 380, 382, 383, 423
Hostile environment 37, 185, 187, 195, 216, 219, 229, 365
Hostility 37, 195, 200, 202-204, 207, 209,
....210, 218-220, 231-233, 242, 244,
.... 316, 365, 386, 391, 394, 395
Hybrid..75, 291, 292
Hypercompetition 233, 280, 399, 400
If-then statements..23, 29, 32, 39, 347
Incentive 2, 17, 63, 71, 87, 89-91, 137, 139, 164,
....182, 183, 209, 216, 222, 226, 238, 239,
.... 264, 349, 350, 388, 401
Incentive system................................2, 17, 63, 71, 87, 89-91, 137, 139,
.... 182, 209, 216, 239, 349, 401
Individualism 236, 238, 240, 241, 343, 350, 351, 352
Inflexible 99, 228, 331, 353
Information overload 15, 55, 170, 214, 215, 226, 334,
....349, 372, 380
Information processing capacity 7, 9, 11, 13, 15, 41, 49, 53,
.... 86, 89, 93, 98, 115, 117, 119, 131,
.... 137, 138, 159, 173, 179, 182, 187,
.... 204, 209, 221, 222, 241, 243, 247,
.... 314, 326, 365, 369, 372, 377, 380
Information processing demand....................... 8, 10, 11, 13, 15, 61, 63, 86,
.... 89, 120, 126, 128, 131, 133,
.... 171, 203, 270, 326, 327, 367,
.... 369, 373, 380, 383, 393, 395
Information processing view 10, 13, 14, 31, 47, 73, 78,
....94, 178, 275
Information richness 221
Information technology..................................... 7, 9, 67, 83, 84, 93, 54, 170,
.... 171, 173, 184, 197, 216, 238, 243,
.... 258, 259, 265-271, 353, 379, 387
Innovation.. 7, 43, 52, 55, 142, 146, 148, 181,
....272, 273, 278, 285, 287-296, 299-301,
....304-312, 315-318, 321, 330-332,
....339-341, 344, 345, 353, 368-370,
.... 373, 375, 376, 384, 387, 388
.... 391-395, 398, 400, 416, 422
Innovative 117, 100, 186, 233, 316, 317, 321,
.... 340, 342, 345, 349, 369, 394, 411
Innovator 297, 316, 388
Integrator.. 224
Interaction 38, 93, 135, 136, 141, 286, 358, 396

Intercontinental...272
Interdivisional 63, 238
Internal process climate 128, 144, 148, 158-164, 319-323,
.... 325, 326, 335, 336, 353, 367, 378
Internal process culture 135, 137, 144
International configuration 47, 75, 77, 78, 315, 314
International dimensions.. 94, 314
Interpersonal.. 95, 110
Jones Company...347, 348, 351
Knowledge base... 7, 11, 13, 18-23, 27,
.... 28-40, 42, 44, 45, 47, 313, 359, 386
Knowledge management 9-11, 14
Knowledge-base... 39, 45, 47
Leader.. 48, 55, 88, 93, 94-98, 101, 102, 110, 111,
.... 114-122, 126, 129, 132, 134, 136, 137,
.... 144-148, 150, 151, 154, 155, 159, 160,
....164, 168, 272, 321, 323, 325, 326, 337,
....338, 343, 346, 349, 350, 353, 354, 362,
.... 371, 373, 374, 382, 389, 390, 391,
....395, 397, 413
Leader credibility..............................137, 144, 145, 147, 150, 151, 154, 155,
....159, 160, 164, 168, 321, 325, 343, 346
Leadership 26, 38, 43, 47
Leadership roles.. 95, 110
Leadership style 26, 38, 88, 92, 94, 95, 98, 99, 101, 102, 110,
.... 111, 113-117, 127, 135, 136, 138, 167,
....236, 312, 323-327, 336-338, 347-354,
.... 359, 362, 363, 365, 366, 368, 371-374,
....377-382, 384-396, 401, 402, 417, 418
Learning.. 7, 11, 14, 132, 420, 421, 422, 423, 427
Liaison 9, 81, 92, 95, 100, 121, 152, 170, 221,
.... 224, 226, 227, 230, 308, 369, 387,
....397, 419, 423
Lifecycle 26, 116, 219, 233, 385, 388-391, 393-395, 400, 418, 424
Linjeflyg 24, 88
Locus of control... 99, 101, 103, 110, 113, 122, 130
Long-term 5, 88, 93, 100, 106, 108, 117-119,
....126, 127, 137, 144, 148, 296, 316, 402, 414
Mack trucks 51, 52, 80, 82, 84, 85, 86, 88, 94
Malevolent.. 200, 202
Management preferences... 20
Manager...88, 94-96, 99, 102, 110-116, 122, 124,
.... 127, 130-135, 163, 177, 213, 242-245,
....259, 260, 296, 298, 323-327, 337-339,
....348, 349, 354, 364-366, 368, 377-379,
....381, 387, 389, 391, 392, 395, 396 403, 418
Manufacturing 4, 50, 52-55, 59, 60, 67, 68, 77, 85,
....94, 175, 251-253, 255, 257, 258, 260
...., 280, 295, 314, 358

Marketing...9, 46, 51-54, 62, 67, 100, 177,
....255, 289, 290, 307, 401, 423
Masculinity 238, 236, 237, 239, 350
Matrix configuration...............................47, 64, 65, 66, 68-70, 76,
.....................................80, 87, 120, 155, 156, 262, 263, 300,
.....................................314, 328-330, 332, 333, 336, 339,
....351-353, 371-376, 387
Matsushita...77, 293, 295, 296
Meetings 7, 24, 71, 92, 93, 121, 125, 129,
....152, 153, 156, 157, 164, 165, 166,
....169, 187, 221, 224, 225, 227-229,
....231, 233, 244-246, 263-265, 267,
....270, 300, 312, 311, 349, 351, 352,
....353, 359, 362, 390, 392, 394, 397,
....412, 419, 420
Metaflexibility.. 233
Metaanalysis 257, 259, 260
Microinvolvement 114, 392
Misfit creation 402, 401
Misfits.............................. 25, 26, 27, 39, 43, 117, 167, 311, 312,
......................................314, 315, 317-330, 332-334, 336-339,
......................................341-353, 354, 385, 387, 388, 392-394,
..........................396-398, 401-409, 413, 414, 417, 424, 427
Mission 15, 19, 127, 272, 314
Moderator 173, 183, 260
Morale..................................... 137, 144-148, 150-152, 154, 155,
........................158, 159, 163, 167, 168, 324, 344, 391
Motivate...88, 102, 113, 118, 119, 127
Multidivisional 94, 181, 182
Multinational 76, 77
Mycin..348, 363, 366
National culture 134, 185, 190, 196, 235, 236, 239,
....243, 339, 341, 350-352
Negotiations 156, 242, 272
Negotiator 96
Network 67, 74, 96, 110, 170, 233, 272, 375, 376, 280
Network organization... 170, 280
Networking.. 96, 110
Non-bureaucratic 116
Non-divisible 238, 262, 333, 369, 394
Non-routine technology 390, 393, 411, 412, 418
Organizational climate........................... 15, 47, 127, 128, 131-134, 137-142,
.... 144, 146, 167, 312, 319, 324, 335,
....346, 353, 397, 412, 413
Organizational complexity 36, 38, 47, 73, 78-85, 120,
.... 128, 152, 156, 160, 173, 176, 177
...............................197, 199, 207, 210-220, 223-228,
.... 233, 242, 244, 260, 261, 271, 307,
.................................327, 329-332, 349-353, 358,

....359, 379-385, 390, 394, 419-422
Organizational configuration 5, 46, 71, 78, 81, 93, 129, 138, 170,
.... 181, 185, 221, 227, 230, 231, 244,
....250, 266-271, 327, 353, 389
Organizational culture.............................. 94, 131, 132, 133, 134, 138-141
Organizational design..........................4, 7, 10, 11, 13-21, 24, 27-33, 36, 37,
.... 42-47, 73, 83, 89-95, 98, 103, 116, 117,
.... 131, 133, 138, 143, 165, 167-169, 181-183,
....185, 205, 207, 209, 218, 222, 233, 238,
....243, 244, 250, 256, 264, 266, 267, 269,
....271, 274, 280, 299, 312, 313, 315, 316,
.... 347, 348, 353, 359, 387-389, 396,
....398-400, 402, 416, 423
Organizational learning 9, 11, 420, 423, 424, 427
Organizational lifecycle... 396
Organizational performance.....................................22, 101, 120, 132, 313
Organizational structure 2, 11, 13, 15, 19, 20, 27-32, 34, 41,
.... 43, 44, 77, 88, 92-94, 98, 99, 102,
.... 115, 117, 131, 133-140, 149, 160,
.... 164-166, 168, 169, 172, 175, 176, 184
.... 185, 186, 196, 198, 199, 209, 217,
.... 235, 238, 241, 243, 244, 266, 271-274
....277, 285, 294, 303, 315, 316, 328, 332,
.... 335, 376, 380, 399, 385, 414
Organizational unit 2, 13, 15, 80, 141, 149, 180
Ownership..16, 20, 62, 139, 280
Panasonic 278, 293
Participation..3, 97, 148, 218, 241
Partnerships.. 73, 280
Personnel.. 43, 52, 54, 77, 79, 84, 85, 88, 94,
.... 127, 128, 137, 163, 167, 168, 177,
....238, 252, 260, 262, 269, 272, 380, 397
Pharmaceutical 53, 59
Planning.. 6, 54, 55, 61, 67, 88, 92, 94-96, 98, 100,
.... 104, 108, 111, 137, 144, 156, 157,
.... 163-166199, 215, 221, 223-226, 229-231,
.... 244, 237, 265, 285, 303, 304, 308,
.... 311, 312, 351-353, 359, 392, 394, 397
Power distance 236, 238-240, 343, 350-352
Private.. 73
Private.. 159, 163
Private.. 177-179
Private..233, 237, 385, 393, 395
Proactive 88, 100, 102, 103, 106, 107, 110, 111, 113,
....118, 119, 122, 126, 127, 138, 280, 398
Procedural.................................... 73, 90, 91, 134, 182, 183, 211, 223, 224,
.... 230, 264, 303, 308, 311, 336, 351-354,
....392, 394, 419
Procedure.. 7, 8, 28, 91, 134, 137, 211, 222, 223,

....225, 226, 240, 244, 258, 263, 264, 274
.... 296, 328, 350, 353, 398, 420
Process innovation 272, 295, 301, 302, 305, 306, 309, 310,
....315, 295, 305
Process of organizational design 388
Process production...241, 243, 251, 254, 258
Product innovation.. 56, 288-290, 295-302,
....305, 306, 309-313, 339,
....341, 344, 345, 344
Productivity...43, 163, 415
Professional bureaucracy.............................. 47, 73, 87, 132, 182, 239, 263,
....300, 377, 378, 379,
Professionalization 72, 80, 87, 173-178, 180, 181, 185
....257, 241, 290, 289, 346, 347, 377, 386, 392
Professionals 47, 72, 87, 93, 175, 180, 214, 245, 247,
....257, 258, 262, 377, 385
Project configuration 64
Prospector............................. 272, 287-300, 305, 307, 309, 313, 316, 290,
.... 291, 296, 298-300, 315-317, 321, 322,
....330-332, 340-342, 344, 345, 349,
.... 350-353, 359, 362, 365, 373-379, 384, 389,
.... 390, 395, 396, 393, 418, 427, 391
Psychological climate.............................. 137, 139, 141, 142, 143, 167
Public.. 53, 54, 60, 65, 88, 150, 177, 178,
....217, 229, 242, 381, 422
Punctuated equilibrium...399, 411, 412
Punishment... 95
Quality... 46, 52, 54, 85, 88, 128, 133, 139, 144
.... 163, 164, 168, 169, 182, 203-205, 237
....238, 246, 257, 272-274, 289-295,
....297, 302, 305-307, 310, 313-316
....335, 347, 348, 385, 394
Rational goal climate 128, 144, 147, 148, 162-166
.... 336, 343-346, 353, 359, 370, 380, 412, 392
Rational goal culture 135
Reactive 102, 103, 106, 107, 111, 113, 118, 122,
....123, 127, 130, 131, 137, 138, 144, 398
Reactor 36, 289, 291, 296, 313, 316, 317, 291, 352, 353, 395, 408
Recognition 142, 146, 340
Recruiting 128, 163
Regulated...................................... 186, 187, 197, 201, 233, 272
Reorganization- 43, 58, 163
Reorganized.. 43, 58, 62
Responsiveness 47, 63
Restructuring... 137, 272
Results based incentives 338, 339, 351, 352, 359, 419, 420, 424
Results-oriented 128, 135, 165, 222, 336
Retail 57, 58, 60, 251-254, 258, 274
Risk 6, 88, 100-104, 106, 109-111, 113,

.... 118-123, 126-131, 137, 138, 141, 142, 148
....205, 213, 219, 292, 298, 316, 320, 339, ,
....348-350, 416, 420, 421
Roles.. 9, 95, 96, 110, 160, 166, 167, 419
Routine technology.................................. 238, 244, 246, 255, 260-264, 274
.... 297, 301, 303, 305, 309, 318, 319,
....322-324, 332, 333, 342-347, 353,
.... 369, 379, 383, 387, 389, 392, 393,
....395-398, 412, 419
Rule-based.. 28, 29
Samsonite.....................................185, 186, 195, 201, 207, 210, 215, 218,
....229, 241, 271, 277
Sensitivity analysis.. 393
Service .... 238, 247, 251-258, 271, 272, 274, 277, 292, ,
....294, 305, 314, 315, 346, 387, 352, 387, 389
Simple configuration 48-50, 82, 88, 128, 181, 182, 219,
....223, 228, 229, 239, 311, 330, 334, 336,
.... 349, 363, 364, 365, 366, 385, 390
Size measure.. 168, 172-176, 183, 184, 185, 259
Skill 47, 70, 83, 89, 163, 171-173, 183, 184,
....227, 238, 242, 243, 245, 248, 255, 259,
.... 326, 333, 334, 346, 347, 393
Skill capabilities 47, 163
Socialization.. 72
Spatial differentiation 47, 78, 79, 84, 85, 380
Specialization.................................. 4, 46, 48, 50, 52, 55, 56, 58, 63, 65, 79,
....80, 81, 116, 133, 164, 168, 177, 182, 197,
.... 212-215, 218, 223, 237, 238, 243, 257,
.... 260, 274, 303, 307, 331-336, 342, 349,
.... 364, 366, 369, 380, 383-385, 421
Spies.. 307
Standardization........................... 15, 35, 55, 72, 85, 87, 180, 192, 218, 243,
....246, 257, 258, 311, 316, 342, 377, 385
Standards 26, 47, 86, 90, 99, 120, 124, 128, 132,
.... 134, 139, 141, 142, 247, 273, 339
Strategic fit 21-23, 25, 27, 233, 274, 312, 313,
....347, 399, 403-405, 407,
Strategic misfits 44, 25, 315, 348, 350, 353, 387, 393,
....394, 402, 417,
Strategical.. 280
Strategy 2, 8, 9, 11, 15, 19-21, 24, 26, 28, 33, 36-40, 43,
....46, 47, 54, 60, 67, 69, 70, 73, 76, 81, 88, 92, 94,
.... 98-101, 112, 115, 116, 124, 128, 134-137, 144,
.... 164, 167, 171, 185, 186, 233, 238, 241, 242, 264,
....271-280, 285-322, 327, 330-332, 339-345, 347,
.... 349-353, 359, 362, 365-381, 384-407, 409,
....411-415, 417, 418, 427
Structuration theory.. 266
Subordinates.. 82, 102

Subsidiary.. 15, 43
Substrategy.. 272
Subtasks.. 46, 55
Subunits 56, 82, 333, 369, 423
Task............................ 4, 8, 9, 11, 24, 30, 31, 45, 48, 49, 52, 55, 60,
.... 79-82, 90, 92, 88, 98, 107, 115, 127, 137, 142,
....152, 159, 161, 170, 173, 180, 183, 198, 199,
.... 209, 212, 223, 243-247, 262, 267, 287, 299,
....303, 334347, 386
Team........................... 10, 11, 43, 70, 71, 96, 128, 272, 280, 401
Teamwork 127, 148
Technology change 397
Telecommunication 201, 232
Theory X 95, 102, 103, 112, 113, 118, 122,
.... 123, 126, 130, 167, 374
Total design fit 22, 27, 315, 347, 350, 351, 388
Transnational...75, 76, 77, 94
Trust.................................. 88, 128, 137, 141-147, 150-160, 163, 167-169
....324, 325, 335, 336,
.... 343-346, 348, 353, 375, 376, 385, 391
Turbulent...17, 199, 280, 286, 312
Uncertainty avoidance...................... 111-116, 118, 120, 122, 126, 130, 132
....236-240, 336-338, 341-344, 350-354, 385,
Unemployment.. 243
Unequivocal 243, 398
Unit of analysis 13, 16, 90, 233
Unit production... 241, 242, 246, 254, 255, 274
Validation.. 34, 40, 46, 47, 135, 172, 175, 244, 350
Validation process 46, 47, 172, 175
Validity 143, 192, 269, 288, 193, 400
Vertical differentiation 47, 79, 81-85, 133, 152, 153, 156, 157
.... 171-173, 177, 212, 215, 242, 243, 260,
....261, 270, 271, 380-383, 422
Viability.. 5, 56, 197, 280, 287, 243, 290, 301
.... 314, 386, 398, 402, 416, 417, 419
Viable..5, 24, 55, 186, 280, 313, 315, 317, 364, 400
.... 386, 389, 395, 396, 402, 412, 420
Video-conferencing 265, 271
Virtual network 47, 74, 85, 132, 134, 152, 155, 157,
.... 161, 280, 300, 338, 339,
....353, 354, 375, 376
Weaknesses... 48
Weaknesses.. 198
Wholesale.. 60
Wholesale...251-254, 258, 274

By purchasing this book the purchaser agrees to abide by the following Terms & Conditions:

1. Agreement. This Agreement consists of these General Terms and Conditions.

2. License. Teknowledge Corporation (Teknowledge) grants to you a non-exclusive, non-transferable license to use the M.4 software ("Product") in concert with the Publication Knowledge Bases included with this book, only. The license is granted to you subject to the following terms and conditions.

3. Ownership/Non-Disclosure. The Product is and shall remain the property of Teknowledge. Teknowledge considers the Product to be and intends that it remain the proprietary information and a trade secret of Teknowledge. You are aware that this Agreement grants you no title or rights of ownership in the Product. You agree to treat Product as the proprietary information and a trade secret of Teknowledge.

4. Restrictions on Use. You agree not to modify or translate all or any part of the Product. You agree not to distribute all or any part of the Product to others except as expressly provided by Sections 7 and 8 or by addenda to this Agreement, if any. You agree to maintain Teknowledge markings, including any copyright notice or legend associated with the Product.

    YOU AGREE NOT TO DISASSEMBLE, REVERSE ENGINEER OR DECOMPILE THE PRODUCT, IN WHOLE OR IN PART. iF YOU DO NOT COMPLY WITH THIS PROVISION, YOUR LICENSE IS AUTOMATICALLY TERMINATED.

5. Limitation of Liability. TEKNOWLEDGE WILL NOT BE LIABLE FOR ANY DAMAGES, INCLUDING ANY LOST PROFITS, LOST SAVINGS OR OTHER INCIDENTAL OR CONSEQUENTIAL DAMAGES ARISING OUT OF THE USE OR INABILITY TO USE THE PRODUCT, EVEN IF TEKNOWLEDGE HAS BEEN ADVISED OF THE POSSIBILITY OF SUCH DAMAGES, OR FOR ANY CLAIM BY ANY OTHER PARTY. SOME STATES DO NOT ALLOW THE LIMITATION OR EXCLUSION OF LIABILITY FOR INCIDENTAL OR CONSEQUENTIAL DAMAGES AND, TO SUCH EXTENT, THE ABOVE LIMITATION OR EXCLUSION MAY NOT APPLY TO YOU.

6. No Warranty. EXCEPT AS OTHERWISE PROVIDED IN THIS AGREEMENT, SHOULD THE PRODUCT PROVE TO BE DEFECTIVE, YOU AGREE TO ASSUME THE ENTIRE COST OF ALL NECESSARY SERVICING, REPAIR OR CORRECTION. YOU AGREE TO ASSUME ALL RESPONSIBILITY FOR THE SELECTION OF THE PRODUCT TO ACHIEVE YOUR INTENDED RESULTS, AND FOR THE INSTALLATION, USE AND RESULTS OBTAINED FROM THE

PRODUCT. TEKNOWLEDGE DOES NOT WARRANT THE PRODUCT WILL MEET YOUR REQUIREMENTS OR THAT THE OPERATION OF THE PRODUCT WILL BE UNINTERRUPTED OR ERROR FREE.

EXCEPT AS OTHERWISE PROVIDED IN THIS AGREEMENT THE PRODUCT IS PROVIDED "AS IS" WITHOUT A WARRANTY OF ANY KIND, EITHER EXPRESSED OR IMPLIED, INCLUDING, BUT NOT LIMITED TO THE IMPLIED WARRANTIES OF MERCHANTABILITY AND FITNESS FOR A PARTICULAR PURPOSE. SOME STATES DO NOT ALLOW THE EXCLUSION OF IMPLIED WARRANTIES AND, TO SUCH EXTENT, THE ABOVE EXCLUSION MAY NOT APPLY TO YOU. YOU MAY ALSO HAVE OTHER RIGHTS WHICH VARY FROM STATE TO STATE.

7. Transfer. You agree not to transfer this Agreement or this license to use all or any part of the Product to any other person or entity except as permitted by this Agreement. You may transfer this license to a third party if (a) such party agrees to accept all of the terms and conditions of this Agreement; (b) the Product, including all related materials, is transferred to the same party or any materials not transferred are destroyed; and (c) such transfer complies with all applicable United States export rules and regulations.

8. No Export. You understand and recognize that the Product may be subject to the Export Administration Regulations of the United States Department of Commerce and other United States government regulations relating to the export of technical data and equipment. You are familiar with and agree to comply with all such regulations, including any future modifications thereof.

9. Term and Termination. This Agreement is effective upon your receipt of the product and will remain effective until terminated. This Agreement may be terminated by (a) you at any time by providing written notice of termination to Teknowledge and a certificate of destruction or return of the Product and all related materials to Licensor, or by (b) Licensor at any time should you fail to comply with any of the terms and conditions of this Agreement. Termination subject to this Section 9 will create no obligation on the part of Teknowledge.

10. General. If any provision of this agreement is legally unenforceable or illegal, such provision shall be severed from this Agreement and the balance of this agreement shall continue in full force and effect. You agree to be responsible for the payment of all state and local taxes (other than California income taxes) incurred on account of this license. This Agreement shall be governed by California law.

PUBLISHER'S NOTE

The Publisher is not liable for any errors in the program. This disclaimer will hold the Publisher harmless against any costs, damages, and expenses caused by the software.

DATE DUE
